TRENDS IN QUANTUM GRAVITY RESEARCH

TRENDS IN QUANTUM GRAVITY RESEARCH

DAVID C. MOORE
EDITOR

Nova Science Publishers, Inc.
New York

LIBRARY OF CONGRESS CATALOGING-IN-PUBLICATION DATA
Trends in quantum gravity research / David C. Moore (editor).
p. cm.
Includes bibliographical references and index.
ISBN 1-59454-670-3
1. Quantum gravity--Research. I. Moore, David C.
QC178.T74 2006
530.14'3--dc22 2005028272

Published by Nova Science Publishers, Inc. ✢ New York

CONTENTS

PREFACE

Quantum gravity is the field of theoretical physics attempting to unify the theory of quantum mechanics, which describes three of the fundamental forces of nature, with general relativity, the theory of the fourth fundamental force: gravity. The ultimate goal is a unified framework for all fundamental forces-a theory of everything. This new book examines state-of-art research in this field.

In Chapter 1, the authors show how the quantum potential arises in various ways and trace its connection to quantum fluctuations and Fisher information along with its realization in terms of Weyl curvature. It represents a genuine quantization factor for certain classical systems as well as an expression for quantum matter in gravity theories of Weyl-Dirac type. Many of the facts and examples are extracted from the literature (with references cited) and we mainly provide connections and interpretation, with a few new observations. We deliberately avoid ontological and epistemological discussion and resort to a collection of contexts where the quantum potential plays a visibly significant role. In particular we sketch some recent results of F. and A. Shojai on Dirac-Weyl action and Bohmian mechanics which connects quantum mass to the Weyl geometry. Connections \`a la Santamato of the quantum potential with Weyl curvature arising from a stochastic geometry, are also indicated for the Schr\"odinger equation (SE) and Klein-Gordon (KG) equation. Quantum fluctuations and quantum geometry are linked with the quantum potential via Fisher information. Derivations of SE and KG from Nottale's scale relativity are sketched along with a variety of approaches to the KG equation. Finally connections of geometry and mass generation via Weyl-Dirac geometry with many cosmological implications are indicated, following M. Israelit and N. Rosen.

Gravitationally bound quantum states of matter were observed for the first time thanks to the unique properties of ultra-cold neutrons (UCN). The neutrons were allowed to fall towards a horizontal mirror which, together with the Earth's gravitational field, provided the necessary confining potential well. In Chapter 2, we discuss the current status of the experiment, as well as possible improvements: the integral and differential measuring modes; the flow-through and storage measuring modes; resonance transitions between the quantum states in the gravitational field or between magnetically split sub-levels of a gravitational quantum state.

This phenomenon and the related experimental techniques could be applied to various domains ranging from the physics of elementary particles and fields (for instance, spin-independent or spin-dependent short-range fundamental forces or the search for a non-zero neutron electric charge) to surface studies (for instance, the distribution of hydrogen in/above the surface of solids or liquids, or thin films on the surface) and the foundations of quantum mechanics (for instance, loss of quantum coherence, quantum-mechanical localization or experiments using the very long path of UCN matter waves in medium and in wave-guides).

In the present article we focus on transitions between the quantum states of neutrons in the gravitational field, consider the characteristic parameters of the problem and examine various methods for producing such transitions. We also analyze the feasibility of experiments with these quantum transitions and their optimization with respect to particular physical goals.

A classical dynamical system in a four-dimensional Euclidean space with universal time is considered in Chapter 3. The space is hypothesized to be originally occupied by a uniform substance, pictured as a liquid, which at some time became supercooled. Our universe began as a nucleation event initiating a liquid to solid transition. The universe we inhabit and are directly aware of consists of only the three-dimensional expanding phase boundary - a crystalline surface. Random energy transfers to the boundary from thermal fluctuations in the adjacent bulk phases are interpreted by us as quantum fluctuations, and give a physical realization to the stochastic quantization technique. Fermionic matter is modeled as screw dislocations; gauge bosons as surface acoustic waves. Minkowski space emerges dynamically through redefining local time to be proportional to the spatial coordinate perpendicular to the boundary. Lorentz invariance is only approximate, and the photon spectrum (now a phonon spectrum) has a maximum energy. Other features include a geometrical quantum gravitational theory based on elasticity theory, and a simple explanation of the quantum measurement process as a spontaneous symmetry breaking. Present, past and future are physically distinct regions, the present being a unique surface where our universe is being continually constructed.

Starting from the action function we have derived a theoretical background that leads to quantization of gravity and the deduction of a correlation between the gravitational and inertial masses, which depends on the *kinetic momentum* of the particle. In Chapter 4, the authors show that there is a reaffirmation of the strong equivalence principle and consequently the Einstein's equations are preserved. In fact such equations are deduced here directly from this kinetic approach to Gravity. Moreover, we have obtained a generalized equation for inertial forces, which incorporates the Mach's principle into Gravitation. Also, we have deduced the equation of Entropy; the Hamiltonian for a particle in an electromagnetic field and the reciprocal fine structure constant. It is possible to deduce the expression of the *Casimir force* and also to explain the *Inflation Period* and the *Missing Matter* without assuming the existence of *vacuum fluctuations*. This new approach for Gravity will allow us to understand some crucial matters in Cosmology. An experiment has been carried out to check the theoretical correlation between the gravitational and inertial masses. The experiment and results are presented on appendix A. The experimental data are in strongly accordance with the theory.

In Chapter 5, it is shown that the inclusion of quantum jumps, i.e., state vector reduction, in the semiclassical gravity construction opens a new avenue for the solution, on the one hand, of the serious difficulties of the construction per se and, on the other hand, of the challenging puzzles of dark energy and dark matter. In the problem of quantum gravity, the simplest and most natural construction is that of semiclassical gravity. In the latter, the energy-momentum tensor entering into the Einstein equation is represented by the expectation value of the corresponding operator. In a conventional treatment, there exists no satisfactory generalization of normal ordering to curved spacetime. The renormalization of the energy-momentum tensor is based on a set of axioms; one of the latter is that the tensor must be four-divergence free. The results of the renormalization suffer from serious difficulties: an ambiguity and a nonlocal dependence on metric. In addition, the conventional treatment denounces the concept of particles and the Hamiltonian. It is commonly accepted that things look even worse when state reduction is involved in dynamics. In fact, the opposite situation occurs. The reduction, being nonlocal and instantaneous, implies a universal time and, as a consequence, the structure of spacetime as the direct product of cosmological time and space. This allows for introducing normal ordering, particles, and the Hamiltonian. The renormalized energy-momentum tensor is unique and involves at most second derivatives of metric. On that basis, semiclassical reductive quantum gravity is constructed---a theory in which metric is treated classically whereas a quantum treatment of matter includes state vector reduction. The theory is assumed to be fundamental. In the theory, the semiclassical Einstein equation is violated due to the following. First, the energy-momentum tensor is not divergence free. Second, the six space components of the Einstein tensor involve the second time derivative of metric, but the other four components involve only the first time derivative. Therefore the latter components must be continuous. The energy-momentum tensor should be complemented by a pseudo energy-momentum tensor with four degrees of freedom which would compensate for the breakdown both of the divergence freedom condition and of the continuity of the four components of the energy-momentum tensor. The compensatory tensor is, by definition, the energy-momentum tensor of pseudomatter. The latter is represented by a pressural dust, i.e., a perfect fluid with a constant pressure, which has four degrees of freedom. The pressural dust comprises both dark energy (cosmological constant) and dark matter. So the presence of dark energy and dark matter in the real world provides an observable evidence of characteristically quantum gravitational effects. That is a challenge to a conventional opinion that there exists no such recognized evidence. The reductive semiclassical Einstein equation is composed of ten equations for six space components of metric and four pseudomatter variables (density and four-velocity). The elimination of the latter variables results in the metric equation. Dark matter is represented by a pseudodust, which implies the fruitlessness of efforts to represent dark matter by any kind of ordinary matter.

As explained in Chapter 6, Black Holes have always played a central role in investigations of quantum gravity. This includes both conceptual issues such as the role of classical singularities and information loss, and technical ones to probe the consistency of candidate theories. Lacking a full theory of quantum gravity, such studies had long been restricted to black hole models which include some aspects of quantization. However, it is then not always clear whether the results are consequences of quantum gravity per se or of the particular steps one had undertaken to bring the system into a treatable form. Over a little more than the last decade loop quantum gravity has emerged as a widely studied candidate for quantum gravity, where it is now possible to introduce black hole models within a quantum theory of gravity. This makes it possible to use only quantum effects which are known to arise also in the full theory, but still work in a rather simple and physically interesting context of black holes. Recent developments have now led to the first physical results about non-rotating quantum black holes obtained in this way. Restricting to the interior inside the Schwarzschild horizon, the resulting quantum model is free of the classical singularity, which is a consequence of discrete quantum geometry taking over for the continuous classical space-time picture. This fact results in a change of paradigm concerning the information loss problem. The horizon itself can also be studied in the quantum theory by imposing horizon conditions at the level of states. Thereby one can illustrate the nature of horizon degrees of freedom and horizon fluctuations. All these developments allow us to study the quantum dynamics explicitly and in detail which provides a rich ground to test the consistency of the full theory.

In: Trends in Quantum Gravity Research
Editor: David C. Moore, pp. 1-63
ISBN 1-59454-670-3

Chapter 1

FLUCTUATIONS, GRAVITY, AND THE QUANTUM POTENTIAL

Robert Carroll*
University of Illinois, Urbana, IL 61801

Abstract

We show how the quantum potential arises in various ways and trace its connection to quantum fluctuations and Fisher information along with its realization in terms of Weyl curvature. It represents a genuine quantization factor for certain classical systems as well as an expression for quantum matter in gravity theories of Weyl-Dirac type. Many of the facts and examples are extracted from the literature (with references cited) and we mainly provide connections and interpretation, with a few new observations. We deliberately avoid ontological and epistemological discussion and resort to a collection of contexts where the quantum potential plays a visibly significant role. In particular we sketch some recent results of F. and A. Shojai on Dirac-Weyl action and Bohmian mechanics which connects quantum mass to the Weyl geometry. Connections à la Santamato of the quantum potential with Weyl curvature arising from a stochastic geometry, are also indicated for the Schrödinger equation (SE) and Klein-Gordon (KG) equation. Quantum fluctuations and quantum geometry are linked with the quantum potential via Fisher information. Derivations of SE and KG from Nottale's scale relativity are sketched along with a variety of approaches to the KG equation. Finally connections of geometry and mass generation via Weyl-Dirac geometry with many cosmological implications are indicated, following M. Israelit and N. Rosen.

1 The Schrödinger Equation

The quantum potential seems to have achieved prominence via the work of L. deBroglie and D. Bohm plus many others on what is often now called Bohmian mechanics. There have been many significant contributions here and we refer to [40, 41, 42, 43, 44, 47, 50, 54]

*E-mail address: rcarroll@math.uiuc.edu

for a reasonably complete list of references. A good picture of the current theory can be obtained from the papers by an American-German-Italian (AGI) group of Allori, Barut, Berndl, Daumer, Dürr, Georgi, Goldstein, Lebowitz, Teufel, Tumulka, and Zanghi (cf. [8, 9, 19, 22, 23, 24, 25, 26, 69, 73, 74, 75, 76, 77, 78, 79, 87, 89, 87, 91, 92, 93, 94, 189, 190, 194]). We refer also to Holland [103, 104, 105, 106], Nikolić [134, 135, 136, 137, 138], Floyd [83, 84], and Bertoldi, Faraggi, and Matone [27, 81, 82] for other approaches and summaries. Other specific references will arise as we go along but we emphasize with apologies that there are many more interesting papers omitted here which hopefully are covered in [54].

First in a simple minded way one can look at a Schrödinger equation (SE)

$$-\frac{\hbar^2}{2m}\psi'' + V\psi = i\hbar\psi_t;\ \psi = Re^{iS/\hbar} \tag{1.1}$$

leading to ($' \sim \partial_x$)

$$S_t + \frac{S_x^2}{2m} + V - \frac{\hbar^2 R''}{R} = 0;\ \partial_t(R^2) + \frac{1}{m}(R^2 S')' = 0 \tag{1.2}$$

$$P = R^2 \sim |\psi|^2;\ Q = -\frac{\hbar^2}{2m}\frac{R''}{R} \Rightarrow S_t + \frac{(S')^2}{2m} + Q + V = 0;\ P_t + \frac{1}{m}(PS')' = 0 \tag{1.3}$$

Here Q is the quantum potential and in 3-dimensions for example one expresses this as $Q = -(\hbar^2/2m)(\Delta\sqrt{\rho})/\sqrt{\rho}\,(R = \sqrt{P},\ P \sim \rho)$.

In a hydrodynamic mode one can write (1-dimension for simplicity and with the proviso that $S \neq const.$) $p = S' = m\dot{q} = mv$ (v a velocity or collective velocity) and $\rho = mP$ (ρ an unspecified mass density) to obtain an Euler type hydrodynamic equation ($\partial \sim \partial_x$)

$$\partial_t(\rho v) + \partial(\rho v^2) + \frac{\rho}{m}\partial V + \frac{\rho}{m}\partial Q = 0 \tag{1.4}$$

REMARK 1.1. Given a wave function ψ with $|\psi|^2$ representing a probability density as in conventional quantum mechanics (QM) it is not unrealistic to imagine an ensemble picture emerging here (as a "cloud" of particles for example). This will be analogous to diffusion or fluid flow of course but can also be modeled on a Bohmian particle picture and this will be discussed later in more detail. We note also that Q appears in the Hamilton-Jacobi (HJ) type equation (1.3) but is not present in the SE (1.1). If one were to interpret ∂V as a hydrodynamical pressure term $-(1/\rho)\partial\mathfrak{P}$ then the SE would be unchanged and the hydrodynamical equation (with no Q term) would be meaningful in the form

$$\partial_t(\rho v) + \partial(\rho v^2) = \frac{1}{m}\partial\mathfrak{P} \tag{1.5}$$

Thinking of Q as a quantization of (1.5) yielding (1.4) leads then to the SE (1.1). ■

REMARK 1.2. The development of the AGI school involves now

$$\dot{q} = v = \frac{\hbar}{m}\Im\frac{\psi^*\psi'}{|\psi|^2} \tag{1.6}$$

and this is derived as the simplest Galilean and time reversal invariant form for velocity transforming correctly under velocity boosts. This is a nice argument and seems to avoid any recourse to Floydian time (cf. [50, 54]). ■

Next we consider relations of diffusion to QM following Nagasawa, Nelson, et al (cf. [131, 132, 133] - see also e.g. [67, 70, 86, 119, 120, 121]) and sketch some formulas for a simple Euclidean metric where $\Delta = \sum(\partial/\partial x^i)^2$. Then $\psi(t,x) = exp[R(t,x) + iS(t,x)]$ satisfies a SE $i\partial_t\psi + (1/2)\Delta\psi + ia(t,x)\cdot\nabla\psi - V(t,x)\psi = 0$ ($\hbar = m = 1$)) if and only if

$$V = -\frac{\partial S}{\partial t} + \frac{1}{2}\Delta R + \frac{1}{2}(\nabla R)^2 - \frac{1}{2}(\nabla S)^2 - a\cdot\nabla S; \tag{1.7}$$

$$0 = \frac{\partial R}{\partial t} + \frac{1}{2}\Delta S + (\nabla S)\cdot(\nabla R) + a\cdot\nabla R$$

in the region $D = \{(s,x) : \psi(s,x) \neq 0\}$. Solutions are often referred to as weak or distributional but we do not belabor this point. From [131] there results

THEOREM 1.1. Let $\psi(t,x) = exp[R(t,x)+iS(t,x)]$ be a solution of the SE above; then $\phi(t,x) = exp[R(t,x) + S(t,x)]$ and $\hat{\phi} = exp[R(t,x) - S(t,x)]$ are solutions of

$$\frac{\partial\phi}{\partial t} + \frac{1}{2}\Delta\phi + a(t,x)\cdot\nabla\phi + c(t,x,\phi)\phi = 0; \tag{1.8}$$

$$-\frac{\partial\hat{\phi}}{\partial t} + \frac{1}{2}\Delta\hat{\phi} - a(t,x)\cdot\nabla\hat{\phi} + c(t,x,\phi)\hat{\phi} = 0$$

where the creation and annihilation term $c(t,x,\phi)$ is given via

$$c(t,x,\phi) = -V(t,x) - 2\frac{\partial S}{\partial t}(t,x) - (\nabla S)^2(t,x) - 2a\cdot\nabla S(t,x) \tag{1.9}$$

Conversely given $(\phi,\hat{\phi})$ as above satisfying (1.8) it follows that ψ satisfies the SE with V as in (1.9) (note $R = (1/2)log(\hat{\phi}\phi)$ and $S = (1/2)log(\phi/\hat{\phi})$ with $exp(R) = (\hat{\phi}\phi)^{1/2}$).■

From this one can conclude that nonrelativistic QM is diffusion theory in terms of Schrödinger processes (described by $(\phi,\hat{\phi})$ - more details later). Further it is shown that key postulates in Nelson's stochastic mechanics or Zambrini's Euclidean QM (cf. [202]) can both be avoided in connecting the SE to diffusion processes (since they are automatically valid). Look now at Theorem 1.1 for one dimension and write $T = \hbar t$ with $X = (\hbar/\sqrt{m})x$; then some simple calculation leads to

COROLLARY 1.1. Equation (1.8), written in the $(X,\,T)$ variables becomes

$$\hbar\phi_T + \frac{\hbar^2}{2m}\phi_{XX} + A\phi_X + \tilde{c}\phi = 0;\; -\hbar\hat{\phi}_T + \frac{\hbar^2}{2m}\hat{\phi}_{XX} - A\hat{\phi}_X + \tilde{c}\hat{\phi} = 0; \tag{1.10}$$

$$\tilde{c} = -\tilde{V}(X,T) - 2\hbar S_T - \frac{\hbar^2}{m}S_X^2 - 2AS_X$$

Thus the diffusion processes pick up factors of $\hbar$ and $\hbar/\sqrt{m}$. ■

Next we sketch a derivation of the SE following scale relativity à la Nottale (cf. [58, 139, 140, 141, 142, 144] and [56, 64, 65, 66] for some refinements and variations); this material is expanded in [40, 54].

REMARK 1.3. One considers quantum paths à la Feynman so that $lim_{t\to t'}[X(t) - X(t')]^2/(t-t')$ exists. This implies $X(t) \in H^{1/2}$ where H^α means $c\epsilon^\alpha \leq |X(t)-X(t')| \leq C\epsilon^\alpha$ and from [80] for example this means $dim_H X[a,b] = 1/2$. Now one "knows" (see e.g. [1]) that quantum and Brownian motion paths (in the plane) have H-dimension 2 and some clarification is needed here. We refer to [125] where there is a paper on Wiener Brownian motion (WBM), random walks, etc. discussing Hausdorff and other dimensions of various sets. Thus given $0 < \lambda < 1/2$ with probability 1 a Browian sample function X satisfies $|X(t+h) - X(t)| \leq b|h|^\lambda$ for $|h| \leq h_0$ where $b = b(\lambda)$. This leads to the result that with probability 1 the graph of a Brownian sample function has Hausdorff and box dimension $3/2$. On the other hand a Browian trail (or path) in 2 dimensions has Hausdorff and box dimension 2 (note a quantum path can have self intersections, etc.). ■

Now fractal spacetime here will mean some kind of continuous nonsmooth pathspace so that a bivelocity structure is defined. One defines first

$$\frac{d_+}{dt}y(t) = lim_{\Delta t\to 0_+}\left\langle\frac{y(t+\Delta t) - y(t)}{\Delta t}\right\rangle; \tag{1.11}$$

$$\frac{d_-}{dt}y(t) = lim_{\Delta t\to 0_+}\left\langle\frac{y(t) - y(t-\Delta t)}{\Delta t}\right\rangle$$

Applied to the position vector x this yields forward and backward mean velocities, namely $(d_+/dt)x(t) = b_+$ and $(d_-/dt)x(t) = b_-$. Here these velocities are defined as the average at a point q and time t of the respective velocities of the outgoing and incoming fractal trajectories; in stochastic QM this corresponds to an average on the quantum state. The position vector $x(t)$ is thus "assimilated" to a stochastic process which satisfies respectively after $(dt > 0)$ and before $(dt < 0)$ the instant t a relation $dx(t) = b_+[x(t)]dt + d\xi_+(t) = b_-[x(t)]dt + d\xi_-(t)$ where $\xi(t)$ is a Wiener process (cf. [133]). It is in the description of ξ that the $D = 2$ fractal character of trajectories is inserted; indeed that ξ is a Wiener process means that the $d\xi$'s are assumed to be Gaussian with mean 0, mutually independent, and such that

$$< d\xi_{+i}(t)d\xi_{+j}(t) >= 2\mathcal{D}\delta_{ij}dt;\ < d\xi_{-i}(t)d\xi_{-j}(t) >= -2\mathcal{D}\delta_{ij}dt \tag{1.12}$$

where $< >$ denotes averaging ($\mathcal{D}$ is now the diffusion coefficient). Nelson's postulate (cf. [133]) is that $\mathcal{D} = \hbar/2m$ and this has considerable justification (cf. [139]). Note also that (1.12) is indeed a consequence of fractal (Hausdorff) dimension 2 of trajectories follows from $< d\xi^2 > /dt^2 = dt^{-1}$, i.e. precisely Feynman's result $< v^2 >^{1/2} \sim \delta t^{-1/2}$. Note that Brownian motion (used in Nelson's postulate) is known to be of fractal (Hausdorff) dimension 2. Note also that any value of $\mathcal{D}$ may lead to QM and for $\mathcal{D} \to 0$ the theory becomes equivalent to the Bohm theory. Now expand any function $f(x,t)$ in a Taylor series up to order 2, take averages, and use properties of the Wiener process ξ to get

$$\frac{d_+f}{dt} = (\partial_t + b_+ \cdot \nabla + \mathcal{D}\Delta)f;\ \frac{d_-f}{dt} = (\partial_t + b_- \cdot \nabla - \mathcal{D}\Delta)f \tag{1.13}$$

Let $\rho(x,t)$ be the probability density of $x(t)$; it is known that for any Markov (hence Wiener) process one has $\partial_t\rho + div(\rho b_+) = \mathcal{D}\Delta\rho$ (forward equation) and $\partial_t\rho + div(\rho b_-) = -\mathcal{D}\Delta\rho$ (backward equation). These are called Fokker-Planck equations and one defines two new average velocities $V = (1/2)[b_+ + b_-]$ and $U = (1/2)[b_+ - b_-]$. Consequently adding and subtracting one obtains $\rho_t + div(\rho V) = 0$ (continuity equation) and $div(\rho U) - \mathcal{D}\Delta\rho = 0$ which is equivalent to $div[\rho(U - \mathcal{D}\nabla log(\rho))] = 0$. One can show, using (1.13) that the term in square brackets in the last equation is zero leading to $U = \mathcal{D}\nabla log(\rho)$. Now place oneself in the (U,V) plane and write $\mathcal{V} = V - iU$. Then write $(d_{\mathcal{V}}/dt) = (1/2)(d_+ + d_-)/dt$ and $(d_{\mathcal{U}}/dt) = (1/2)(d_+ - d_-)/dt$. Combining the equations in (1.13) one defines $(d_{\mathcal{V}}/dt) = \partial_t + V\cdot\nabla$ and $(d_{\mathcal{U}}/dt) = \mathcal{D}\Delta + U\cdot\nabla$; then define a complex operator $(d'/dt) = (d_{\mathcal{V}}/dt) - i(d_{\mathcal{U}}/dt)$ which becomes

$$\frac{d'}{dt} = \left(\frac{\partial}{\partial t} - i\mathcal{D}\Delta\right) + \mathcal{V}\cdot\nabla \tag{1.14}$$

One now postulates that the passage from classical mechanics to a new nondifferentiable process considered here can be implemented by the unique prescription of replacing the standard d/dt by d'/dt. Thus consider $\mathfrak{S} = \left\langle \int_{t_1}^{t_2} \mathcal{L}(x,\mathcal{V},t)dt \right\rangle$ yielding by least action $(d'/dt)(\partial\mathcal{L}/\partial\mathcal{V}_i) = \partial\mathcal{L}/\partial x_i$. Define then $\mathcal{P}_i = \partial\mathcal{L}/\partial\mathcal{V}_i$ leading to $\mathcal{P} = \nabla\mathfrak{S}$ (recall the classical action principle with $dS = pdq - Hdt$). Now for Newtonian mechanics write $L(x,v,t) = (1/2)mv^2 - \mathbf{U}$ which becomes $\mathcal{L}(x,\mathcal{V},t) = (1/2)m\mathcal{V}^2 - \mathfrak{U}$ leading to $-\nabla\mathfrak{U} = m(d'/dt)\mathcal{V}$. One separates real and imaginary parts of the complex acceleration $\gamma = (d'\mathcal{V}/dt$ to get

$$d'\mathcal{V} = (d_{\mathcal{V}} - id_{\mathcal{U}})(V - iU) = (d_{\mathcal{V}}V - d_{\mathcal{U}}U) - i(d_{\mathcal{U}}V + d_{\mathcal{V}}U) \tag{1.15}$$

The force $F = -\nabla\mathfrak{U}$ is real so the imaginary part of the complex acceleration vanishes; hence

$$\frac{d_{\mathcal{U}}}{dt}V + \frac{d_{\mathcal{V}}}{dt}U = \frac{\partial U}{\partial t} + U\cdot\nabla V + V\cdot\nabla U + \mathcal{D}\Delta V = 0 \tag{1.16}$$

from which $\partial U/\partial t$ may be obtained. This is a weak point in the derivation since one has to assume e.g. that $U(x,t)$ has certain smoothness properties. Now considerable calculation leads to the SE $i\hbar\psi_t = -(\hbar^2/2m)\Delta\psi + \mathfrak{U}\psi$ and this suggests an interpretation of QM as mechanics in a nondifferentiable (fractal) space. In fact (using one space dimension for convenience) we see that if $\mathfrak{U} = 0$ then the free motion $m(d'/dt)\mathcal{V} = 0$ yields the SE $i\hbar\psi_t = -(\hbar^2/2m)\psi_{xx}$ as a geodesic equation in "fractal" space. Further from $U = (\hbar/m)(\partial\sqrt{\rho}/\sqrt{\rho})$ and $Q = -(\hbar^2/2m)(\Delta\sqrt{\rho}/\sqrt{\rho})$ one arrives at a lovely relation, namely

PROPOSITION 1.1. The quantum potential Q can be written in the form $Q = -(m/2)U^2 -(\hbar/2)\partial U$. Hence the quantum potential arises directly from the fractal nonsmooth nature of the quantum paths. Since Q can be thought of as a quantization of a classical motion we see that the quantization corresponds exactly to the existence of nonsmooth paths. Consequently smooth paths imply no quantum mechanics.

REMARK 1.4. In [5] one writes again $\psi = Rexp(iS/\hbar)$ with field equations in the hydrodynamical picture (1-D for convenience)

$$d_t(m_0\rho v) = \partial_t(m_0\rho v) + \nabla(m_0\rho v) = -\rho\nabla(u + Q);\ \partial_t\rho + \nabla\cdot(\rho v) = 0 \tag{1.17}$$

where $Q = -(\hbar^2/2m_0)(\Delta\sqrt{\rho}/\sqrt{\rho})$. The Nottale approach is used as above with $d_v \sim d_{\mathcal{V}}$ and $d_u \sim d_{\mathcal{U}}$. One assumes that the velocity field from the hydrodynamical model agrees with the real part v of the complex velocity $V = v - iu$ so $v = (1/m_0)\nabla s \sim 2\mathcal{D}\partial s$ and $u = -(1/m_0)\nabla\sigma \sim \mathcal{D}\partial log(\rho)$ where $\mathcal{D} = \hbar/2m_0$. In this context the quantum potential $Q = -(\hbar^2/2m_0)\Delta\mathcal{D}\sqrt{\rho}/\sqrt{\rho}$ becomes

$$Q = -m_0\mathcal{D}\nabla\cdot u - (1/2)m_0u^2 \sim -(\hbar/2)\partial u - (1/2)m_0u^2 \tag{1.18}$$

Consequently Q arises from the fractal derivative and the nondifferentiability of spacetime again, as in Proposition 1.1. Further one can relate u (and hence Q) to an internal stress tensor whereas the v equations correspond to systems of Navier-Stokes type.

REMARK 1.5. We note that it is the presence of $\pm$ derivatives that makes possible the introduction of a complex plane to describe velocities and hence QM; one can think of this as the motivation for a complex valued wave function and the nature of the SE. ■

REMARK 1.6. In [56] one extends ideas of Nottale and Ord (cf. [148, 149, 150, 151]) in order to derive an interesting nonlinear Schrödinger equation (NLSE) using a complex diffusion coefficient and a hydrodynamic model.

1.1 The Schrödinger Equation in Weyl Space

We go now to Santamato [171] and derive the SE from classical mechanics in Weyl space (i.e. from Weyl geometry - cf. also [18, 42, 43, 55, 108, 172, 199]). The idea is to relate the quantum force (arising from the quantum potential) to geometrical properties of spacetime; the Klein-Gordon (KG) equation is also treated in this spirit in [55, 172]. One wants to show how geometry acts as a guidance field for matter (as in general relativity). Initial positions are assumed random (as in the Madelung approach) and thus the theory is statistical and is really describing the motion of an ensemble. Thus assume that the particle motion is given by some random process $q^i(t,\omega)$ in a manifold M (where ω is the sample space tag) whose probability density $\rho(q,t)$ exists and is properly normalizable. Assume that the process $q^i(t,\omega)$ is the solution of differential equations

$$\dot{q}^i(t,\omega) = (dq^i/dt)(t,\omega) = v^i(q(t,\omega),t) \tag{1.19}$$

with random initial conditions $q^i(t_0, \omega) = q_0^i(\omega)$. Once the joint distribution of the random variables $q_0^i(\omega)$ is given the process $q^i(t, \omega)$ is uniquely determined by (1.19). One knows that in this situation $\partial_t \rho + \partial_i(\rho v^i) = 0$ (continuity equation) with initial Cauchy data $\rho(q,t) = \rho_0(q)$. The natural origin of v^i arises via a least action principle based on a Lagrangian $L(q, \dot{q}, t)$ with

$$L^*(q, \dot{q}, t) = L(q, \dot{q}, t) - \Phi(q, \dot{q}, t);\ \Phi = \frac{dS}{dt} = \partial_t S + \dot{q}^i \partial_i S \tag{1.20}$$

Then $v^i(q,t)$ arises by minimizing

$$I(t_0, t_1) = E[\int_{t_0}^{t_1} L^*(q(t,\omega), \dot{q}(t,\omega), t)dt] \tag{1.21}$$

where t_0, t_1 are arbitrary and E denotes the expectation (cf. [40, 41, 131, 132, 133] for stochastic ideas). The minimum is to be achieved over the class of all random motions $q^i(t, \omega)$ obeying (1.20) with arbitrarily varied velocity field $v^i(q,t)$ but having common initial values. One proves first

$$\partial_t S + H(q, \nabla S, t) = 0;\ v^i(q,t) = \frac{\partial H}{\partial p_i}(q, \nabla S(q,t), t) \tag{1.22}$$

Thus the value of I in (1.21) along the random curve $q^i(t, q_0(\omega))$ is

$$I(t_1, t_0, \omega) = \int_{t_0}^{t_1} L^*(q(, q_0(\omega)), \dot{q}(t, q_0(\omega)), t)dt \tag{1.23}$$

Let $\mu(q_0)$ denote the joint probability density of the random variables $q_0^i(\omega)$ and then the expectation value of the random integral is

$$I(t_1, t_0) = E[I(t_1, t_0, \omega)] = \int_{\mathbf{R}^n} \int_{t_0}^{t_1} \mu(q_0) L^*(q(t, q_0), \dot{q}(t, q_0), t) d^n q_0 dt \tag{1.24}$$

Standard variational methods give then

$$\delta I = \int_{\mathbf{R}^n} d^n q_0 \mu(_0) \left[\frac{\partial L^*}{\partial \dot{q}^i}(q(t_1, q_0), \partial_t q(t_1, q_0), t) \delta q^i(t_1, q_0) - \right. \tag{1.25}$$

$$\left. - \int_{t_0}^{t_1} dt \left(\frac{\partial}{\partial t} \frac{\partial L^*}{\partial \dot{q}^i}(q(t, q_0), \partial_t q)t, q_0), t) - \frac{\partial L^*}{\partial q^i}(q(t, q_0), \partial_t q(t, q_0), t) \right) \delta q^i(t, q_0) \right]$$

where one uses the fact that $\mu(q_0)$ is independent of time and $\delta q^i(t_0, q_0) = 0$ (recall common initial data is assumed). Therefore

$$(\mathbf{A})\ (\partial L^* / \partial \dot{q}^i)(q(t, q_0), \partial_t q(t, q_0), t) = 0; \tag{1.26}$$

$$(\mathbf{B})\ \frac{\partial}{\partial t} \frac{\partial L^*}{\partial \dot{q}^i}(q(t, q_0), \partial_t q(t, q_0, t) - \frac{\partial L^*}{\partial q^i}(q(t, q_0), \partial_t q(t, q_0), t) = 0$$

are the necessary conditions for obtaining a minimum of I. Conditions **(B)** are the usual Euler-Lagrange (EL) equations whereas **(A)** is a consequence of the fact that in the most general case one must retain varied motions with $\delta q^i(t_1, q_0)$ different from zero at the final time t_1. Note that since L^* differs from L by a total time derivative one can safely replace L^* by L in **(B)** and putting (1.20) into **(A)** one obtains the classical equations

$$p_i = (\partial L/\partial \dot{q}^i)(q(t, q_0), \dot{q}(t, q_0), t) = \partial_i S(q(t, q_0), t) \quad (1.27)$$

It is known now that if $det[(\partial^2 L/\partial \dot{q}^i \partial \dot{q}^j] \neq 0$ then the second equation in (1.22) is a consequence of the gradient condition (1.27) and of the definition of the Hamiltonian function $H(q, p, t) = p_i \dot{q}^i - L$. Moreover **(B)** in (1.26) and (1.27) entrain the HJ equation in (1.63), (1.33). In order to show that the average action integral (1.24) actually gives a minimum one needs $\delta^2 I > 0$ but this is not necessary for Lagrangians whose Hamiltonian H has the form

$$H_C(q, p, t) = \frac{1}{2m} g^{ik}(p_i - A_i)(p_k - A_k) + V \quad (1.28)$$

with arbitrary fields A_i and V (particle of mass m in an EM field A) which is the form for nonrelativistic applications; given positive definite g_{ik} such Hamiltonians involve sufficiency conditions $det[\partial^2 L/\partial \dot{q}^i \partial \dot{q}^k] = mg > 0$. Finally **(B)** in (1.26) with L^* replaced by L) shows that along particle trajectories the EL equations are satisfied, i.e. the particle undergoes a classical motion with probability one. Notice here that in (1.22) no explicit mention of generalized momenta is made; one is dealing with a random motion entirely based on position. Moreover the minimum principle (1.21) defines a 1-1 correspondence between solutions $S(q, t)$ in (1.22) and minimizing random motions $q^i(t, \omega)$. Provided v^i is given via (1.22) the particle undergoes a classical motion with probability one. Thus once the Lagrangian L or equivalently the Hamiltonian H is given, $\partial_t \rho + \partial_i(\rho v^i) = 0$ and (1.22) uniquely determine the stochastic process $q^i(t, \omega)$. Now suppose that some geometric structure is given on M so that the notion of scalar curvature $R(q, t)$ of M is meaningful. Then we assume (ad hoc) that the actual Lagrangian is

$$L(q, \dot{q}, t) = L_C(q, \dot{q}, t) + \gamma(\hbar^2/m)R(q, t) \quad (1.29)$$

where $\gamma = (1/6)(n-2)/(n-1)$ with $n = dim(M)$. Since both L_C and R are independent of $\hbar$ we have $L \to L_C$ as $\hbar \to 0$.

Now for a differential manifold with $ds^2 = g_{ik}(q)dq^i dq^k$ it is standard that in a transplantation $q^i \to q^i + \delta q^i$ one has $\delta A^i = \Gamma^i_{k\ell} A^\ell dq^k$ with $\Gamma^i_{k\ell}$ general affine connection coefficients on M (Riemannian structure is not assumed). In [171] it is assumed that for $\ell = (g_{ik} A^i A^k)^{1/2}$ one has $\delta \ell = \ell \phi_k dq^k$ where the ϕ_k are covariant components of an arbitrary vector (Weyl geometry). Then the actual affine connections $\Gamma^i_{k\ell}$ can be found by comparing this with $\delta \ell^2 = \delta(g_{ik} A^i A^k)$ and using $\delta A^i = \Gamma^i_{k\ell} A^\ell dq^k$. A little linear algebra gives then

$$\Gamma^i_{k\ell} = -\left\{ \begin{array}{c} i \\ k\ \ell \end{array} \right\} + g^{im}(g_{mk}\phi_\ell + g_{m\ell}\phi_k - g_{k\ell}\phi_m) \quad (1.30)$$

Thus we may prescribe the metric tensor g_{ik} and ϕ_i and determine via (1.30) the connection coefficients. Note that $\Gamma^i_{k\ell} = \Gamma^i_{\ell k}$ and for $\phi_i = 0$ one has Riemannian geometry. Covariant derivatives are defined via

$$A^k_{,i} = \partial_i A^k - \Gamma^{k\ell} A^\ell; \; A_{k,i} = \partial_i A_k + \Gamma^\ell_{ki} A_\ell \tag{1.31}$$

for covariant and contravariant vectors respectively (where $S_{,i} = \partial_i S$). Note Ricci's lemma no longer holds (i.e. $g_{ik,\ell} \neq 0$) so covariant differentiation and operations of raising or lowering indices do not commute. The curvature tensor $R^i_{k\ell m}$ in Weyl geometry is introduced via $A^i_{,k,\ell} - A^i_{,\ell,k} = F^i_{mk\ell} A^m$ from which arises the standard formula of Riemannian geometry

$$R^i_{mk\ell} = -\partial_\ell \Gamma^i_{mk} + \partial_k \Gamma^i_{m\ell} + \Gamma^i_{n\ell}\Gamma^n_{mk} - \Gamma^i_{nk}\Gamma^n_{m\ell} \tag{1.32}$$

where (1.30) is used in place of the Christoffel symbols. The tensor $R^i_{mk\ell}$ obeys the same symmetry relations as the curvature tensor of Riemann geometry as well as the Bianchi identity. The Ricci symmetric tensor R_{ik} and the scalar curvature R are defined by the same formulas also, viz. $R_{ik} = R^\ell_{i\ell k}$ and $R = g^{ik} R_{ik}$. For completeness one derives here

$$R = \dot{R} + (n-1)[(n-2)\phi_i\phi^i - 2(1/\sqrt{g})\partial_i(\sqrt{g}\phi^i)] \tag{1.33}$$

where $\dot{R}$ is the Riemannian curvature built by the Christoffel symbols. Thus from (1.30) one obtains

$$g^{k\ell}\Gamma^i_{k\ell} = -g^{k\ell}\left\{ \begin{array}{c} i \\ k\;\ell \end{array} \right\} - (n-2)\phi^i; \; \Gamma^i_{k\ell} = -\left\{ \begin{array}{c} i \\ k\;\ell \end{array} \right\} + n\phi_k \tag{1.34}$$

Since the form of a scalar is independent of the coordinate system used one may compute R in a geodesic system where the Christoffel symbols and all $\partial_\ell g_{ik}$ vanish; then (1.30) reduces to $\Gamma^i_{k\ell} = \phi_k \kappa^i_\ell + \phi_\ell \delta^i_k - g_{k\ell}\phi^i$ and hence

$$R = -g^{km}\partial_m \Gamma^i_{k\ell} + \partial_i(g^{k\ell}\Gamma^i_{k\ell}) + g^{\ell m}\Gamma^i_{n\ell}\Gamma^n_{mi} - g^{m\ell}\Gamma^i_{n\ell}\Gamma^n_{m\ell} \tag{1.35}$$

Further one has $g^{\ell m}\Gamma^i_{n\ell}\Gamma^n_{mi} = -(n-2)(\phi_k\phi^k)$ at the point in consideration. Putting all this in (1.35) one arrives at

$$R = \dot{R} + (n-1)(n-2)(\phi_k\phi^k) - 2(n-1)\partial_k\phi^k \tag{1.36}$$

which becomes (1.33) in covariant form. Now the geometry is to be derived from physical principles so the ϕ_i cannot be arbitrary but must be obtained by the same averaged least action principle (1.21) giving the motion of the particle. The minimum in (1.21) is to be evaluated now with respect to the class of all Weyl geometries having arbitrarily varied gauge vectors but fixed metric tensor. Note that once (1.29) is inserted in (1.20) the only term in (1.21) containing the gauge vector is the curvature term. Then observing that $\gamma > 0$ when $n \geq 3$ the minimum principle (1.21) may be reduced to the simpler form $E[R(q(t,\omega),t)] = min$ where only the gauge vectors ϕ_i are varied. Using (1.33) this is easily done. First a little argument shows that $\hat{\rho}(q,t) = \rho(q,t)/\sqrt{g}$ transforms as a scalar in a coordinate change and this will be called the scalar probability density of the random motion of the particle (statistical determination of geometry). Starting from $\partial_t\rho + \partial_i(\rho v^i) = 0$ a manifestly covariant equation for $\hat{\rho}$ is found to be $\partial_t\hat{\rho} + (1/\sqrt{g})\partial_i(\sqrt{g}v^i\hat{\rho}) = 0$. Now return to the minimum problem $E[R(q(t,\omega),t)] = min$; from (1.33) and $\hat{\rho} = \rho/\sqrt{g}$ one obtains

$$E[R(q(t,\omega),t)] = E[\dot{R}(q(t,\omega),t)]+ \tag{1.37}$$

$$+(n-1)\int_M [(n-2)\phi_i\phi^i - 2(1/\sqrt{g})\partial_i(\sqrt{g}\phi^i)]\hat{\rho}(q,t)\sqrt{g}d^nq$$

Assuming fields go to 0 rapidly enough on ∂M and integrating by parts one gets then

$$E[R] = E[\dot{R}] - \frac{n-1}{n-2}E[g^{ik}\partial_i(log(\hat{\rho})\partial_k(log(\hat{\rho})]+ \tag{1.38}$$

$$+\frac{n-1}{n-2}E\{g^{ik}[(n-2)\phi_i + \partial_i(log(\hat{\rho})][(n-2)\phi_k + \partial_k(log(\hat{\rho})]\}$$

Since the first two terms on the right are independent of the gauge vector and g^{ik} is positive definite $E[R]$ will be a minimum when

$$\phi_i(q,t) = -[1/(n-2)]\partial_i[log(\hat{\rho})(q,t)] \tag{1.39}$$

This shows that the geometric properties of space are indeed affected by the presence of the particle and in turn the alteration of geometry acts on the particle through the quantum force $f_i = \gamma(\hbar^2/m)\partial_i R$ which according to (1.33) depends on the gauge vector and its derivatives. It is this peculiar feedback between the geometry of space and the motion of the particle which produces quantum effects.

In this spirit one goes now to a geometrical derivation of the SE. Thus inserting (1.39) into (1.33) one gets

$$R = \dot{R} + (1/2\gamma\sqrt{\hat{\rho}})[1/\sqrt{g})\partial_i(\sqrt{g}g^{ik}\partial_k\sqrt{\hat{\rho}})] \tag{1.40}$$

where the value $(n-2)/6(n-1)$ for γ is used. On the other hand the HJ equation (1.20) can be written as

$$\partial_t S + H_C(q, \nabla S, t) - \gamma(\hbar^2/m)R = 0 \tag{1.41}$$

where (1.29) has been used. When (1.40) is introduced into (1.41) the HJ equation and the continuity equation $\partial_t\hat{\rho} + (1/\sqrt{g})(\sqrt{g}v^i\hat{\rho}) = 0$, with velocity field given by (1.22), form a set of two nonlinear PDE which are coupled by the curvature of space. Therefore self consistent random motions of the particle (i.e. random motions compatible with (1.35)) are obtained by solving (1.41) and the continuity equation simultaneously. For every pair of solutions $S(q,t,\hat{\rho}(q,t))$ one gets a possible random motion for the particle whose invariant probability density is $\hat{\rho}$. The present approach is so different from traditional QM that a proof of equivalence is needed and this is only done for Hamiltonians of the form (1.28) (which is not very restrictive). The HJ equation corresponding to (1.28) is

$$\partial_t S + \frac{1}{2m}g^{ik}(\partial_i S - A_i)(\partial_k S - A_k) + V - \gamma\frac{\hbar^2}{m}R = 0 \tag{1.42}$$

with R given by (1.40). Moreover using (1.22) as well as (1.33) the continuity equation becomes

$$\partial_t \hat{\rho} + (1/m\sqrt{g})\partial_i[\hat{\rho}\sqrt{g}g^{ik}(\partial_k S - A_k)] = 0 \tag{1.43}$$

Owing to (1.40), (1.42) and (1.43) form a set of two nonlinear PDE which must be solved for the unknown functions S and $\hat{\rho}$. Now a straightforward calculations shows that, setting

$$\psi(q,t) = \sqrt{\hat{\rho}(q,t)}exp](i/\hbar)S(q,t)], \tag{1.44}$$

the quantity ψ obeys a linear PDE (corrected from [171])

$$i\hbar\partial_t\psi = \frac{1}{2m}\left\{\left[\frac{i\hbar\partial_i\sqrt{g}}{\sqrt{g}} + A_i\right]g^{ik}(i\hbar\partial_k + A_k)\right\}\psi + \left[V - \gamma\frac{\hbar^2}{m}\dot{R}\right]\psi \tag{1.45}$$

where only the Riemannian curvature $\dot{R}$ is present (any explicit reference to the gauge vector ϕ_i having disappeared). (1.45) is of course the SE in curvilinear coordinates whose invariance under point transformations is well known. Moreover (1.44) shows that $|\psi|^2 = \hat{\rho}(q,t)$ is the invariant probability density of finding the particle in the volume element $d^n q$ at time t. Then following Nelson's arguments that the SE together with the density formula contains QM the present theory is physically equivalent to traditional nonrelativistic QM. One sees also from (1.44) and (1.45) that the time independent SE is obtained via $S = S_0(q) - Et$ with constant E and $\hat{\rho}(q)$. In this case the scalar curvature of space becomes time independent; since starting data at t_0 is meaningless one replaces the continuity equation with a condition $\int_M \hat{\rho}(q)\sqrt{g}d^n q = 1$.

REMARK 1.6. We recall that in the nonrelativistic context the quantum potential has the form $Q = -(\hbar^2/2m)(\partial^2\sqrt{\rho}/\sqrt{\rho})$ ($\rho \sim \hat{\rho}$ here) and in more dimensions this corresponds to $Q = -(\hbar^2/2m)(\Delta\sqrt{\rho}/\sqrt{\rho})$. Here we have a SE involving $\psi = \sqrt{\rho}exp[(i/\hbar)S]$ with corresponding HJ equation (1.42) which corresponds to the flat space 1-D $S_t + (s')^2/2m + V + Q = 0$ with continuity equation $\partial_t\rho + \partial(\rho S'/m) = 0$ (take $A_k = 0$ here). The continuity equation in (1.43) corresponds to $\partial_t\rho + (1/m\sqrt{g})\partial_i[\rho\sqrt{g}g^{ik}(\partial_k S)] = 0$. For $A_k = 0$ (1.42) becomes

$$\partial_t S + (1/2m)g^{ik}\partial_i S\partial_k S + V - \gamma(\hbar^2/m)R = 0 \tag{1.46}$$

This leads to an identification $Q \sim -\gamma(\hbar^2/m)R$ where R is the Ricci scalar in the Weyl geometry (related to the Riemannian curvature built on standard Christoffel symbols via (1.33)). Here $\gamma = (1/6)[(n-2)(n-2)]$ as above which for $n = 3$ becomes $\gamma = 1/12$; further the Weyl field $\phi_i = -\partial_i log(\rho)$. Consequently (see below).

PROPOSITION 1.2. For the SE (1.45) in Weyl space the quantum potential is $Q = -(\hbar^2/12m)R$ where R is the Weyl-Ricci scalar curvature. For Riemannian flat space $\dot{R} = 0$ this becomes via (1.40)

$$R = \frac{1}{2\gamma\sqrt{\rho}}\partial_i g^{ik}\partial_k\sqrt{\rho} \sim \frac{1}{2\gamma}\frac{\Delta\sqrt{\rho}}{\sqrt{\rho}} \Rightarrow Q = -\frac{\hbar^2}{2m}\frac{\Delta\sqrt{\rho}}{\sqrt{\rho}} \tag{1.47}$$

as is should and the SE (1.45) reduces to the standard SE in the form $i\hbar\partial_t\psi = -(\hbar^2/2m)\Delta\psi + V\psi$ ($A_k = 0$). ■

REMARK 1.7. In [172] (first paper) one begins with a generic 4-dimensional manifold with torsion free connections and a metric tensor $g_{\mu\nu}$ ($\hbar = c = 1$ for convenience). Then working with an average action principle based on [95] the particle motion and (Weyl) spacetime geometry are derived in a gauge invariant manner (cf. Section 3.2). Thus an integrable Weyl geometry is produced from a stochastic background via an extremization procedure (see Section 3). An effective particle mass is taken as $m^2 - (R/6) \sim m^2(1+Q) \approx m^2exp(Q)$ corresponding to $R/6 = -m^2Q = -\Box\sqrt{\rho}/\sqrt{\rho}$ (here $\hbar = c = 1$ and one has signature $(-,+,+,+)$ while the term $exp(Q)$ arises from [182]). We refer to [42, 43, 54, 55, 172] and Section 2 for details (for various other approaches see [18, 199]).■

1.2 Fisher Information Revisited

We recall first that the classical Fisher information associated with translations of a 1-D observable X with probability density $P(x)$ (related to a quantum geometry probability measure $ds^2 = \sum[(dp_j)^2/p_j]$) is

$$F_X = \int dx\, P(x)([log(P(x)]')^2 > 0 \quad (1.48)$$

(cf. [40, 43, 85, 96, 97, 98, 99, 160, 161]). One has a well known Cramer-Rao inequality $Var(X) \geq F_X^{-1}$ where $Var(X) \sim$ variance of X. A Fisher length for X is defined via $\delta X = F_X^{-1/2}$ and this quantifies the length scale over which $p(x)$ (or better $log(p(x))$) varies appreciably. Then the root mean square deviation ΔX satisfies $\Delta X \geq \delta X$. Let now P be the momentum observable conjugate to X, and P_{cl} a classical momentum observable corresponding to the state ψ given via $p_{cl}(x) = (\hbar/2i)[(\psi'/\psi) - (\bar{\psi}'/\bar{\psi})]$. One has then the identity $<p>_\psi = <p_{cl}>_\psi$ following via integration by parts. Now define the nonclassical momentum by $p_{nc} = p - p_{cl}$ and one shows then

$$\Delta X \Delta p \geq \delta X \Delta p \geq \delta X \Delta p_{nc} = \hbar/2 \quad (1.49)$$

Then consider a classical ensemble of n-dimensional particles of mass m moving under a potential V. The motion can be described via the HJ and continuity equations

$$\frac{\partial s}{\partial t} + \frac{1}{2m}|\nabla s|^2 + V = 0; \quad \frac{\partial P}{\partial t} + \nabla\cdot\left[P\frac{\nabla s}{m}\right] = 0 \quad (1.50)$$

for the momentum potential s and the position probability density P (note that there is no quantum potential and this will be supplied by the information term). These equations follow from the variational principle $\delta L = 0$ with Lagrangian

$$L = \int dt\, d^nx\, P\left[(\partial s/\partial t) + (1/2m)|\nabla s|^2 + V\right] \quad (1.51)$$

It is now assumed that the classical Lagrangian must be modified due to the existence of random momentum fluctuations. The nature of such fluctuations is immateria and one can assume that the momentum associated with position x is given by $p = \nabla s + N$ where the fluctuation term N vanishes on average at each point x. Thus s changes to being an average momentum potential. It follows that the average kinetic energy $< |\nabla s|^2 > /2m$ appearing in the Lagrangian above should be replaced by $< |\nabla s + N|^2 > /2m$ giving rise to

$$L' = L + (2m)^{-1} \int dt < N \cdot N >= L + (2m)^{-1} \int dt (\Delta N)^2 \tag{1.52}$$

where $\Delta N =< N \cdot N >^{1/2}$ is a measure of the strength of the quantum fluctuations. The additional term is specified uniquely, up to a multiplicative constant, by the three assumptions

1. Action principle: L' is a scalar Lagrangian with respect to the fields P and s where the principle $\delta L' = 0$ yields causal equations of motion. Thus

$$(\Delta N)^2 = \int d^n x p f(P, \nabla P, \partial P/\partial t, s, \nabla s, \partial s/\partial t, x, t)$$

for some scalar function f.

2. Additivity: If the system comprises two independent noninteracting subsystems with $P = P_1 P_2$ then the Lagrangian decomposes into additive subsystem contributions; thus $f = f_1 + f_2$ for $P = P_1 P_2$.

3. Exact uncertainty: The strength of the momentum fluctuation at any given time is determined by and scales inversely with the uncertainty in position at that time. Thus $\Delta N \to k\Delta N$ for $x \to x/k$. Moreover since position uncertainty is entirely characterized by the probability density P at any given time the function f cannot depend on s, nor explicitly on t, nor on $\partial P/\partial t$.

This leads to the result that (cf. [40, 54, 96])

$$(\Delta N)^2 = c \int d^n x \, P |\nabla log(P)|^2 \tag{1.53}$$

where c is a positive universal constant. Further for $\hbar = 2\sqrt{c}$ and $\psi = \sqrt{P} exp(is/\hbar)$ the equations of motion for p and s arising from $\delta L' = 0$ are $i\hbar \frac{\partial \psi}{\partial t} = -\frac{\hbar^2}{2m} \nabla^2 \psi + V\psi$.

A second derivation is given in [161, 161]. Thus let $P(y^i)$ be a probability density and $P(y^i + \Delta y^i)$ be the density resulting from a small change in the y^i. Calculate the cross entropy via

$$J(P(y^i + \Delta y^i) : P(y^i)) = \int P(y^i + \Delta y^i) log \frac{P(y^i + \Delta y^i)}{P(y^i)} d^n y \simeq \tag{1.54}$$

$$\simeq \left[\frac{1}{2}\int \frac{1}{P(y^i)}\frac{\partial P(y^i)}{\partial y^i}\frac{\partial P(y^i)}{\partial y^k)}d^n y\right]\Delta y^i \Delta y^k = I_{jk}\Delta y^i \Delta y^k$$

The I_{jk} are the elements of the Fisher information matrix. The most general expression has the form

$$I_{jk}(\theta^i) = \frac{1}{2}\int \frac{1}{P(x^i|\theta^i)}\frac{\partial P(x^i|\theta^i)}{\partial \theta^j}\frac{\partial P(x^i|\theta^i)}{\partial \theta^k}d^n x \tag{1.55}$$

where $P(x^i|\theta^i)$ is a probability distribution depending on parameters θ^i in addition to the x^i. For $P(x^i|\theta^i) = P(x^i + \theta^i)$ one recovers (1.54). If P is defined over an n-dimensional manifold with positive inverse metric g^{ik} one obtains a natural definition of the information associated with P via

$$I = g^{ik}I_{ik} = \frac{g^{ik}}{2}\int \frac{1}{P}\frac{\partial P}{\partial y^i}\frac{\partial P}{\partial y^k}d^n y \tag{1.56}$$

Now in the HJ formulation of classical mechanics the equation of motion takes the form

$$\frac{\partial S}{\partial t} + \frac{1}{2}g^{\mu\nu}\frac{\partial S}{\partial x^\mu}\frac{\partial S}{\partial x^\nu} + V = 0 \tag{1.57}$$

where $g^{\mu\nu} = diag(1/m, \cdots, 1/m)$. The velocity field u^μ is given by $u^\mu = g^{\mu\nu}(\partial S/\partial x^\nu)$. When the exact coordinates are unknown one can describe the system by means of a probability density $P(t, x^\mu)$ with $\int Pd^n x = 1$ and

$$(\partial P/\partial t) + (\partial/\partial x^\mu)(Pg^{\mu\nu}(\partial S/\partial x^\nu) = 0 \tag{1.58}$$

These equations completely describe the motion and can be derived from the Lagrangian

$$L_{CL} = \int P\left\{(\partial S/\partial t) + (1/2)g^{\mu\nu}(\partial S/\partial x^\mu)(\partial S/\partial x^\nu) + V\right\}dtd^n x \tag{1.59}$$

using fixed endpoint variation in S and P. Quantization is obtained by adding a term proportional to the information I defined in (1.56). This leads to

$$L_{QM} = L_{CL} + \lambda I = \int P\left\{\frac{\partial S}{\partial t} + \frac{1}{2}g^{\mu\nu}\left[\frac{\partial S}{\partial x^\mu}\frac{\partial S}{\partial x^\nu} + \frac{\lambda}{P^2}\frac{\partial P}{\partial x^\mu}\frac{\partial P}{\partial x^\nu}\right] + V\right\}dtd^n x \tag{1.60}$$

Fixed endpoint variation in S leads again to (1.58) while variation in P leads to

$$\frac{\partial S}{\partial t} + \frac{1}{2}g^{\mu\nu}\left[\frac{\partial S}{\partial x^\mu}\frac{\partial S}{\partial x^\nu} + \lambda\left(\frac{1}{P^2}\frac{\partial P}{\partial x^\mu}\frac{\partial P}{\partial x^\nu} - \frac{2}{P}\frac{\partial^2 P}{\partial x^\mu \partial x^\nu}\right)\right] + V = 0 \tag{1.61}$$

These equations are equivalent to the SE if $\psi = \sqrt{P}exp(iS/\hbar)$ with $\lambda = (2\hbar)^2$.

REMARK 1.8. Following ideas in [55, 56, 139] we note in (1.60) for $\phi_\mu \sim A_\mu = \partial_\mu log(P)$ (cf. (1.39)) and $P_\mu = \partial_u S$, a complex velocity can be envisioned leading to

$$|P_\mu + i\sqrt{\lambda}A_\mu|^2 = P_\mu^2 + \lambda A_\mu^2 \sim g^{\mu\nu}\left(\frac{\partial S}{\partial x^\mu}\frac{\partial S}{\partial x^\nu} + \frac{\lambda}{P^2}\frac{\partial P}{\partial x^\mu}\frac{\partial P}{\partial x^\nu}\right) \tag{1.62}$$

Further I in (1.56) is exactly known from ϕ_μ so one has a direct connection between Fisher information and the Weyl field ϕ_μ, along with motivation for a complex velocity. ■

REMARK 1.9. Comparing now with [43] and quantum geometry in the form $ds^2 = \sum(dp_j^2/p_j)$ on a space of probability distributions we can define (1.56) as a Fisher information metric in the present context. This should be positive definite in view of its relation to $(\Delta N)^2$ in (1.53) for example. Now for $\psi = Rexp(iS/\hbar)$ one has ($\rho \sim \hat{\rho}$ here)

$$-\frac{\hbar^2}{2m}\frac{R''}{R} \equiv -\frac{\hbar^2}{2m}\frac{\partial^2\sqrt{\rho}}{\sqrt{\rho}} = -\frac{\hbar^2}{8m}\left[\frac{2\rho''}{\rho} - \left(\frac{\rho'}{\rho}\right)^2\right] \tag{1.63}$$

in 1-D while in more dimensions we have a form ($\rho \sim P$)

$$Q \sim -2\hbar^2 g^{\mu\nu}\left[\frac{1}{P^2}\frac{\partial P}{\partial x^\mu}\frac{\partial P}{\partial x^\nu} - \frac{2}{P}\frac{\partial^2 P}{\partial x^\mu \partial x^\nu}\right] \tag{1.64}$$

as in (1.63) (arising from the Fisher metric I of (1.56) upon variation in P in the Lagrangian). It can also be related to an osmotic velocity field $u = D\nabla log(\rho)$ via $Q = (1/2)u^2 + D\nabla\cdot u$ connected to Brownian motion where D is a diffusion coefficient (cf. [56, 67, 86, 139]). For $\phi_\mu = -\partial_\mu log(P)$ we have then $u = -D\phi$ with $Q = D^2((1/2)(|u|^2 - \nabla\cdot\phi)$, expressing Q directly in terms of the Weyl vector. This enforces the idea that QM is built into Weyl geometry! ■

2 Bohmian Mechanics and Weyl Geometry

From Chapters 1 and 2 we know something about Bohmian mechanics and the quantum potential and we go now to the papers [178, 179, 180, 181, 182, 183, 185, 186] by A. and F. Shojai to begin the present discussion (cf. also [2, 28, 29, 68, 129, 130, 163, 164, 165, 166, 167, 168, 169, 170, 173, 174, 175, 176, 177, 184, 187, 188]). for related work from the Tehran school and [43, 55, 124, 147, 171, 172, 181] for linking of dBB theory with Weyl geometry). In nonrelativistic deBroglie-Bohm theory the quantum potential is $Q = -(\hbar^2/2m)(\nabla^2|\Psi|/|\Psi|)$. The particles trajectory can be derived from Newton's law of motion in which the quantum force $-\nabla Q$ is present in addition to the classical force $-\nabla V$. The enigmatic quantum behavior is attributed here to the quantum force or quantum potential (with Ψ determining a "pilot wave" which guides the particle motion). Setting $\Psi = \sqrt{\rho}exp[iS/\hbar]$ one has

$$\frac{\partial S}{\partial t} + \frac{|\nabla S|^2}{2m} + V + Q = 0; \;\; \frac{\partial\rho}{\partial t} + \nabla\cdot\left(\rho\frac{\nabla S}{m}\right) = 0 \tag{2.1}$$

The first equation in (2.1) is a Hamilton-Jacobi (HJ) equation which is identical to Newton's law and represents an energy condition $E = (|p|^2/2m) + V + Q$ (recall from HJ theory $-(\partial S/\partial t) = E(= H)$ and $\nabla S = p$). The second equation represents a continuity equation for a hypothetical ensemble related to the particle in question. For the relativistic extension one could simply try to generalize the relativistic energy equation $\eta_{\mu\nu}P^{\mu}P^{\nu} = m^2c^2$ to the form

$$\eta_{\mu\nu}P^{\mu}P^{\nu} = m^2c^2(1+\mathcal{Q}) = \mathcal{M}^2c^2;\ \mathcal{Q} = (\hbar^2/m^2c^2)(\Box|\Psi|/|\Psi|) \tag{2.2}$$

$$\mathcal{M}^2 = m^2\left(1+\alpha\frac{\Box|\Psi|}{|\Psi|}\right);\ \alpha = \frac{\hbar^2}{m^2c^2} \tag{2.3}$$

This could be derived e.g. by setting $\Psi = \sqrt{\rho}exp(iS/\hbar)$ in the Klein-Gordon (KG) equation and separating the real and imaginary parts, leading to the relativistic HJ equation $\eta_{\mu\nu}\partial^{\mu}S\partial^{\nu}S = \mathfrak{M}^2c^2$ (as in (2.1) - note $P^{\mu} = -\partial^{\mu}S$) and the continuity equation is $\partial_{\mu}(\rho\partial^{\mu}S) = 0$. The problem of $\mathcal{M}^2$ not being positive definite here (i.e. tachyons) is serious however and in fact (2.2) is not the correct equation (see e.g. [180, 182, 185]). One must use the covariant derivatives ∇_{μ} in place of ∂_{μ} and for spin zero in a curved background there results

$$\nabla_{\mu}(\rho\nabla^{\mu}S) = 0;\ g^{\mu\nu}\nabla_{\mu}S\nabla_{\nu}S = \mathfrak{M}^2c^2; \tag{2.4}$$

To see this one must require that a correct relativistic equation of motion should not only be Poincaré invariant but also it should have the correct nonrelativistic limit. Thus for a relativistic particle of mass $\mathfrak{M}$ (which is a Lorentz invariant quantity)

$$\mathfrak{A} = \int d\lambda(1/2)\mathfrak{M}(r)(dr_{\mu}/d\lambda)(dr^{\nu}/d\lambda) \tag{2.5}$$

is the action functional where λ is any scalar parameter parametrizing the path $r_{\mu}(\lambda)$ (it could e.g. be the proper time τ). Varying the path via $r_{\mu} \to r'_{\mu} = r_{\mu} + \epsilon_{\mu}$ one gets

$$\mathfrak{A} \to \mathfrak{A}' = \mathfrak{A} + \delta\mathfrak{A} = \mathfrak{A} + \int d\lambda\left[\frac{dr_{\mu}}{d\lambda}\frac{d\epsilon^{\mu}}{d\lambda} + \frac{1}{2}\frac{dr_{\mu}}{d\lambda}\frac{dr^{\mu}}{d\lambda}\epsilon_{\nu}\partial^{\nu}\mathfrak{M}\right] \tag{2.6}$$

By least action the correct path satisfies $\delta\mathfrak{A} = 0$ with fixed boundaries so the equation of motion is

$$(d/d\lambda)(\mathfrak{M}u_{\mu}) = (1/2)u_{\nu}u^{\nu}\partial_{\mu}\mathfrak{M}; \tag{2.7}$$

$$\mathfrak{M}(du_{\mu}/d\lambda) = ((1/2)\eta_{\mu\nu}u_{\alpha}u^{\alpha} - u_{\mu}u_{\nu})\partial^{\nu}\mathfrak{M}$$

where $u_{\mu} = dr_{\mu}/d\lambda$. Now look at the symmetries of the action functional via $\lambda \to \lambda + \delta$. The conserved current is then the Hamiltonian $\mathfrak{H} = -\mathfrak{L} + u_{\mu}(\partial\mathfrak{L}/\partial u_{\mu}) = (1/2)\mathfrak{M}u_{\mu}u^{\mu} = E$. This can be seen by setting $\delta\mathfrak{A} = 0$ where

$$0 = \delta\mathfrak{A} = \mathfrak{A}' - \mathfrak{A} = \int d\lambda\left[\frac{1}{2}u_{\mu}u^{\mu}u^{\nu}\partial_{\nu}\mathfrak{M} + \mathfrak{M}u_{\mu}\frac{du^{\mu}}{d\lambda}\right]\delta \tag{2.8}$$

which means that the integrand is zero, i.e. $(d/d\lambda)[(1/2)\mathfrak{M}u_{\mu}u^{\mu}] = 0$. Since the proper time is defined as $c^2d\tau^2 = dr_{\mu}dr^{\mu}$ this leads to $(d\tau/d\lambda) = \sqrt{(2E/\mathfrak{M}c^2)}$ and the equation of motion becomes

$$\mathfrak{M}(dv_{\mu}/d\tau) = (1/2)(c^2\eta_{\mu\nu} - v_{\mu}v_{\nu})\partial^{\nu}\mathfrak{M} \tag{2.9}$$

where $v_\mu = dr_\mu/d\tau$. The nonrelativistic limit can be derived by letting the particles velocity be ignorable with respect to light velocity. In this limit the proper time is identical to the time coordinate $\tau = t$ and the result is that the $\mu = 0$ component is satisfied identically via $(r \sim \vec{r})$

$$\mathfrak{M}\frac{d^2r}{dt^2} = -\frac{1}{2}c^2\nabla\mathfrak{M} \Rightarrow m\left(\frac{d^2r}{dt^2}\right) = -\nabla\left[\frac{mc^2}{2}log\left(\frac{\mathfrak{M}}{\mu}\right)\right] \tag{2.10}$$

where μ is an arbitrary mass scale. In order to have the correct limit the term in parenthesis on the right side should be equal to the quantum potential so

$$\mathfrak{M} = \mu exp[-(\hbar^2/m^2c^2)(\nabla^2|\Psi|/|\Psi|)] \tag{2.11}$$

The relativistic quantum mass field (manifestly invariant) is $\mathfrak{M} = \mu exp[(\hbar^2/2m)(\Box|\Psi|/|\Psi|)]$ and setting $\mu = m$ we get

$$\mathfrak{M} = m exp[(\hbar^2/m^2c^2)(\Box|\Psi|/|\Psi|)] \tag{2.12}$$

If one starts with the standard relativistic theory and goes to the nonrelativistic limit one does not get the correct nonrelativistic equations; this is a result of an improper decomposition of the wave function into its phase and norm in the KG equation (cf. also [27] for related procedures). One notes here also that (2.12) leads to a positive definite mass squared. Also from [180] this can be extended to a many particle version and to a curved spacetime. In summary, for a particle in a curved background one has (cf. [182] which we continue to follow)

$$\nabla_\mu(\rho\nabla^\mu S) = 0;\; g^{\mu\nu}\nabla_\mu S\nabla_\nu S = \mathfrak{M}^2c^2;\; \mathfrak{M}^2 = m^2e^{\mathfrak{Q}};\; \mathfrak{Q} = \frac{\hbar^2}{m^2c^2}\frac{\Box_g|\Psi|}{|\Psi|} \tag{2.13}$$

Since, following deBroglie, the quantum HJ equation (QHJE) in (2.13) can be written in the form $(m^2/\mathfrak{M}^2)g^{\mu\nu}\nabla_\mu S\nabla_\nu S = m^2c^2$ the quantum effects are identical to a change of spacetime metric

$$g_{\mu\nu} \to \tilde{g}_{\mu\nu} = (\mathfrak{M}^2/m^2)g_{\mu\nu} \tag{2.14}$$

which is a conformal transformation. The QHJE becomes then $\tilde{g}^{\mu\nu}\tilde{\nabla}_\mu S\tilde{\nabla}_\nu S = m^2c^2$ where $\tilde{\nabla}_\mu$ represents covariant differentiation with respect to the metric $\tilde{g}_{\mu\nu}$ and the continuity equation is then $\tilde{g}_{\mu\nu}\tilde{\nabla}_\mu(\rho\tilde{\nabla}_\nu S) = 0$. The important conclusion here is that the presence of the quantum potential is equivalent to a curved spacetime with its metric given by (2.14). This is a geometrization of the quantum aspects of matter and it seems that there is a dual aspect to the role of geometry in physics. The spacetime geometry sometimes looks like "gravity" and sometimes reveals quantum behavior. The curvature due to the quantum potential may have a large influence on the classical contribution to the curvature of spacetime. The particle trajectory can now be derived from the guidance relation via differentiation of (2.13) leading to the Newton equations of motion

$$\mathfrak{M}\frac{d^2x^\mu}{d\tau^2} + \mathfrak{M}\Gamma^\mu_{\nu\kappa}u^\nu u^\kappa = (c^2g^{\mu\nu} - u^\mu u^\nu)\nabla_\nu\mathfrak{M} \tag{2.15}$$

Using the conformal transformation above (2.15) reduces to the standard geodesic equation.

Now a general "canonical" relativistic system consisting of gravity and classical matter (no quantum effects) is determined by the action

$$\mathcal{A} = \frac{1}{2\kappa}\int d^4x\sqrt{-g}\mathcal{R} + \int d^4x\sqrt{-g}\frac{\hbar^2}{2m}\left(\frac{\rho}{\hbar^2}\mathcal{D}_\mu S\mathcal{D}^\mu S - \frac{m^2}{\hbar^2}\rho\right) \tag{2.16}$$

where $\kappa = 8\pi G$ and $c = 1$ for convenience. It was seen above that via deBroglie the introduction of a quantum potential is equivalent to introducing a conformal factor $\Omega^2 = \mathfrak{M}^2/m^2$ in the metric. Hence in order to introduce quantum effects of matter into the action (2.16) one uses this conformal transformation to get $(1 + Q \sim exp(Q))$

$$\mathfrak{A} = \frac{1}{2\kappa}\int d^4x\sqrt{-\bar{g}}(\bar{\mathcal{R}}\Omega^2 - 6\bar{\nabla}_\mu\Omega\bar{\nabla}^\mu\Omega) + \tag{2.17}$$

$$+\int d^4x\sqrt{-\bar{g}}\left(\frac{\rho}{m}\Omega^2\bar{\nabla}_\mu S\bar{\nabla}^\mu S - m\rho\Omega^4\right) + \int d^4x\sqrt{-\bar{g}}\lambda\left[\Omega^2 - \left(1 + \frac{\hbar^2}{m^2}\frac{\Box\sqrt{\rho}}{\sqrt{\rho}}\right)\right]$$

where a bar over any quantity means that it corresponds to the nonquantum regime. Here only the first two terms of the expansion of $\mathfrak{M}^2 = m^2 exp(\mathfrak{Q})$ in (2.13) have been used, namely $\mathfrak{M}^2 \sim m^2(1 + \mathfrak{Q})$. No physical change is involved in considering all the terms. λ is a Lagrange multiplier introduced to identify the conformal factor with its Bohmian value. One uses here $\bar{g}_{\mu\nu}$ to raise of lower indices and to evaluate the covariant derivatives; the physical metric (containing the quantum effects of matter) is $g_{\mu\nu} = \Omega^2\bar{g}_{\mu\nu}$. By variation of the action with respect to $\bar{g}_{\mu\nu}$, Ω, ρ, S, and λ one arrives at the following quantum equations of motion:

1. The equation of motion for Ω

$$\bar{\mathcal{R}}\Omega + 6\Box\Omega + \frac{2\kappa}{m}\rho\Omega(\bar{\nabla}_\mu S\bar{\nabla}^\mu S - 2m^2\Omega^2) + 2\kappa\lambda\Omega = 0 \tag{2.18}$$

2. The continuity equation for particles $\bar{\nabla}_\mu(\rho\Omega^2\bar{\nabla}^\mu S) = 0$

3. The equations of motion for particles (here $a' \equiv \bar{a}$)

$$(\bar{\nabla}_\mu S\bar{\nabla}^\mu S - m^2\Omega^2)\Omega^2\sqrt{\rho} + \frac{\hbar^2}{2m}\left[\Box'\left(\frac{\lambda}{\sqrt{\rho}}\right) - \lambda\frac{\Box'\sqrt{\rho}}{\rho}\right] = 0 \tag{2.19}$$

4. The modified Einstein equations for $\bar{g}_{\mu\nu}$

$$\Omega^2\left[\bar{\mathcal{R}}_{\mu\nu} - \frac{1}{2}\bar{g}_{\mu\nu}\bar{\mathcal{R}}\right] - [\bar{g}_{\mu\nu}\Box' - \bar{\nabla}_\mu\bar{\nabla}_\nu]\Omega^2 - 6\bar{\nabla}_\mu\Omega\bar{\nabla}_\nu\Omega + 3\bar{g}_{\mu\nu}\bar{\nabla}_\alpha\Omega\bar{\nabla}^\alpha\Omega + \tag{2.20}$$

$$+\frac{2\kappa}{m}\rho\Omega^2\bar{\nabla}_\mu S\bar{\nabla}_\nu S - \frac{\kappa}{m}\rho\Omega^2\bar{g}_{\mu\nu}\bar{\nabla}_\alpha S\bar{\nabla}^\alpha S + \kappa m\rho\Omega^4\bar{g}_{\mu\nu} +$$

$$+\frac{\kappa\hbar^2}{m^2}\left[\bar{\nabla}_\mu\sqrt{\rho}\bar{\nabla}_\nu\left(\frac{\lambda}{\sqrt{\rho}}\right) + \bar{\nabla}_\nu\sqrt{\rho}\bar{\nabla}_\mu\left(\frac{\lambda}{\sqrt{\rho}}\right)\right] - \frac{\kappa\hbar^2}{m^2}\bar{g}_{\mu\nu}\bar{\nabla}_\alpha\left[\lambda\frac{\bar{\nabla}^\alpha\sqrt{\rho}}{\sqrt{\rho}}\right] = 0$$

5. The constraint equation $\Omega^2 = 1 + (\hbar^2/m^2)[(\Box\sqrt{\rho})/\sqrt{\rho}]$

Thus the back reaction effects of the quantum factor on the background metric are contained in these highly coupled equations. A simpler form of (2.8) can be obtained by taking the trace of (2.20) and using (2.18) which produces $\lambda = (\hbar^2/m^2)\bar{\nabla}_\mu[\lambda(\bar{\nabla}^\mu\sqrt{\rho})/\sqrt{\rho}]$. A solution of this via perturbation methods using the small parameter $\alpha = \hbar^2/m^2$ yields the trivial solution $\lambda = 0$ so the above equations reduce to

$$\bar{\nabla}_\mu(\rho\Omega^2\bar{\nabla}^\mu S) = 0;\ \bar{\nabla}_\mu S\bar{\nabla}^\mu S = m^2\Omega^2;\ \mathfrak{G}_{\mu\nu} = -\kappa\mathfrak{T}^{(m)}_{\mu\nu} - \kappa\mathfrak{T}^{(\Omega)}_{\mu\nu} \tag{2.21}$$

where $\mathfrak{T}^{(m)}_{\mu\nu}$ is the matter energy-momentum (EM) tensor and

$$\kappa\mathfrak{T}^{(\Omega)}_{\mu\nu} = \frac{[g_{\mu\nu}\Box - \nabla_\mu\nabla_\nu]\Omega^2}{\Omega^2} + 6\frac{\nabla_\mu\Omega\nabla_\nu\Omega}{\omega^2} - 2g_{\mu\nu}\frac{\nabla_\alpha\Omega\nabla^\alpha\Omega}{\Omega^2} \tag{2.22}$$

with $\Omega^2 = 1+\alpha(\Box\sqrt{\rho}/\sqrt{\rho})$. Note that the second relation in (2.21) is the Bohmian equation of motion and written in terms of $g_{\mu\nu}$ it becomes $\nabla_\mu S\nabla^\mu S = m^2c^2$.

REMARK 2.1. In the preceeding one has tacitly assumed that there is an ensemble of quantum particles so what about a single particle? One translates now the quantum potential into purely geometrical terms without reference to matter parameters so that the original form of the quantum potential can only be deduced after using the field equations. Thus the theory will work for a single particle or an ensemble. One notes that the use of $\psi\psi^*$ automatically suggests or involves an ensemble if (or its square root) it is to be interpreted as a probability density. Thus the idea that a particle has only a probability of being at or near x seems to mean that some paths take it there but others don't and this is consistent with Feynman's use of path integrals for example. This seems also to say that there is no such thing as a particle, only a collection of versions or cloud connected to the particle idea. Bohmian theory on the other hand for a fixed energy gives a one parameter family of trajectories associated to ψ (see here [47, 50, 54] for details). This is because the trajectory arises from a third order differential while fixing the solution ψ of the second order stationary Schrödinger equation involves only two "boundary" conditions. As was shown in [50] this automatically generates a Heisenberg inequality $\Delta x\Delta p \geq c\hbar$; i.e. the uncertainty is built in when using the wave function ψ and amazingly can be expressed by the operator theoretical framework of quantum mechanics. Thus a one parameter family of paths can be associated with the use of $\psi\psi^*$ and this generates the cloud or ensemble automatically associated with the use of ψ. In fact (cf. Remark 3.2) one might conjecture that upon using a wave function discription of quantum particle motion, one opens the door to a cloud of particles, all of whose motions are incompletely governed by the SE, since one determining condition for particle motion is ignored. Thus automatically the quantum potential will give rise to a force acting on any such particular trajectory and the "ensemble" idea naturally applies to a cloud of identical particles. ■

REMARK 2.2. Now first ignore gravity and look at the geometrical properties of the conformal factor given via

$$g_{\mu\nu} = e^{4\Sigma}\eta_{\mu\nu};\ e^{4\Sigma} = \frac{\mathfrak{M}^2}{m^2} = exp\left(\alpha\frac{\Box_\eta\sqrt{\rho}}{\sqrt{\rho}}\right) = exp\left(\alpha\frac{\Box_\eta\sqrt{|\mathfrak{T}|}}{\sqrt{|\mathfrak{T}|}}\right) \tag{2.23}$$

where $\mathfrak{T}$ is the trace of the EM tensor and is substituted for ρ (true for dust). The Einstein tensor for this metric is

$$\mathfrak{G}_{\mu\nu} = 4g_{\mu\nu}\Box_\eta exp(-\Sigma) + 2exp(-2\Sigma)\partial_\mu\partial_\nu exp(2\Sigma);\ \Sigma = \frac{\alpha}{4}\frac{\Box_\eta\sqrt{\rho}}{\sqrt{\rho}} \tag{2.24}$$

Hence as an Ansatz one can suppose that in the presence of gravitational effects the field equation would have a form

$$\mathcal{R}_{\mu\nu} - \frac{1}{2}\mathcal{R}g_{\mu\nu} = \kappa\mathfrak{T}_{\mu\nu} + 4g_{\mu\nu}e^{\Sigma}\Box e^{-\Sigma} + 2e^{-2\Sigma}\nabla_\mu\nabla_\nu e^{2\Sigma} \tag{2.25}$$

This is written in a manner such that in the limit $\mathfrak{T}_{\mu\nu} \to 0$ one will obtain (2.23). Taking the trace of the last equation one gets $-\mathcal{R} = \kappa\mathfrak{T} - 12\Box\Sigma + 24(\nabla\Sigma)^2$ which has the iterative solution $\kappa\mathfrak{T} = -\mathcal{R} + 12\alpha\Box[(\Box\sqrt{\mathcal{R}})/\sqrt{\mathcal{R}}]$ leading to

$$\Sigma \propto \alpha[(\Box\sqrt{|\mathfrak{T}|}/\sqrt{|\mathfrak{T}|})] \simeq \alpha[(\Box\sqrt{|\mathcal{R}|})/\sqrt{|\mathcal{R}|})] \tag{2.26}$$

to first order in α. One goes now to the field equations for this toy model. First from the above one sees that $\mathfrak{T}$ can be replaced by $\mathcal{R}$ in the expression for the quantum potential or for the conformal factor of the metric. This is important since the explicit reference to ensemble density is removed and the theory works for a single particle or an ensemble. So from (1.32), (1.24) for a toy quantum gravity theory one assumes the following field equations

$$\mathfrak{G}_{\mu\nu} - \kappa\mathfrak{T}_{\mu\nu} - \mathfrak{Z}_{\mu\nu\alpha\beta}exp\left(\frac{\alpha}{2}\Phi\right)\nabla^\alpha\nabla^\beta exp\left(-\frac{\alpha}{2}\Phi\right) = 0 \tag{2.27}$$

where $\mathfrak{Z}_{\mu\nu\alpha\beta} = 2[g_{\mu\nu}g_{\alpha\beta} - g_{\mu\alpha}g_{\nu\beta}]$ and $\Phi = (\Box\sqrt{|\mathcal{R}|}/\sqrt{|\mathcal{R}|})$. The number 2 and the minus sign of the second term are chosen so that the energy equation derived later will be correct. Note that the trace of (2.27) is

$$\mathcal{R} + \kappa\mathfrak{T} + 6exp(\alpha\Phi/2)\Box exp(-\alpha\Phi/2) = 0 \tag{2.28}$$

and this represents the connection of the Ricci scalar curvature of space time and the trace of the matter EM tensor. If a perturbative solution is admitted one can expand in powers of α to find $\mathcal{R}^{(0)} = -\kappa\mathfrak{T}$ and $\mathcal{R}^{(1)} = -\kappa\mathfrak{T} - 6exp(\alpha\Phi^0/2)\Box exp(-\alpha\Phi^0/2)$ where $\Phi^{(0)} = \Box\sqrt{|\mathfrak{T}|}/\sqrt{|\mathfrak{T}|}$. The energy relation can be obtained by taking the four divergence of the field equations and since the divergence of the Einstein tensor is zero one obtains

$$\kappa\nabla^\nu\mathfrak{T}_{\mu\nu} = \alpha\mathcal{R}_{\mu\nu}\nabla^\nu\Phi - \frac{\alpha^2}{4}\nabla_\mu(\nabla\Phi)^2 + \frac{\alpha^2}{2}\nabla_\mu\Phi\Box\Phi \tag{2.29}$$

For a dust with $\mathfrak{T}_{\mu\nu} = \rho u_\mu u_\nu$ and u_μ the velocity field, the conservation of mass law is $\nabla^\nu(\rho\mathfrak{M}u_\nu) = 0$ so one gets to first order in α $\nabla_\mu\mathfrak{M}/\mathfrak{M} = -(\alpha/2)\nabla_\mu\Phi$ or $\mathfrak{M}^2 = m^2 exp(-\alpha\Phi)$ where m is an integration constant. This is the correct relation of mass and quantum potential. ■

In [182] there is then some discussion about making the conformal factor dynamical via a general scalar tensor action (cf. also [176]) and subsequently one makes both the conformal factor and the quantum potential into dynamical fields and creates a scalar tensor theory with two scalar fields. Thus first start with a general action

$$\mathfrak{A} = \int d^4x\sqrt{-g}\left[\phi\mathcal{R} - \omega\frac{\nabla_\mu\phi\nabla^\mu\phi}{\phi} - \frac{\nabla_\mu Q\nabla^\mu Q}{\phi} + 2\Lambda\phi + \mathfrak{L}_m\right] \tag{2.30}$$

The cosmological constant generally has an interaction term with the scalar field and here one uses an ad hoc matter Lagrangian

$$\mathfrak{L}_m = \frac{\rho}{m}\phi^a\nabla_\mu S\nabla^\mu S - m\rho\phi^b - \Lambda(1+Q)^c + \alpha\rho(e^{\ell Q} - 1) \tag{2.31}$$

(only the first two terms $1 + Q$ from $exp(Q)$ are used for simplicity in the third term). Here a, b, c are constants to be fixed later and the last term is chosen (heuristically) in such a manner as to have an interaction between the quantum potential field and the ensemble density (via the equations of motion); further the interaction is chosen so that it vanishes in the classical limit but this is ad hoc. Variation of the above action yields

1. The scalar fields equation of motion

$$\mathcal{R} + \frac{2\omega}{\phi}\Box\phi - \frac{\omega}{\phi^2}\nabla^\mu\phi\nabla_\mu\phi + 2\Lambda + \tag{2.32}$$

$$+\frac{1}{\phi^2}\nabla^\mu Q\nabla_\mu Q + \frac{a}{m}\rho\phi^{a-1}\nabla^\mu S\nabla_\mu S - mb\rho\phi^{b-1} = 0$$

2. The quantum potential equations of motion

$$(\Box Q/\phi) - (\nabla_\mu Q\nabla^\mu\phi/\phi^2) - \Lambda c(1+Q)^{c-1} + \alpha\ell\rho exp(\ell Q) = 0 \tag{2.33}$$

3. The generalized Einstein equations

$$\mathfrak{G}^{\mu\nu} - \Lambda g^{\mu\nu} = -\frac{1}{\phi}\mathfrak{T}^{\mu\nu} - \frac{1}{\phi}[\nabla^\mu\nabla^\nu - g^{\mu\nu}\Box]\phi + \frac{\omega}{\phi^2}\nabla^\mu\phi\nabla^\nu\phi - \tag{2.34}$$

$$-\frac{\omega}{2\phi^2}g^{\mu\nu}\nabla^\alpha\phi\nabla_\alpha\phi + \frac{1}{\phi^2}\nabla^\mu Q\nabla^\nu Q - \frac{1}{2\phi^2}g^{\mu\nu}\nabla^\alpha Q\nabla_\alpha Q$$

4. The continuity equation $\nabla_\mu(\rho\phi^a\nabla^\mu S) = 0$

5. The quantum Hamilton Jacobi equation

$$\nabla^\mu S\nabla_\mu S = m^2\phi^{b-a} - \alpha m\phi^{-a}(e^{\ell Q}-1) \tag{2.35}$$

In (2.32) the scalar curvature and the term $\nabla^\mu S\nabla_\mu S$ can be eliminated using (2.34) and (2.35); further on using the matter Lagrangian and the definition of the EM tensor one has

$$(2\omega-3)\Box\phi = (a+1)\rho\alpha(e^{\ell Q}-1) - 2\Lambda(1+Q)^c + 2\Lambda\phi - \frac{2}{\phi}\nabla_\mu Q\nabla^\mu Q \tag{2.36}$$

(where $b = a+1$). Solving (2.33) and (2.36) with a perturbation expansion in α one finds

$$Q = Q_0 + \alpha Q_1 + \cdots\,;\ \phi = 1 + \alpha Q_1 + \cdots\,;\ \sqrt{\rho} = \sqrt{\rho_0} + \alpha\sqrt{\rho_1} + \cdots \tag{2.37}$$

where the conformal factor is chosen to be unity at zeroth order so that as $\alpha \to 0$ (2.35) goes to the classical HJ equation. Further since by (2.35) the quantum mass is $m^2\phi + \cdots$ the first order term in ϕ is chosen to be Q_1 (cf. (2.13)). Also we will see that $Q_1 \sim \Box\sqrt{\rho}/\sqrt{\rho}$ plus corrections which is in accord with Q as a quantum potential field. In any case after some computation one obtains $a = 2\omega k,\ b = a+1$, and $\ell = (1/4)(2\omega k+1) = (1/4)(a+1) = b/4$ with $Q_0 = [1/c(2c-3)]\{[-(2\omega k+1)/2\Lambda]k\sqrt{\rho_0} - (2c^2-c+1)\}$ while ρ_0 can be determined (cf. [182] for details). Thus heuristically the quantum potential can be regarded as a dynamical field and perturbatively one gets the correct dependence of quantum potential upon density, modulo some corrective terms.

REMARK 2.3. The gravitational effects determine the causal structure of spacetime as long as quantum effects give its conformal structure. This does not mean that quantum effects have nothing to do with the causal structure; they can act on the causal structure through back reaction terms appearing in the metric field equations. The conformal factor of the metric is a function of the quantum potential and the mass of a relativistic particle is a field produced by quantum corrections to the classical mass. One has shown that the presence of the quantum potential is equivalent to a conformal mapping of the metric. Thus in different conformally related frames one feels different quantum masses and different curvatures. In particular there are two frames with one containing the quantum mass field and the classical metric while the other contains the classical mass and the quantum metric. In general frames both the spacetime metric and the mass field have quantum properties so one can state that different conformal frames are identical pictures of the gravitational and quantum phenomena. We feel different quantum forces in different conformal frames. The question then arises of whether the geometrization of quantum effects implies conformal invariance just as gravitational effects imply general coordinate invariance. One sees here that Weyl geometry provides additional degrees of freedom which can be identified with quantum effects and seems to create a unified geometric framework for understanding both gravitational and quantum forces. Some features here are: (i) Quantum effects appear independent of any preferred length scale. (ii) The quantum mass of a particle is a field. (iii) The gravitational constant is also a field depending on the matter distribution via the quantum potential (cf. [176, 183]). (iv) A local variation of matter field distribution changes the quantum potential acting on the geometry and alters it globally; the nonlocal character is forced by the quantum potential (cf. [177]). ∎

2.1 Dirac-Weyl Action

Next (still following [182]) one goes to Weyl geometry based on the Weyl-Dirac action

$$\mathfrak{A} = \int d^4x\sqrt{-g}(F_{\mu\nu}F^{\mu\nu} - \beta^2\ {}^W\mathcal{R} + (\sigma+6)\beta_{;\mu}\beta^{;\mu} + \mathfrak{L}_{matter} \tag{2.38}$$

Here $F_{\mu\nu}$ is the curl of the Weyl 4-vector ϕ_μ, σ is an arbitrary constant and β is a scalar field of weight -1. The symbol ";" represents a covariant derivative under general coordinate and conformal transformations (Weyl covariant derivative) defined as $X_{;\mu} = {}^W\nabla_\mu X - \mathcal{N}\phi_\mu X$ where $\mathcal{N}$ is the Weyl weight of X. The equations of motion are then

$$\mathfrak{G}^{\mu\nu} = -\frac{8\pi}{\beta^2}(\mathfrak{T}^{\mu\nu} + M^{\mu\nu}) + \frac{2}{\beta}(g^{\mu\nu\,W}\nabla^{\alpha\,W}\nabla_\alpha\beta - {}^W\nabla^{\mu\,W}\nabla^\nu\beta) + \tag{2.39}$$

$$+\frac{1}{\beta^2}(4\nabla^\mu\beta\nabla^\nu\beta - g^{\mu\nu}\nabla^\alpha\beta\nabla_\alpha\beta) + \frac{\sigma}{\beta^2}(\beta^{;\mu}\beta^{;\nu} - \frac{1}{2}g^{\mu\nu}\beta^{;\alpha}\beta_{;\alpha});$$

$${}^W\nabla_\mu F^{\mu\nu} = \frac{1}{2}\sigma(\beta^2\phi^\mu + \beta\nabla^\mu\beta) + 4\pi J^\mu;$$

$$\mathcal{R} = -(\sigma+6)\frac{{}^W\Box\beta}{\beta} + \sigma\phi_\alpha\phi^\alpha - \sigma^W\nabla^\alpha\phi_\alpha + \frac{\psi}{2\beta}$$

where

$$M^{\mu\nu} = (1/4\pi)[(1/4)g^{\mu\nu}F^{\alpha\beta}F_{\alpha\beta} - F^\mu_\alpha F^{\nu\alpha} \tag{2.40}$$

and

$$8\pi\mathfrak{T}^{\mu\nu} = \frac{1}{\sqrt{-g}}\frac{\delta\sqrt{-g}\mathfrak{L}_{matter}}{\delta g_{\mu\nu}};\ 16\pi J^\mu = \frac{\delta\mathfrak{L}_{matter}}{\delta\phi_\mu};\ \psi = \frac{\delta\mathfrak{L}_{matter}}{\delta\beta} \tag{2.41}$$

For the equations of motion of matter and the trace of the EM tensor one uses invariance of the action under coordinate and gauge transformations, leading to

$${}^W\nabla_\nu\mathfrak{T}^{\mu\nu} - \mathfrak{T}\frac{\nabla^\mu\beta}{\beta} = J_\alpha\phi^{\alpha\mu} - \left(\phi^\mu + \frac{\nabla^\mu\beta}{\beta}\right){}^W\nabla_\alpha J^\alpha; \tag{2.42}$$

$$16\pi\mathfrak{T} - 16\pi^W\nabla_\mu J^\mu - \beta\psi = 0$$

The first relation is a geometrical identity (Bianchi identity) and the second shows the mutual dependence of the field equations. Note that in the Weyl-Dirac theory the Weyl vector does not couple to spinors so ϕ_μ cannot be interpreted as the EM potential; the Weyl vector is used as part of the spacetime geometry and the auxillary field (gauge field) β represents the quantum mass field. The gravity fields $g_{\mu\nu}$ and ϕ_μ and the quantum mass field determine the spacetime geometry. Now one constructs a Bohmian quantum gravity which is conformally invariant in the framework of Weyl geometry. If the model has mass this must be a field (since mass has non-zero Weyl weight). The Weyl-Dirac action is a general Weyl invariant action as above and for simplicity now assume the matter Lagrangian does not depend on the Weyl vector so that $J_\mu = 0$. The equations of motion are then

$$\mathfrak{G}^{\mu\nu} = -\frac{8\pi}{\beta^2}(\mathfrak{T}^{\mu\nu} + M^{\mu\nu}) + \frac{2}{\beta}(g^{\mu\nu W}\nabla^{\alpha W}\nabla_\alpha\beta - {}^W\nabla^{\mu W}\nabla^\nu\beta) + \tag{2.43}$$

$$+\frac{1}{\beta^2}(4\nabla^\mu\beta\nabla^\nu\beta - g^{\mu\nu}\nabla^\alpha\beta\nabla_\alpha\beta) + \frac{\sigma}{\beta^2}\left(\beta^{;\mu}\beta^{;\nu} - \frac{1}{2}g^{\mu\nu}\beta^{;\alpha}\beta_{;\alpha}\right);$$

$${}^W\nabla_\nu F^{\mu\nu} = \frac{1}{2}\sigma(\beta^2\phi^\mu + \beta\nabla^\mu\beta);\ \mathcal{R} = -(\sigma+6)\frac{{}^W\Box\beta}{\beta} + \sigma\phi_\alpha\phi^\alpha - \sigma{}^W\nabla^\alpha\phi_\alpha + \frac{\psi}{2\beta}$$

The symmetry conditions are

$${}^W\nabla_\nu\mathfrak{T}^{\mu\nu} - \mathfrak{T}(\nabla^\mu\beta/\beta) = 0;\ 16\pi\mathfrak{T} - \beta\psi = 0 \tag{2.44}$$

(recall $\mathfrak{T} = \mathfrak{T}^{\mu\nu}_{\mu\nu}$). One notes that from (2.43) results ${}^W\nabla_\mu(\beta^2\phi^\mu + \beta\nabla^\mu\beta) = 0$ so ϕ_μ is not independent of β. To see how this is related to the Bohmian quantum theory one introduces a quantum mass field and shows it is proportional to the Dirac field. Thus using (2.43) and (2.44) one has

$$\Box\beta + \frac{1}{6}\beta\mathcal{R} = \frac{4\pi}{3}\frac{\mathfrak{T}}{\beta} + \sigma\beta\phi_\alpha\phi^\alpha + 2(\sigma-6)\phi^\gamma\nabla_\gamma\beta + \frac{\sigma}{\beta}\nabla^\mu\beta\nabla_\mu\beta \tag{2.45}$$

This can be solved iteratively via

$$\beta^2 = (8\pi\mathfrak{T}/\mathcal{R}) - \{1/[(\mathcal{R}/6) - \sigma\phi_\alpha\phi^\alpha]\}\beta\Box\beta + \cdots \tag{2.46}$$

Now assuming $\mathfrak{T}^{\mu\nu} = \rho u^\mu u^\nu$ (dust with $\mathfrak{T} = \rho$) we multiply (2.44) by u_μ and sum to get

$${}^W\nabla_\nu(\rho u^\nu) - \rho(u_\mu\nabla^\mu\beta/\beta) = 0 \tag{2.47}$$

Then put (2.44) into (2.47) which yields

$$u^{\nu W}\nabla_\nu u^\mu = (1/\beta)(g^{\mu\nu} - u^\mu u^\nu)\nabla_\nu\beta \tag{2.48}$$

To see this write (assuming $g^{\mu\nu}\nabla_\nu\beta = \nabla^\mu\beta$)

$${}^W\nabla_\nu(\rho u^\mu u^\nu) = u^{\mu W}\nabla_\nu\rho u^\mu + \rho u^{\nu W}\nabla_\nu u^\mu \Rightarrow \tag{2.49}$$

$$\Rightarrow u^\mu\left(\frac{u_\mu\nabla^\mu\beta}{\beta}\right) + u^{\nu W}\nabla_\nu u^\mu - \frac{\nabla^\mu\beta}{\beta} = 0 \Rightarrow u^{\nu W}\nabla_\nu u^\mu = (1 - u^\mu u_\mu)\frac{\nabla^\mu\beta}{\beta} =$$

$$(g^{\mu\nu} - u^\mu u_\mu g^{\mu\nu})\frac{\nabla_\nu\beta}{\beta} = (g^{\mu\nu} - u^\mu u^\nu)\frac{\nabla_\nu\beta}{\beta}$$

which is (2.47). Then from (2.46)

$$\beta^{2(1)} = \frac{8\pi\mathfrak{T}}{\mathcal{R}};\ \beta^{2(2)} = \frac{8\pi\mathfrak{T}}{\mathcal{R}}\left(1 - \frac{1}{(\mathcal{R}/6) - \sigma\phi_\alpha\phi^\alpha}\frac{\Box\sqrt{\mathfrak{T}}}{\sqrt{\mathfrak{T}}}\right);\cdots \tag{2.50}$$

Comparing with (2.15) and (2.3) shows that we have the correct equations for the Bohmian theory provided one identifies

$$\beta \sim \mathfrak{M};\ \frac{8\pi\mathfrak{T}}{\mathcal{R}} \sim m^2;\ \frac{1}{\sigma\phi_\alpha\phi^\alpha - (\mathcal{R}/6)} \sim \alpha \tag{2.51}$$

Thus β is the Bohmian quantum mass field and the coupling constant α (which depends on $\hbar$) is also a field, related to geometrical properties of spacetime. One notes that the quantum effects and the length scale of the spacetime are related. To see this suppose one is in a gauge in which the Dirac field is constant; apply a gauge transformation to change this to a general spacetime dependent function, i.e.

$$\beta = \beta_0 \to \beta(x) = \beta_0 exp(-\Xi(x));\ \phi_\mu \to \phi_\mu + \partial_\mu\Xi \tag{2.52}$$

Thus the gauge in which the quantum mass is constant (and the quantum force is zero) and the gauge in which the quantum mass is spacetime dependent are related to one another via a scale change. In particular ϕ_μ in the two gauges differ by $-\nabla_\mu(\beta/\beta_0)$ and since ϕ_μ is a part of Weyl geometry and the Dirac field represents the quantum mass one concludes that the quantum effects are geometrized (cf. also (2.43) which shows that ϕ_μ is not independent of β so the Weyl vector is determined by the quantum mass and thus the geometrical aspect of the manifold is related to quantum effects).

3 More on Klein Gordon Equations

We give several approaches here, from various points of view.

3.1 Bertoldi-Faraggi-Matone Theory

The equivalence principle (EP) of Faraggi-Matone (cf. [27, 44, 46, 51, 82]) is based on the idea that all physical systems can be connected by a coordinate transformation to the free situation with vanishing energy (i.e. all potentials are equivalent under coordinate transformations). This automatically leads to the quantum stationary Hamilton-Jacobi equation (QSHJE) which is a third order nonlinear differential equation providing a trajectory representation of quantum mechanics (QM). The theory transcends in several respects the Bohm theory and in particular utilizes a Floydian time (cf. [83, 84]) leading to $\dot{q} = p/m_Q \neq p/m$ where $m_Q = m(1 - \partial_E Q)$ is the "quantum mass" and Q the "quantum potential". Thus the EP is reminscient of the Einstein equivalence of relativity theory. This latter served as a midwife to the birth of relativity but was somewhat inaccurate in its original form. It is better put as saying that all laws of physics should be invariant under general coordinate transformations (cf. [146]). This demands that not only the form but also the content of the equations be unchanged. More precisely the equations should be covariant and all absolute constants in the equations are to be left unchanged (e.g. c, $\hbar$, e, m and $\eta_{\mu\nu}$ = Minkowski tensor). Now for the EP, the classical picture with $S^{cl}(q, Q^0, t)$ the Hamilton principal function ($p = \partial S^{cl}/\partial q$) and P^0, Q^0 playing the role of initial conditions involves the classical HJ equation (CHJE) $H(q, p = (\partial S^{cl}/\partial q), t) + (\partial S^{cl}/\partial t) = 0$. For time independent V one writes $S^{cl} = S_0^{cl}(q, Q^0) - Et$ and arrives at the classical stationary HJ equation (CSHJE) $(1/2m)(\partial S_0^{cl}/\partial q)^2 + \mathfrak{W} = 0$ where $\mathfrak{W} = V(q) - E$. In the Bohm theory one looked at Schrödinger equations $i\hbar\psi_t = -(\hbar^2/2m)\psi'' + V\psi$ with $\psi = \psi(q)exp(-iEt/\hbar)$ and $\psi(q) = R(qexp(i\hat{W}/\hbar)$ leading to

$$\left(\frac{1}{2m}\right)(\hat{W}')^2 + V - E - \frac{\hbar^2 R''}{2mR} = 0; \ (R^2\hat{W}')' = 0 \tag{3.1}$$

where $\hat{Q} = -\hbar^2 R''/2mR$ was called the quantum potential; this can be written in the Schwartzian form $\hat{Q} = (\hbar^2/4m)\{\hat{W}; q\}$ (via $R^2\hat{W}' = c$). Here $\{f; q\} = (f'''/f') - (3/2)(f''/f')^2$. Writing $\mathfrak{W} = V(q) - E$ as in above we have the quantum stationary HJ equation (QSHJE)

$$(1/2m)(\partial\hat{W}'/\partial q)^2 + \mathfrak{W}(q) + \hat{Q}(q) = 0 \equiv \mathfrak{W} = -(\hbar^2/4m)\{exp(2iS_0/\hbar); q\} \tag{3.2}$$

This was worked out in the Bohm school (without the Schwarzian connections) but $\psi = Rexp(i\hat{W}/\hbar)$ is not appropriate for all situations and care must be taken ($\hat{W} = constant$ must be excluded for example - cf. [82, 83, 84]). The technique of Faraggi-Matone (FM) is completely general and with only the EP as guide one exploits the relations between Schwarzians, Legendre duality, and the geometry of a second order differential operator $D_x^2 + V(x)$ (Möbius transformations play an important role here) to arrive at the QSHJE in the form

$$\frac{1}{2m}\left(\frac{\partial S_0^v(q^v)}{\partial q^v}\right)^2 + \mathfrak{W}(q^v) + \mathfrak{Q}^v(q^v) = 0 \tag{3.3}$$

where $v : q \to q^v$ represents an arbitrary locally invertible coordinate transformation. Note in this direction for example that the Schwarzian derivative of the the ratio of two linearly independent elements in $ker(D_x^2 + V(x))$ is twice $V(x)$. In particular given an arbitrary system with coordinate q and reduced action $S_0(q)$ the system with coordinate q^0 corresponding to $V - E = 0$ involves $\mathfrak{W}(q) = (q^0; q)$ where (q^0, q) is a cocycle term which has the form $(q^a; q^b) = -(\hbar^2/4m)\{q^a; q^b\}$. In fact it can be said that the essence of the EP is the cocycle conditon

$$(q^a; q^c) = (\partial_{q^c} q^b)^2[(q^a; q^b) - (q^c; q^b)] \tag{3.4}$$

In addition FM developed a theory of (x, ψ) duality (cf. [81])) which related the space coordinate and the wave function via a prepotential (free energy) in the form $\mathfrak{F} = (1/2)\psi\bar{\psi} + iX/\epsilon$ for example. A number of interesting philosophical points arise (e.g. the emergence of space from the wave function) and we connected this to various features of dispersionless KdV in [44, 51] in a sort of extended WKB spirit. One should note here that although a form $\psi = Rexp(i\hat{W}/\hbar)$ is not generally appropriate it is correct when one is dealing with two independent solutions of the Schrödinger equation ψ and $\bar{\psi}$ which are not proportional. In this context we utilized some interplay between various geometric properties of KdV which involve the Lax operator $L^2 = D_x^2 + V(x)$ and of course this is all related to Schwartzians, Virasoro algebras, and vector fields on S^1 (see e.g. [44, 45, 51, 52, 53]). Thus the simple presence of the Schrödinger equation (SE) in QM automatically incorporates a host of geometrical properties of $D_x = d/dx$ and the circle S^1. In fact since the FM theory exhibits the fundamental nature of the SE via its geometrical properties connected to the QSHJE one could speculate about trivializing QM (for 1-D) to a study of S^1 and ∂_x.

We import here some comments based on [27] concerning the Klein-Gordon (KG) equation and the equivalence principle (EP) (details are in [27] and cf. also [72]). One starts with the relativistic classical Hamilton-Jacobi equation (RCHJE) with a potential $V(q,t)$ given as

$$\frac{1}{2m}\sum_1^D(\partial_k S^{cl}(q,t))^2 + \mathfrak{W}_{rel}(q,t) = 0; \tag{3.5}$$

$$\mathfrak{W}_{rel}(q,t) = \frac{1}{2mc^2}[m^2c^4 - (V(q,t) + \partial_t S^{cl}(q,t))^2]$$

In the time-independent case one has $S^{cl}(q,t) = S_0^{cl}(q) - Et$ and (3.3) becomes

$$\frac{1}{2m}\sum_1^D(\partial_k S_0^{cl})^2 + \mathfrak{W}_{rel} = 0;\ \mathfrak{W}_{rel}(q) = \frac{1}{2mc^2}[m^2c^4 - (V(q) - E)^2] \tag{3.6}$$

In the latter case one can go through the same steps as in the nonrelativistic case and the relativistic quantum HJ equation (RQHJE) becomes

$$(1/2m)(\nabla S_0)^2 + \mathfrak{W}_{rel} - (\hbar^2/2m)(\Delta R/R) = 0;\ \nabla\cdot(R^2\nabla S_0) = 0 \tag{3.7}$$

these equations imply the stationary KG equation

$$-\hbar^2c^2\Delta\psi + (m^2c^4 - V^2 + 2EV - E^2)\psi = 0 \tag{3.8}$$

where $\psi = Rexp(iS_0/\hbar)$. Now in the time dependent case the (D+1)-dimensional RCHJE is ($\eta^{\mu\nu} = diag(-1,1,\cdots,1)$)

$$(1/2m)\eta^{\mu\nu}\partial_\mu S^{cl}\partial_\nu S^{cl} + \mathfrak{W}'_{rel} = 0; \tag{3.9}$$

$$\mathfrak{W}'_{rel} = (1/2mc^2)[m^2c^4 - V^2(q) - 2cV(q)\partial_0 S^{cl}(q)]$$

with $q = (ct, q_1, \cdots, q_D)$. Thus (3.9) has the same structure as (3.6) with Euclidean metric replaced by the Minkowskian one. We know how to implement the EP by adding Q via

$(1/2m)(\partial S)^2 + \mathfrak{W}_{rel} + Q = 0$ (cf. [82] and remarks above). Note now that $\mathfrak{W}'_{rel}$ depends on S^{cl} requires an identification

$$\mathfrak{W}_{rel} = (1/2mc^2)[m^2c^4 - V^2(q) - 2cV(q)\partial_0 S(q)] \tag{3.10}$$

(S replacing S^{cl}) and implementation of the EP requires that for an arbitrary $\mathfrak{W}^a$ state ($q \sim q^a$) one must have

$$\mathfrak{W}^b_{rel}(q^b) = (p^b|p^a)\mathfrak{W}^a_{rel}(q^a) + (q^q; q^b);\ Q^b(q^b) = (p^b|p^a)Q(q^a) - (q^a; q^b) \tag{3.11}$$

where

$$(p^b|p) = [\eta^{\mu\nu}p^b_\mu p^b_\nu/\eta^{\mu\nu}p_\mu p_\nu] = p^T J\eta J^T p/p^T\eta p;\ J^\mu_\nu = \partial q^\mu/\partial q^{b^\nu} \tag{3.12}$$

(J is a Jacobian and these formulas are the natural multidimensional generalization - see [27] for details). Furthermore there is a cocycle condition $(q^a; q^c) = (p^c|p^b)[(q^a; q^b) - (q^c; q^b)]$.

Next one shows that $\mathfrak{W}_{rel} = (\hbar^2/2m)[\Box(Rexp(iS/\hbar))/Rexp(iS/\hbar)]$ and hence the corresponding quantum potential is $Q_{rel} = -(\hbar^2/2m)[\Box R/R]$. Then the RQHJE becomes $(1/2m)(\partial S)^2 + \mathfrak{W}_{rel} + Q = 0$ with $\partial \cdot (R^2\partial S) = 0$ (here $\Box R = \partial_\mu\partial^\mu R$) and this reduces to the standard SE in the classical limit $c \to \infty$ (note $\partial \sim (\partial_0, \partial_1, \cdots, \partial_D)$ with $q_0 = ct$, etc. - cf. (3.9)). To see how the EP is simply implemented one considers the so called minimal coupling prescription for an interaction with an electromagnetic four vector A_μ. Thus set $P^{cl}_\mu = p^{cl}_\mu + eA_\mu$ where p^{cl}_μ is a particle momentum and $P^{cl}_\mu = \partial_\mu S^{cl}$ is the generalized momentum. Then the RCHJE reads as $(1/2m)(\partial S^{cl} - eA)^2 + (1/2)mc^2 = 0$ where $A_0 = -V/ec$. Then $\mathfrak{W} = (1/2)mc^2$ and the critical case $\mathfrak{W} = 0$ corresponds to the limit situation where $m = 0$. One adds the standard Q correction for implementation of the EP to get $(1/2m)(\partial S - eA)^2 + (1/2)mc^2 + Q = 0$ and there are transformation properties (here $(\partial S - eA)^2 \sim \sum(\partial_\mu S - eA_\mu)^2$)

$$\mathfrak{W}(q^b) = (p^b|p^a)\mathfrak{W}^a(q^a) + (q^a; q^b);\ Q^b(q^b) = (p^q|p^a)Q^a(q^a) - (q^a; q^b) \tag{3.13}$$

$$(p^b|p) = \frac{(p^b - eA^b)^2}{(p - eA)^2} = \frac{(p - eA)^T J\eta J^T(p - eA)}{(p - eA)^T\eta(p - eA)}$$

Here J is a Jacobian $J^\mu_\nu = \partial q^\mu/\partial q^{b^\nu}$ and this all implies the cocycle condition again. One finds now that (recall $\partial \cdot (R^2(\partial S - eA)) = 0$ - continuity equation)

$$(\partial S - eA)^2 = \hbar^2\left(\frac{\Box R}{R} - \frac{D^2(Re^{iS/\hbar})}{Re^{iS/\hbar}}\right);\ D_\mu = \partial_\mu - \frac{i}{\hbar}eA_\mu \tag{3.14}$$

and it follows that

$$\mathfrak{W} = \frac{\hbar^2}{2m}\frac{D^2(Re^{iS/\hbar})}{Re^{iS/\hbar}};\ Q = -\frac{\hbar^2}{2m}\frac{\Box R}{R};\ D^2 = \Box - \frac{2ieA\partial}{\hbar} - \frac{e^2A^2}{\hbar^2} - \frac{ie\partial A}{\hbar} \tag{3.15}$$

$$(\partial S - eA)^2 + m^2c^2 - \hbar^2 \frac{\Box R}{R} = 0;\ \partial \cdot (R^2(\partial S - eA)) = 0 \tag{3.16}$$

Note also that (3.9) agrees with $(1/2m)(\partial S^{cl} - eA)^2 + (1/2)mc^2 = 0$ after setting $\mathfrak{W}_{rel} = mc^2/2$ and replacing $\partial_\mu S^{cl}$ by $\partial_\mu S^{cl} - eA_\mu$. One can check that (3.16) implies the KG equation $(i\hbar\partial + eA)^2\psi + m^2c^2\psi = 0$ with $\psi = Rexp(iS/\hbar)$.

REMARK 3.1. We extract now a remark about mass generation and the EP from [19]. Thus a special property of the EP is that it cannot be implemented in classical mechanics (CM) because of the fixed point corresponding to $\mathfrak{W} = 0$. One is forced to introduce a uniquely determined piece to the classical HJ equation (namely a quantum potential Q). In the case of the RCHJE the fixed point $\mathfrak{W}(q^0) = 0$ corresponds to $m = 0$ and the EP then implies that all the other masses can be generated by a coordinate transformation. Consequently one concludes that masses correspond to the inhomogeneous term in the transformation properties of the $\mathfrak{W}^0$ state, i.e. $(1/2)mc^2 = (q^0; q)$. Furthermore by (3.13) masses are expressed in terms of the quantum potential $(1/2)mc^2 = (p|p^0)Q^0(q^0) - Q(q)$. In particular in [82] the role of the quantum potential was seen as a sort of intrinsic self energy which is reminiscent of the relativistic self energy and this provides a more explicit evidence of such an interpretation. ■

REMARK 3.2. In a previous paper [47] (working with stationary states and ψ satisfying the Schrödinger equation (SE) $-(\hbar^2/2m)\psi'' + V\psi = E\psi$) we suggested that the notion of uncertainty in quantum mechanics (QM) could be phrased as incomplete information. The background theory here is taken to be the trajectory theory of Bertoldi-Faraggi-Matone-Floyd as above and the idea in [47] goes as follows. First recall that Floydian microstates satisfy a third order quantum stationary Hamilton-Jacobi equation (QSHJE)

$$\frac{1}{2m}(S_0')^2 + \mathfrak{W}(q) + Q(q) = 0;\ Q(q) = \frac{\hbar^2}{4m}\{S_0; q\}; \tag{3.17}$$

$$\mathfrak{W}(q) = -\frac{\hbar^2}{4m}\{exp(2iS_0/\hbar); q\} \sim V(q) - E$$

where $\{f;q\} = (f'''/f') - (3/2)(f''/f')^2$ is the Schwarzian and S_0 is the Hamilton principle function. Also one recalls that the EP of Faraggi-Matone can only be implemented when $S_0 \neq const$; thus consider $\psi = Rexp(iS_0/\hbar)$ with $Q = -\hbar^2 R''/2mR$ and $(R^2 S_0')' = 0$ where $S_0' = p$ and $m_Q\dot{q} = p$ with $m_Q = m(1 - \partial_E Q)$ and $t \sim \partial_E S_0$. Thus microstates require three initial or boundary conditions in general to determine S_0 whereas the SE involves only two such conditions. Hence in dealing with the SE in the standard QM Hilbert space formulation one is not using complete information about the "particles" described by microstate trajectories. The price of underdetermination is then uncertainty in q, p, t for example. In the present note we will make this more precise and add further discussion. Following [50] we now make this more precise and add further discussion. For the stationary SE $-(\hbar^2/2m)\psi'' + V\psi = E\psi$ it is shown in [82] that one has a general formula

$$e^{2iS_0(\delta)/\hbar} = e^{i\alpha}\frac{w + i\bar{\ell}}{w - i\ell} \tag{3.18}$$

($\delta \sim (\alpha, \ell)$) with three integration constants, α, ℓ_1, ℓ_2 where $\ell = \ell_1 + i\ell_2$ and $w \sim \psi^D/\psi \in \mathbf{R}$. Note ψ and ψ^D are linearly independent solutions of the SE and one can arrange that $\psi^D/\psi \in \mathbf{R}$ in describing any situation. Here p is determined by the two constants in ℓ and has a form

$$p = \frac{\pm\hbar\Omega\ell_1}{|\psi^D - i\ell\psi|^2} \tag{3.19}$$

(where $w \sim \psi^D/\psi$ above and $\Omega = \psi'\psi^D - \psi(\psi^D)'$). Now let p be determined exactly with $p = p(q, E)$ via the Schrödinger equation and S_0'. Then $\dot{q} = (\partial_E p)^{-1}$ is also exact so $\Delta q = (\partial_E p)^{-1}(\tau)\Delta t$ for some τ with $0 \leq \tau \leq t$ is exact (up to knowledge of τ). Thus given the wave function ψ satisfying the stationary SE with two boundary conditions at $q = 0$ say to fix uniqueness, one can create a probability density $|\psi|^2(q, E)$ and the function S_0'. This determines p uniquely and hence $\dot{q}$. The additional constant needed for S_0 appears in (3.18) and we can write $S_0 = S_0(\alpha, q, E)$ since from (3.18) one has

$$S_0 - (\hbar/2)\alpha = -(i\hbar/2)log(\beta) \tag{3.20}$$

and $\beta = (w+i\bar{\ell})/(w-i\ell)$ with $w = \psi^D/\psi$ is to be considered as known via a determination of suitable ψ, ψ^D. Hence $\partial_\alpha S_0 = -\hbar/2$ and consequently $\Delta S_0 \sim \partial_\alpha S_0 \delta\alpha = -(\hbar/2)\Delta\alpha$ measures the indeterminacy in S_0.

Let us expand upon this as follows. Note first that the determination of constants necessary to fix S_0 from the QSHJE is not usually the same as that involved in fixing ℓ, $\bar{\ell}$ in (3.18). In paricular differentiating in q one gets

$$S_0' = -\frac{i\hbar\beta'}{\beta};\ \beta' = -\frac{2i\Re\ell w'}{(w - i\ell)^2} \tag{3.21}$$

Since $w' = -\Omega/\psi^2$ where $\Omega = \psi'\psi^D - \psi(\psi^D)'$ we get $\beta' = -2i\ell_1\Omega/(\psi^D - i\ell\psi)^2$ and consequently

$$S_0' = -\frac{\hbar\ell_1\Omega}{|\psi^D - i\ell\psi|^2} \tag{3.22}$$

which agrees with p in (3.19) ($\pm\hbar$ simply indicates direction). We see that e.g. $S_0(x_0) = i\hbar\ell_1\Omega/|\psi^D(x_0) - i\ell\psi(x_0)|^2 = f(\ell_1, \ell_2, x_0)$ and $S_0'' = g(\ell_1, \ell_2, x_0)$ determine the relation between $(p(x_0), p'(x_0))$ and $(\ell_1,\ \ell_2)$ but they are generally different numbers. In any case, taking α to be the arbitrary unknown constant in the determination of S_0, we have $S_0 = S_0(q, E, \alpha)$ with $q = q(S_0, E, \alpha)$ and $t = t(S_0, E, \alpha) = \partial_E S_0$ (emergence of time from the wave function). One can then write e.g.

$$\Delta q = (\partial q/\partial S_0)(\hat{S}_0, E, \alpha)\Delta S_0 = (1/p)(\hat{q}, E)\Delta S_0 = -(1/p)(\hat{q}, E)(\hbar/2)\Delta\alpha \tag{3.23}$$

(for intermediate values ($\hat{S}_0$, $\hat{q}$)) leading to

THEOREM 3.1. With p determined uniquely by two "initial" conditions so that Δp is determined and q given via (3.18) we have from (3.23) the inequality $\Delta p \Delta q = O(\hbar)$ which resembles the Heisenberg uncertainty relation.

COROLLARY 3.1. Similarly $\Delta t = (\partial t/\partial S_0)(\hat{S}_0, E, \alpha)\Delta S_0$ for some intermediate value $\hat{S}_0$ and hence as before $\Delta E\Delta t = O(\hbar)$ (ΔE being precise).

Note that there is no physical argument here; one is simply looking at the number of conditions necessary to fix solutions of a differential equation. In fact (based on some corresondence with E. Floyd) it seems somewhat difficult to produce a physical argument. We refer also to Remark 3.1.2 for additional discussion. ∎

REMARK 3.3. In order to get at the time dependent SE from the BFM (Bertoldi-Faraggi-Matone) theory we proceed as follows. From the previous discussion on the KG equation one sees that (dropping the A terms) in the time independent case one has $S^{cl}(q,t) = S_0^{cl}(q) - Et$

$$(1/2m)\sum_1^D(\partial_k S_0^{cl})^2 + \mathfrak{W}_{rel} = 0;\ \mathfrak{W}_{rel}(q) = (1/2mc^2)[m^2c^4 - (V(q) - E)^2] \quad (3.24)$$

leading to a stationary RQHJE

$$(1/2m)(\nabla S_0)^2 + \mathfrak{W}_{rel} - (\hbar^2/2m)(\Delta R/R) = 0;\ \nabla\cdot(R^2\nabla S_0) = 0 \quad (3.25)$$

This implies also the stationary KG equation

$$-\hbar^2c^2\Delta\psi + (m^2c^4 - V^2 + 2VE - E^2)\psi = 0 \quad (3.26)$$

Now in the time dependent case one can write $(1/2m)\eta^{\mu\nu}\partial_\mu S^{cl}\partial_\nu S^{cl} + \mathfrak{W}'_{rel} = 0$ where $\eta \sim diag(-1, 1, \cdots, 1)$ and

$$\mathfrak{W}'_{rel}(q) = (1/2mc^2)[m^2c^4 - V^2(q) - 2cV(q)\partial_0 S^{cl}(q)] \quad (3.27)$$

with $q \equiv (ct, q_1, \cdots, q_D)$. Thus we have the same structure as (3.24) with Euclidean metric replaced by a Minkowskian one. To implement the EP we have to modify the classical equation by adding a function to be determined, namely $(1/2m)(\partial S)^2 + \mathfrak{W}_{rel} + Q = 0$ ($(\partial S)^2 \sim \sum(\partial_\mu S)^2$ etc.). Observe that since $\mathfrak{W}'_{rel}$ depends on S^{cl} we have to make the identification $\mathfrak{W}_{rel} = (1/2mc^2)[m^2c^4 - V^2(q) - 2cV(q)\partial_0 S(q)]$ which differs from $\mathfrak{W}'_{rel}$ since S now appears instead of S^{cl}. Implementation of the EP requires that for an arbitrary $\mathfrak{W}^a$ state

$$\mathfrak{W}^b_{rel}(q^b) = (p^b|p^a)\mathfrak{W}^a_{rel}(q^a) + (q^q; q^b);\ Q^b(q^b) = (p^b|p^a)Q^a(q^a) - (q^a; q^b) \quad (3.28)$$

where now $(p^b|p) = \eta^{\mu\nu}p^b_\mu p^b_\nu/\eta^{\mu\nu}p_\mu p_\nu = p^TJ\eta J^Tp/p^T\eta p$ and $J^\mu_\nu = \partial q^\mu/\partial(q^b)^\nu$. This leads to the cocycle condition $(q^a; q^c) = (p^c|p^b)[(q^q; q^b) - (q^c; q^b)]$ as before. Now consider the identity

$$\alpha^2(\partial S)^2 = \Box(Rexp(\alpha S))/Rexp(\alpha S) - (\Box R/R) - (\alpha\partial\cdot(R^2\partial S)/R^2) \quad (3.29)$$

and if R satisfies the continuity equation $\partial\cdot(R^2\partial S)=0$ one sets $\alpha=i/\hbar$ to obtain

$$\frac{1}{2m}(\partial S)^2=-\frac{\hbar^2}{2m}\frac{\Box(Re^{iS/\hbar})}{Re^{iS/\hbar}}+\frac{\hbar^2}{2m}\frac{\Box R}{R} \tag{3.30}$$

Then it is shown that $\mathfrak{W}_{rel}=(\hbar^2/2m)(\Box(Rexp(iS/\hbar))/Rexp(iS/\hbar)$ so there results $Q_{rel}=-(\hbar^2/2m)(\Box R/R)$. Thus the RQHJE has the form (cf. (3.14) - (3.16))

$$\frac{1}{2m}(\partial S)^2+\mathfrak{W}_{rel}-\frac{\hbar^2}{2m}\frac{\Box R}{R}=0;\ \partial\cdot(R^2\partial S)=0 \tag{3.31}$$

Now for the time dependent SE one takes the nonrelativistic limit of the RQHJE. For the classical limit one makes the usual substitution $S=S'-mc^2t$ so as $c\to\infty$ $\mathfrak{W}_{rel}\to(1/2)mc^2+V$ and $-(1/2m)(\partial_0 S)^2\to\partial_t S'-(1/2)mc^2$ with $\partial(R^2\partial S)=0\to m\partial_t(R')^2+\nabla\cdot((R')^2\nabla S')=0$. Therefore (removing the primes) (3.31) becomes $(1/2m)(\nabla S)^2+V+\partial_t S-(\hbar^2/2m)(\Delta R/R)=0$ with the time dependent nonrelativistic continuity equation being $m\partial_t R^2+\nabla\cdot(R^2\nabla S)=0$. This leads then (for $\psi\sim Rexp(iS/\hbar)$) to the SE

$$i\hbar\partial_t\psi=\left(-\frac{\hbar^2}{2m}\Delta+V\right)\psi \tag{3.32}$$

One sees from all this that the BFM theory is profoundly governed by the equivalence principle and produces a usable framework for computation. It is surprising that it has not attracted more adherents. ■

3.2 Klein Gordon à La Santamato

The derivation of the SE in [171] (treated in Section 1.1) was modified in [172] to a derivation of the Klein-Gordon (KG) equation via a somewhat different average action principle. Recall that the spacetime geometry in [171] was obtained from the average action principle to obtain Weyl connections with a gauge field ϕ_μ (thus the geometry had a statistical origin). The Riemann scalar curvature $\dot{R}$ was then related to the Weyl scalar curvature R via an equation

$$R=\dot{R}-3[(1/2)g^{\mu\nu}\phi_\mu\phi_\nu+(1/\sqrt{-g})\partial_\mu(\sqrt{-g}g^{\mu\nu}\phi_\nu)] \tag{3.33}$$

Explicit reference to the underlying Weyl structure disappears in the resulting SE and we refer to Remark 1.7 for a few comments (cf. also [55] for an incisive review). We recall also here from [156, 157, 158, 159] (cf. [42, 43, 54]) that in the conformal geometry the particles do not follow geodesics of the conformal metric alone; further the work in [156, 157, 158, 159] is absolutely fundamental in exhibiting a correct framework for general relativity via the conformal (Weyl) version. Summarizing from [171] and the second paper in [172] one can say that traditional QM is equivalent (in some sense) to classical statistical mechanics in Weyl spaces. The moral seems to be (loosely) that quantum mechanics in Riemannian spacetime is the same as classical statistical mechanics in a Weyl space. In particular one wants to establish that traditional QM, based on wave equations and ad hoc probability calculus is merely a convenient mathematical construction to overcome the complications arising from a nontrivial spacetime geometric structure. Here one works from first principles and includes gauge invariance (i.e. invariance with respect to an arbitrary choice of the spacetime calibration). The spacetime is supposed to be a generic 4-dimensional differential manifold with torsion free connections $\Gamma^{\lambda}_{\mu\nu} = \Gamma^{\lambda}_{\nu\mu}$ and a metric tensor $g_{\mu\nu}$ with signature $(+,-,-,-)$ (one takes $\hbar = c = 1$). Here the (restrictive) hypothesis of assuming a Weyl geometry from the beginning is released, both the particle motion and the spacetime geometric structure are derived from a single average action principle. A result of this approach is that the spacetime connections are forced to be integrable Weyl connections by the extremization principle.

The particle is supposed to undergo a motion in spacetime with deterministic trajectories and random initial conditions taken on an arbitrary spacelike 3-dimensional hypersurface; thus the theory describes a relativistic Gibbs ensemble of particles (cf. [95, 172] for all this in detail and see also [54]). Both the particle motion and the spacetime connections can be obtained from the average stationary action principle

$$\delta\left[E\left(\int_{\tau_1}^{\tau_2} L(x(\tau,\dot{x}(\tau))d\tau\right)\right] = 0 \tag{3.34}$$

This action integral must be parameter invariant, coordinate invariant, and gauge invariant. All of these requirements are met if L is positively homogeneous of the first degree in $\dot{x}^{\mu} = dx^{\mu}/d\tau$ and transforms as a scalar of Weyl type $w(L) = 0$. The underlying probabiity measure must also be gauge invariant. A suitable Lagrangian is then

$$L(x,dx) = (m^2 - (R/6))^{1/2}ds + A_{\mu}dx^{\mu} \tag{3.35}$$

where $ds = (g_{\mu\nu}\dot{x}^{\mu}\dot{x}^{\nu})^{1/2}d\tau$ is the arc length and R is the space time scalar curvature; m is a parameterlike scalar field of Weyl type (or weight) $w(m) = -(1/2)$. The factor 6 is essentially arbitrary and has been chosen for future convenience. The vector field A_{μ} can be interpreted as a 4-potential due to an externally applied EM field and the curvature dependent factor in front of ds is an effective particle mass. This seems a bit ad hoc but some feeling for the nature of the Lagrangian can be obtained from Section 1.1 (cf. also [18]). The Lagrangian will be gauge invariant provided the A_{μ} have Weyl type $w(A_{\mu}) = 0$. Now one can split A_{μ} into its gradient and divergence free parts $A_{\mu} = \bar{A}_{\mu} - \partial_{\mu}S$, with both S and $\bar{A}_{\mu}$ having Weyl type zero, and with $\bar{A}_{\mu}$ interpreted as and EM 4-potential in the Lorentz gauge. Due to the nature of the action principle regarding fixed endpoints in variation one notes that the average action principle is not invariant under EM gauge transformations $A_{\mu} \to A_{\mu} + \partial_{\mu}S$; but one knows that QM is also not invariant under EM gauge transformations (cf. [7]) so there is no incompatability with QM here.

Now the set of all spacetime trajectories accessible to the particle (the particle path space) may be obtained from (3.34) by performing the variation with respect to the particle trajectory with fixed metric tensor, connections, and an underlying probability measure.

Thus (cf. [54, 95, 172]) the solution is given by the so-called Carathéodory complete figure associated with the Lagrangian

$$\bar{L}(x, dx) = (m^2 - (R/6))^{1/2} ds + \bar{A}_\mu dx^\mu \tag{3.36}$$

(note this leads to the same equations as (3.35) since the Lagrangians differ by a total differential dS). The resulting complete figure is a geometric entity formed by a one parameter family of hypersurfaces $S(x) = const.$ where S satisfies the HJ equation

$$g^{\mu\nu}(\partial_\mu S - \bar{A}_\mu)(\partial_\nu S - \bar{A}_\nu) = m^2 - \frac{R}{6} \tag{3.37}$$

and by a congruence of curves intersecting this family given by

$$\frac{dx^\mu}{ds} = \frac{g^{\mu\nu}(\partial_\nu S - \bar{A}_\nu)}{[g^{\rho\sigma}(\partial_\rho S - \bar{A}_\rho)(\partial_\sigma S - \bar{A}_\sigma)]^{1/2}} \tag{3.38}$$

The congruence yields the actual particle path space and the underlying probability measure on the path space may be defined on an arbitrary 3-dimensional hypersurface intersecting all of the members of the congruence without tangencies (cf. [95]). The measure will be completely identified by its probability current density j^μ (see [54, 172]). Moreover, since the measure is independent of the arbitrary choice of the hypersurface, j^μ must be conserved, i.e. $\partial_\mu j^\mu = 0$. Since the trajectories are deterministically defined by (3.38), j^μ must be parallel to the particle 4-velocity (3.38), and hence

$$j^\mu = \rho\sqrt{-g} g^{\mu\nu}(\partial_\nu S - \bar{A}_\nu) \tag{3.39}$$

with some $\rho > 0$. Now gauge invariance of the underlying measure as well as of the complete figure requires that j^μ transforms as a vector density of Weyl type $w(j^\mu) = 0$ and S as a scalar of Weyl type $w(S) = 0$. From (3.39) one sees then that ρ transforms as a scalar of Weyl type $w(\rho) = -1$ and ρ is called the scalar probability density of the particle random motion.

The actual spacetime affine connections are obtained from (3.34) by performing the variation with respect to the fields $\Gamma^\lambda_{\mu\nu}$ for a fixed metric tensor, particle trajectory, and probability measure. It is expedient to tranform the average action principle to the form of a 4-volume integral

$$\delta\left[\int_\Omega d^4x[(m^2 - (R/6))(g_{\mu\nu}j^\mu j^\nu]^{1/2} + A_\mu j^\mu\right] = 0 \tag{3.40}$$

where Ω is the spacetime region occupied by the congruence (3.38) and j^μ is given by (3.39) (cf. [54, 172] for proofs). Since the connection fields $\Gamma^\lambda_{\mu\nu}$ are contained only in the curvature term R the variational problem (3.40) can be further reduced to

$$\delta\left[\int_\Omega \rho R\sqrt{-g} d^4x\right] = 0 \tag{3.41}$$

(here the HJ equation (3.37) has been used). This states that the average spacetime curvature must be stationary under a variation of the fields $\Gamma^{\lambda}_{\mu\nu}$ (principle of stationary average curvature). The extremal connections $\Gamma^{\lambda}_{\mu\nu}$ arising from (3.41) are derived in [172] using standard field theory techniques and the result is

$$\Gamma^{\lambda}_{\mu\nu} = \left\{ \begin{array}{c} \lambda \\ \mu\ \nu \end{array} \right\} + \frac{1}{2}(\phi_{\mu}\delta^{\lambda}_{\nu} + \phi_{\nu}\delta^{\lambda}_{\mu} - g_{\mu\nu}g^{\lambda\rho}\phi_{\rho});\ \phi_{\mu} = \partial_{\mu}log(\rho) \tag{3.42}$$

This shows that the resulting connections are integrable Weyl connections with a gauge field ϕ_{μ} (cf. [171] and Sections 1.1-1.2). The HJ equation (3.37) and the continuity equation $\partial_{\mu}j^{\mu} = 0$ can be consolidated in a single complex equation for S, namely

$$e^{iS}g^{\mu\nu}(iD_{\mu} - \bar{A}_{\mu})(iD_{\nu} - \bar{A}_{\nu})e^{-iS} - (m^2 - (R/6)) = 0;\ D_{\mu}\rho = 0 \tag{3.43}$$

Here D_{μ} is (doubly covariant - i.e. gauge and coordinate invariant) Weyl derivative given by (cf. [18])

$$D_{\mu}T^{\alpha}_{\ \beta} = \partial_{\mu}T^{\alpha}_{\ \beta} + \Gamma^{\alpha}_{\mu\epsilon}T^{\epsilon}_{\ \beta} - \Gamma^{\epsilon}_{\ \mu\beta}T^{\alpha}_{\ \epsilon} + w(T)\phi_{\mu}T^{\alpha}_{\ \beta} \tag{3.44}$$

It is to be noted that the probability density (but not the rest mass) remains constant relative to D_{μ}. When written out (3.43) for a set of two coupled partial differential equations for ρ and S. To any solution corresponds a particular random motion of the particle.

Next one notes that (3.43) can be cast in the familiar KG form, i.e.

$$[(i/\sqrt{-g})\partial_{\mu}\sqrt{-g} - \bar{A}_{\mu}]g^{\mu\nu}(i\partial_{\nu} - \bar{A}_{\nu})\psi - (m^2 - (\dot{R}/6))\psi = 0 \tag{3.45}$$

where $\psi = \sqrt{\rho}exp(-iS)$ and $\dot{R}$ is the Riemannian scalar curvature built out of $g_{\mu\nu}$ only. We have the (by now) familiar formula

$$R = \dot{R} - 3[(1/2)g^{\mu\nu}\phi_{\mu}\phi_{\nu} + (1/\sqrt{-g})\partial_{\mu}(\sqrt{-g}g^{\mu\nu}\phi_{\nu})] \tag{3.46}$$

According to point of view **(A)** above in the KG equation (3.45) any explicit reference to the underlying spacetime Weyl structure has disappeared; thus the Weyl structure is hidden in the KG theory. However we note that no physical meaning is attributed to ψ or to the KG equation. Rather the dynamical and statistical behavior of the particle, regarded as a classical particle, is determined by (3.43), which, although completely equivalent to the KG equation, is expressed in terms of quantities having a more direct physical interpretation.

REMARK 3.4. The formula (3.46) goes back to Weyl [198] and the connection of matter to geometry arises from (3.42). The time variable is treated in a special manner here related to a Gibbs ensemble and $\rho > 0$ is built into the theory. ∎

3.3 Klein Gordon via Scale Relativity

In [40, 54] and Section 1.1 we sketched a few developments in the theory of scale relativity. This is by no means the whole story and we want to give a taste of some further main ideas while deriving the KG equation in this context (cf. [3, 58, 64, 65, 66, 139, 140, 141, 142, 143, 144, 145]). A main idea here is that the Schrödinger, Klein-Gordon, and Dirac equations are all geodesic equations in the fractal framework. They have the form $D^2/ds^2 = 0$ where D/ds represents the appropriate covariant derivative. The complex nature of the SE and KG equaton arises from a discrete time symmetry breaking based on nondifferentiability. For the Dirac equation further discrete symmetry breakings are needed on the spacetime variables in a biquaternionic context (cf. here [58]). First we go back to [139, 140, 144] and sketch some of the fundamentals of scale relativity. This is a very rich and beautiful theory extending in both spirit and generality the relativity theory of Einstein (cf. also [57] for variations involving Clifford theory). The basic idea here is that (following Einstein) the laws of nature apply whatever the state of the system and hence the relevant variables can only be defined relative to other states. Standard scale laws of power-law type correspond to Galilean scale laws and from them one actually recovers quantum mechanics (QM) in a nondifferentiable space. The quantum behavior is a manifestation of the fractal geometry of spacetime. In particular the quantum potential is a manifestation of fractality in the same way as the Newton potential is a manifestation of spacetime curvature. In this spirit one can also conjecture (cf. [144]) that this quantum potential may explain various dynamical effects presently attributed to dark matter (cf. also [6]). Now for the KG equation via scale relativity the derivation in the first paper of [58] seems the most concise and we follow that at first (cf. also [140]). All of the elements of the approach for the SE remain valid in the motion relativistic case with the time replaced by the proper time s, as the curvilinear parameter along the geodesics. Consider a small increment dX^μ of a nondifferentiable four coordinate along one of the geodesics of the fractal spacetime. One can decompose this in terms of a large scale part $\overline{LS} < dX^\mu >= dx^\mu = v_\mu ds$ and a fluctuation $d\xi^\mu$ such that $\overline{LS} < d\xi^\mu >= 0$. One is led to write the displacement along a geodesic of fractal dimension $D = 2$ via

$$dX^\mu_\pm = d_\pm x^\mu + d\xi^\mu_\pm = v^\mu_\pm ds + u^\mu_\pm \sqrt{2\mathcal{D}} ds^{1/2} \tag{3.47}$$

Here $u^\mu_\pm$ is a dimensionless fluctuation andd the length scale $2\mathcal{D}$ is introduced for dimensional purposes. The large scale forward and backward derivatives d/ds_+ and d/ds_- are defined via

$$\frac{d}{ds_\pm} f(s) = lim_{s\to 0_\pm} \overline{LS} \left\langle \frac{f(s+\delta s) - f(s)}{\delta s} \right\rangle \tag{3.48}$$

Applied to x^μ one obtains the forward and backward large scale four velocities of the form

$$(d/dx_+)x^\mu(s) = v^\mu_+; \ (d/ds_-)x^\mu = v^\mu_- \tag{3.49}$$

Combining yields

$$\frac{d'}{ds} = \frac{1}{2}\left(\frac{d}{ds_+} + \frac{d}{ds_-}\right) - \frac{i}{2}\left(\frac{d}{ds_+} - \frac{d}{ds_-}\right); \tag{3.50}$$

$$\mathcal{V}^\mu = \frac{d'}{ds}x^\mu = V^\mu - iU^\mu = \frac{v_+^\mu + v_-^\mu}{2} - i\frac{v_+^\mu - v_-^\mu}{2}$$

For the fluctuations one has

$$\overline{LS} < d\xi_\pm^\mu d\xi_\pm^\nu >= \mp 2\mathcal{D}\eta^{\mu\nu}ds \tag{3.51}$$

One chooses here $(+,-,-,-)$ for the Minkowski signature for $\eta^{\mu\nu}$ and there is a mild problem because the diffusion (Wiener) process makes sense only for positive definite metrics. Various solutions have been given and they are all basically equivalent, amounting to the transformatin a Laplacian into a D'Alembertian. Thus the two forward and backward differentials of $f(x,s)$ should be written as

$$(df/ds_\pm) = (\partial_s + v_\pm^\mu\partial_\mu \mp \mathcal{D}\partial^\mu\partial_\mu)f \tag{3.52}$$

One considers now only stationary functions f, not depending explicitly on the proper time s, so that the complex covariant derivative operator reduces to

$$(d'/ds) = (\mathcal{V}^\mu + i\mathcal{D}\partial^\mu)\partial_\mu \tag{3.53}$$

Now assume that the large scale part of any mechanical system can be characterized by a complex action $\mathfrak{S}$ leading one to write

$$\delta\mathfrak{S} = -mc\delta\int_a^b ds = 0;\ ds = \overline{LS} < \sqrt{dX^\nu dX_\nu} > \tag{3.54}$$

This leads to $\delta\mathfrak{S} = -mc\int_a^b \mathcal{V}_\nu d(\delta x^\nu)$ with $\delta x^\nu = \overline{LS} < dX^\nu >$. Integrating by parts yields

$$\delta\mathfrak{S} = -[mc\delta x^\nu]_a^b + mc\int_a^b \delta x^\nu(d\mathcal{V}_\mu/ds)ds \tag{3.55}$$

To get the equations of motion one has to determine $\delta\mathfrak{S} = 0$ between the same two points, i.e. at the limits $(\delta x^\nu)_a = (\delta x^\nu)_b = 0$. From (3.55) one obtains then a differential geodesic equation $d\mathcal{V}/ds = 0$. One can also write the elementary variation of the action as a functional of the coordinates. So consider the point a as fixed so $(\delta x^\nu)_a = 0$ and consider b as variable. The only admissable solutions are those satisfying the equations of motion so the integral in (3.55) vanishes and writing $(\delta x^\nu)_b$ as δx^ν gives $\delta\mathfrak{S} = -mc\mathcal{V}_\nu\delta x^\nu$ (the minus sign comes from the choice of signature). The complex momentum is now

$$\mathcal{P}_\nu = mc\mathcal{V}_\nu = -\partial_\nu\mathfrak{S} \tag{3.56}$$

and the complex action completely characterizes the dynamical state of the particle. Hence introduce a wave function $\psi = exp(i\mathfrak{S}/\mathfrak{S}_0)$ and via (3.56) one gets

$$\mathcal{V}_\nu = (i\mathfrak{S}_0/mc)\partial_\nu log(\psi) \tag{3.57}$$

Now for the scale relativistic prescription replace the derivative in d/ds by its covariant expression d'/ds. Using (3.57) one transforms $d\mathcal{V}/ds = 0$ into

$$-\frac{\mathfrak{S}_0^2}{m^2c^2}\partial^\mu log(\psi)\partial_\mu\partial_\nu log(\psi) - \frac{\mathfrak{S}_0\mathcal{D}}{mc}\partial^\mu\partial_\mu\partial_\nu log(\psi) = 0 \tag{3.58}$$

The choice $\mathfrak{S}_0 = \hbar = 2mc\mathcal{D}$ allows a simplification of (3.58) when one uses the identity

$$\frac{1}{2}\left(\frac{\partial_\mu\partial^\mu\psi}{\psi}\right) = \left(\partial_\mu log(\psi) + \frac{1}{2}\partial_\mu\right)\partial^\mu\partial^\nu log(\psi) \tag{3.59}$$

Dividing by $\mathcal{D}^2$ one obtains the equation of motion for the free particle $\partial^\nu[\partial^\mu\partial_\mu\psi/\psi] = 0$. Therefore the KG equation (no electromagnetic field) is

$$\partial^\mu\partial_\mu\psi + (m^2c^2/\hbar^2)\psi = 0 \tag{3.60}$$

and this becomes an integral of motion of the free particle provided the integration constant is chosen in terms of a squared mass term $m^2c^2/\hbar^2$. Thus the quantum behavior described by this equation and the probabilistic interpretation given to ψ is reduced here to the description of a free fall in a fractal spacetime, in analogy with Einstein's general relativity. Moreover these equations are covariant since the relativistic quantum equation written in terms of d'/ds has the same form as the equation of a relativistic macroscopic and free particle using d/ds. One notes that the metric form of relativity, namely $V^\mu V_\mu = 1$ is not conserved in QM and it is shown in [155] that the free particle KG equation expressed in terms of $\mathcal{V}$ leads to a new equality

$$\mathcal{V}^\mu\mathcal{V}_\mu + 2i\mathcal{D}\partial^\mu\mathcal{V}_\mu = 1 \tag{3.61}$$

In the scale relativistic framework this expression defines the metric that is induced by the internal scale structures of the fractal spacetime. In the absence of an electromagnetic field $\mathcal{V}^\mu$ and $\mathfrak{S}$ are related by (3.56) which can be writen as $\mathcal{V}_\mu = -(1/mc)\partial_\mu\mathfrak{S}$ so (3.61) becomes

$$\partial^\mu\mathfrak{S}\partial_\mu\mathfrak{S} - 2imc\mathcal{D}\partial^\mu\partial_\mu\mathfrak{S} = m^2c^2 \tag{3.62}$$

which is the new form taken by the Hamilton-Jacobi equation.

REMARK 3.5. We go back to [140, 155] now and repeat some of their steps in a perhaps more primitive but revealing form. Thus one omits the $\overline{LS}$ notation and uses $\lambda \sim 2\mathcal{D}$; equations (3.47) - (3.53) and (3.50) are the same and one writes now $\mathfrak{d}/ds$ for d'/ds. Then $\mathfrak{d}/ds = \mathcal{V}^\mu\partial_\mu + (i\lambda/2)\partial^\mu\partial_\mu$ plays the role of a scale covariant derivative and one simply takes the equation of motion of a free relativistic quantum particle to be given as $(\mathfrak{d}/ds)\mathcal{V}^\nu = 0$, which can be interpreted as the equations of free motion in a fractal spacetime or as geodesic equations. In fact now $(\mathfrak{d}/ds)\mathcal{V}^\nu = 0$ leads directly to the KG equation upon writing $\psi = exp(i\mathfrak{S}/mc\lambda)$ and $\mathfrak{P}^\mu = -\partial^\mu\mathfrak{S} = mc\mathcal{V}^\mu$ so that $i\mathfrak{S} = mc\lambda log(\psi)$ and $\mathcal{V}^\mu = i\lambda\partial^\mu log(\psi)$. Then

$$\left(\mathcal{V}^\mu\partial_\mu + \frac{i\lambda}{2}\partial^\mu\partial_\mu\right)\partial^\nu log(\psi) = 0 = i\lambda\left(\frac{\partial^\mu\psi}{\psi}\partial_\mu + \frac{1}{2}\partial^\mu\partial_\mu\right)\partial^\nu \log(\psi) \tag{3.63}$$

Now some identities are given in [155] for aid in calculation here, namely

$$\frac{\partial^\mu \psi}{\psi}\partial_\mu \frac{\partial^\nu \psi}{\psi} = \frac{\partial^\mu \psi}{\psi}\partial^\nu \left(\frac{\partial_\mu \psi}{\psi}\right) = \tag{3.64}$$

$$= \frac{1}{2}\partial^\nu \left(\frac{\partial^\mu \psi}{\psi}\frac{\partial_\mu \psi}{\psi}\right);\ \partial_\mu \left(\frac{\partial^\mu \psi}{\psi}\right) + \frac{\partial^\mu \psi}{\psi}\frac{\partial_\mu \psi}{\psi} = \frac{\partial^\mu \partial_\mu \psi}{\psi}$$

The first term in the last equation of (3.63) is then $(1/2)[(\partial^\mu \psi/\psi)(\partial_\mu \psi/\psi)]$ and the second is

$$(1/2)\partial^\mu \partial_\mu \partial^\nu log(\psi) = (1/2)\partial^\mu \partial^\nu \partial_\mu log(\psi) = \tag{3.65}$$

$$= (1/2)\partial^\nu \partial^\mu \partial_\mu log(\psi) = (1/2)\partial^\nu \left(\frac{\partial^\mu \partial_\mu \psi}{\psi} - \frac{\partial^\mu \psi \partial_\mu \psi}{\psi^2}\right)$$

Combining we get $(1/2)\partial^\nu(\partial^\mu \partial_\mu \psi/\psi) = 0$ which integrates then to a KG equation

$$-(\hbar^2/m^2c^2)\partial^\mu \partial_\mu \psi = \psi \tag{3.66}$$

for suitable choice of integration constant (note $\hbar/mc$ is the Compton wave length).

Now in this context or above we refer back to Section 3.1 for example and write $Q = -(1/2m)(\Box R/R)$ ($\hbar = c = 1$ for convenience here). Then recall $\psi = exp(i\mathfrak{S}/m\lambda)$ and $\mathfrak{P}_\mu = m\mathcal{V}_\mu = -\partial_\mu \mathfrak{S}$ with $i\mathfrak{S} = m\lambda log(\psi)$. Also $\mathcal{V}_\mu = -(1/m)\partial_\mu \mathfrak{S} = i\lambda \partial_\mu log(\psi)$ with $\psi = Rexp(iS/m\lambda)$ so $log(\psi) = i\mathfrak{S}/m\lambda = log(R) + iS/m\lambda$, leading to

$$\mathcal{V}_\mu = i\lambda[\partial_\mu log(R) + (i/m\lambda)\partial_\mu S] = -\frac{1}{m}\partial_\mu S + i\lambda \partial_\mu log(R) = V_\mu + iU_\mu \tag{3.67}$$

Then $\Box = \partial^\mu \partial_\mu$ and $U_\mu = \lambda \partial_\mu log(R)$ leads to

$$\partial^\mu U_\mu = \lambda \partial^\mu \partial_\mu log(R) = \lambda \Box log(R) \tag{3.68}$$

Further $\partial^\mu \partial_\nu log(R) = (\partial^\mu \partial_\nu R/R) - (R_\nu R_\mu/R^2)$ so

$$\Box log(R) = \partial^\mu \partial_\mu log(R) = (\Box R/R) - (\sum R_\mu^2/R^2) = \tag{3.69}$$

$$= (\Box R/R) - \sum(\partial_\mu R/R)^2 = (\Box R/R) - |U|^2$$

for $|U|^2 = \sum U_\mu^2$. Hence via $\lambda = 1/2m$ for example one has

$$Q = -(1/2m)(\Box R/R) = -\frac{1}{2m}\left[|U|^2 + \frac{1}{\lambda}\Box log(R)\right] = \tag{3.70}$$

$$= -\frac{1}{2m}\left[|U|^2 + \frac{1}{\lambda}\partial^\mu U_\mu\right] = -\frac{1}{2m}|U|^2 - \frac{1}{2}div(\vec{U})$$

(cf. Proposition 1.1). ∎

3.4 Field Theoretic Methods

In trying to imagine particle trajectories of a fractal nature or in a fractal medium we are tempted to abandon (or rather relax) the particle idea and switch to quantum fields (QF). Let the fields sense the bumps and fractality; if one can think of fields as operator valued distributions for example then fractal supports for example are quite reasonable. There are other reasons of course since the notion of particle in quantum field theory (QFT) has a rather fuzzy nature anyway. Then of course there are problems with QFT itself (cf. [197]) as well as arguments that there is no first quantization (except perhaps in the Bohm theory - cf. [134]). Some aspects of particles arising from QF and QFT methods, especially in a Bohmian spirit are reviewed in [41, 54] and here we only briefly indicate one approach due to Nikolić for bosonic fields (cf. [134, 135, 136, 137, 138] (cf. also [37, 103, 104, 105, 106, 107] for other field aspects of KG). We refer also to [100, 197] for interesting philosophical discussion about particles and localized objects in a QFT. Many details are omitted and standard QFT techniques are assumed to be known (see e.g. [101]) and we will concentrate here on derivations of KG type equations. First note that the papers [136] are impressive in producing a local operator describing the particle density current for scalar and spinor fields in an arbitrary gravitational and electromagnetic background. This enables one to describe particles in a local, general covariant, and gauge invariant manner and this is reviewed in [54]. We follow here [135] concerning Bohmian particle trajectories in relativistic bosonic and fermionic QFT. First we recall that there is no objection to a Bohmian type theory for QFT and no contradiction to Bell's theorems etc. (see e.g. [30, 75]). Without discussing philosophical aspects of such a theory we simply construct one following Nikolic. Thus consider first particle trajectories in relativistic QM and posit a real scalar field $\phi(x)$ satisfying the Klein-Gordon equation in a Minkowski metric $\eta_{\mu\nu} = diag(1, -1, -1, -1)$ written as $(\partial_0^2 - \nabla^2 + m^2)\phi = 0$. Let $\psi = \phi^+$ with $\psi^* = \phi^-$ correspond to positive and negative frequency parts of $\phi = \phi^+ + \phi^-$. The particle current is $j_\mu = i\psi^* \overleftrightarrow{\partial_\mu}\psi$ and $N = \int d^3x j_0$ is the positive definite number of particles (not the charge). This is most easily seen from the plane wave expansion $\phi^+(x) = \int d^3k a(\kappa) exp(-ikx)/\sqrt{(2\pi)^3 2k_0}$ since then $N = \int d^3k a^\dagger(\kappa)a(\kappa)$ (see above and [134, 136] where it is shown that the particle current and the decomposition $\phi = \phi^+ + \phi^-$ make sense even when a background gravitational field or some other potential is present). One can write also $j_0 = i(\phi^-\pi^+ - \phi^+\pi^-)$ where $\pi = \pi^+ + \pi^-$ is the canonical momentum (cf. [103]). Alternatively ϕ may be interpreted not as a field containg an arbitrary number of particles but rather as a one particle wave function. Here we note that contrary to a field a wave function is not an observable and so doing we normalize ϕ here so that $N = 1$. The current j_μ is conserved via $\partial_\mu j^\mu = 0$ which implies that $N = \int d^3x j_0$ is also conserved, i.e. $dN/dt = 0$. In the causal interpretation one postulates that the particle has the trajectory determined by $dx^\mu/d\tau = j^\mu/2m\psi^*\psi$. The affine parameter τ can be eliminated by writing the trajectory equation as $d\mathbf{x}/dt = \mathbf{j}(t, \mathbf{x})/j_0(t, \mathbf{x})$ where $t = x^0$, $\mathbf{x} = (x^1, x^2, x^3)$ and $\mathbf{j} = (j^1, j^2, j^3)$. By writing $\psi = Rexp(iS)$ where R, S) are real one arrives at a Hamilton-Jacobi (HJ) form $dx^\mu/d\tau = -(1/m)\partial^{mu} S$ and the KG equation is equivalent to

$$\partial^\mu(R^2\partial_\mu S) = 0;\quad \frac{(\partial^\mu S)(\partial_\mu S)}{2m} - \frac{m}{2} + Q = 0 \tag{3.71}$$

Here $Q = -(1/2m)(\partial^\mu\partial_\mu R/R$ is the quantum potential ($c = \hbar = 1$). From the HJ form and (3.71) plus the identity $d/d\tau = (dx^\mu/dt)\partial_\mu$ one arrives at the equations of motion $m(d^2x^\mu/d\tau^2) = \partial^\mu Q$. A typical trajectory arising from $d\mathbf{x}/dt = \mathbf{j}/j_0$ could be imagined as an S shaped curve in the $t - x$ plane (with t horizontal) and cut with a vertical line through the middle of the S. The velocity may be superluminal and may move backwards in time (at points where $j_0 < 0$). There is no paradox with backwards in time motion since it is physically indistinguishable from a motion forwards with negative energy. One introduces a physical number of particles via $N_{phys} = \int d^3x|j_0|$. Contrary to $N = \int d^3x j_0$ the physical number of particles is not conserved. A pair of particles one with positive and the other with negative energy may be created or annihilated; this resembles the behavior of virtual particles in convential QFT.

Now go to relativistic QFT where in the Heisenberg picture the Hermitian field operator $\hat{\phi}(x)$ satisfies

$$(\partial_0^2 - \nabla^2 + m^2)\hat{\phi} = J(\hat{\phi}) \tag{3.72}$$

where J is a nonlinear function describing the interaction. In the Schrödinger picture the time evolution is determined via the Schrödinger equation (SE) $H[\phi, -i\delta/\delta\phi]\Psi[\phi, t] = i\partial_t\Psi[\phi, t]$ where Ψ is a functional with respect to $\phi(\mathbf{x})$ and a function of t. A normalized solution of this can be expanded as $\Psi[\phi, t] = \sum_{-\infty}^{\infty}\tilde{\Psi}_n[\phi, t]$ where the $\tilde{\Psi}_n$ are unnormalized n-particle wave functionals and the analysis proceeds from there (cf. [135]). In the deBroglie-Bohm (dBB) interpretation the field $\phi(x)$ has a causal evolution determined by

$$(\partial_0^2 - \nabla^2 + m^2)\phi(x) = J(\phi(x)) - \left(\frac{\delta Q[\phi, t]}{\delta\phi(\mathbf{x})}\right)_{\phi(\mathbf{x})=\phi(x)}; \tag{3.73}$$

$$Q = -\frac{1}{2|\Psi|}\int d^3x \frac{\delta^2|\Psi|}{\delta\phi^2(\mathbf{x})}$$

where Q is the quantum potential again. However the n particles attributed to the wave function ψ_n also have causal trajectories determined by a generalization of $d\mathbf{x}/dt = \mathbf{j}/j_0$ as

$$\frac{d\mathbf{x}_{n,j}}{dt} = \left(\frac{\psi_n^*(x^{(n)})\overleftrightarrow{\nabla_j}\psi_n(x^{(n)})}{\psi_n^*(x^{(n)})\overleftrightarrow{\partial_{t_j}}\psi_n(x^{(n)})}\right)_{t_1=\cdots=t_n=t} \tag{3.74}$$

where the n-particle wave function is

$$\psi_n(\mathbf{x}^{(n)}, t) =< 0|\hat{\phi}(t, \mathbf{x}_1)\cdots\hat{\phi}(t, \mathbf{x}_n)|\Psi > \tag{3.75}$$

These n-particles have well defined trajectories even when the probability (in the conventional interpretation of QFT) of the experimental detection is equal to zero. In the dBB interpretation of QFT we can introduce a new causally evolving parameter $e_n[\phi, t]$ defined as

$$e_n[\phi, t] = |\tilde{\Psi}_n[\phi, t]|^2 / \sum_{n'}^{\infty} |\tilde{\Psi}_{n'}[\phi, t]|^2 \tag{3.76}$$

The evolution of this parameter is determined by the evolution of ϕ given via (3.73) and by the solution $\Psi = \sum \tilde{\Psi}$ of the SE. This parameter might be interpreted as a probability that there are n particles in the system at time t if the field is equal (but not measured!) to be $\phi(\mathbf{x})$ at that time. However in the dBB theory one does not want a stochastic interpretation. **Hence assume that e_n is an actual property of the particles guided by the wave function ψ_n and call it the effectivity of these n particles.** This is a nonlocal hidden variable attributed to the particles and it is introduced to provide a deterministic description of the creation and destruction of particles (see [41, 54, 135] for more on this).

REMARK 3.6. In [134] an analogous fermionic theory is developed but it is even more technical and we refer to [54] for a sketch. ■

REMARK 3.7. In [138] one addresses the question of statistical transparency. Thus the probabilitistic interpretation of the nonrelativistic SE does not work for the relativistic KG equation $(\partial^\mu \partial_\mu + m^2)\psi = 0$ (where $x = (\mathbf{x}, t)$ and $\hbar = c = 1$) since $|\psi|^2$ does not correspond to a probability density. There is a conserved current $j^\mu = i\psi^* \overleftrightarrow{\partial^\mu} \psi$ (where $a \overleftrightarrow{\partial^\mu} b = a\partial^\mu b - b\partial^\mu a$) but the time component j^0 is not positive definite. In [134, 135] the equations that determine the Bohmain trajectories of relativistic quantum particles described by many particle wave functions were written in a form requiring a preferred time coordinate. However a preferred Lorentz frame is not necessary (cf. [25]) and this is developed in [138] following [25, 135]. First note that as in [25, 135] it appears that particles may be superluminal and the principle of Lorentz covariance does not forbid superluminal velocities and conversly superluminal velocities do not lead to causal paradoxes (cf. [25, 138]). As noted in [25] the Lorentz-covariant Bohmian interprtation of the many particle KG equation is not statistically transparent. This means that the statistical distribution of particle positions cannot be calculated in a simple way from the wave function alone without the knowledge of particle trajectories. One knows that classcal QM is statistically transparent of course and this perhaps helps to explain why Bohmian mechanics has not attracted more attention. However statistical transparency (ST) may not be a fundamental property of nature as suggested by looking at standard theories (cf. [138]) The upshot is that since statistical probabilities can be calculated via Bohmian trajectories that theory is more powerful than other interpretations of general QM and we refer to [138] for discussion on this, on the KG equation, and on Lorentz covariance. ■

3.5 DeDonder-Weyl and Kg

We go here to a paper [137] which gives a manifestly covariant canonical method of field quantization based on the classical DeDonder-Weyl formulation of field theory. The Bohmian formulation is not postulated for intepretational purposes here but derived from purely technical requirements, namely covariance and consistency with standard QM. It arises automatically as a part of the formalism without which the theory cannot be formulated consistently. This together with the results of [134, 138] suggest that it is Bohmian mechanics that might be the missing bridge between QM and relativity; further it should play an important role in cosmology (cf. [54, 110, 119, 112, 113, 114, 115, 116, 117, 162, 191, 192, 193]). The classical covariant canonical DeDonder-Weyl formalism is given first following the excellent development in [122] and for simplicity one real scalar field in Minkowski spacetime is used. Thus (classical formulation) let $\phi(x)$ be a real scalar field described by

$$\mathfrak{A} = \int d^4x\mathfrak{L};\ \mathfrak{L} = \frac{1}{2}(\partial^\mu\phi)(\partial_\mu\phi) - V(\phi) \tag{3.77}$$

As usual one has

$$\pi^\mu = \frac{\partial\mathfrak{L}}{\partial(\partial_\mu\phi)} = \partial^\mu\phi;\ \partial_\mu\phi = \frac{\partial\mathfrak{H}}{\partial\pi^\mu};\ \partial_\mu\pi^\mu = -\frac{\partial\mathfrak{H}}{\partial\phi} \tag{3.78}$$

where the scalar DeDonder-Weyl (DDW) Hamilonian (not related to the energy density) is given by the Legendre transform $\mathfrak{H}(\pi^\mu,\phi) = \pi^\mu\partial_\mu\phi - \mathfrak{L} = (1/2)\pi^\mu\pi_\mu + V$. The equations (3.78) are equivalent to the standard Euler-Lagrange (EL) equations and by introducing the local vector $S^\mu(\phi(x),x)$ the dynamics can also be described by the covariant DDW HJ equation and equations of motion

$$\mathfrak{H}\left(\frac{\partial S^\alpha}{\partial\phi},\phi\right) + \partial_\mu S^\mu = 0;\ \partial^\mu\phi = \pi^\mu = \frac{\partial S^\mu}{\partial\phi} \tag{3.79}$$

Note here ∂_μ is the partial derivative acting only on the second argument of $S^\mu(\phi(x),x)$; the corresonding total derivative is $d_\mu = \partial_\mu + (\partial_\mu\phi)(\partial/\partial\phi)$. Further the first equation in (3.79) is a single equation for four quantities S^μ so there is a lot of freedom in finding solutions. Nevertheless the theory is equivalent to other formulations of classical field theory. Now following [118] one considers the relation between the covariant HJ equation and the conventional HJ equation; the latter can be derived from the former as follows. Using (3.78), (3.79) takes the form $(1/2)\partial_\phi S_\mu\partial_\phi S^\mu + V + \partial_\mu S^\mu = 0$. Then using the equation of motion in (3.79) write the first term as

$$\frac{1}{2}\frac{\partial S_\mu}{\partial\phi}\frac{\partial S^\mu}{\partial\phi} = \frac{1}{2}\frac{\partial S^0}{\partial\phi}\frac{\partial S^0}{\partial\phi} + \frac{1}{2}(\partial_i\phi)(\partial^i\phi) \tag{3.80}$$

Similarly using (3.79) the last term is $\partial_\mu S^\mu = \partial_0 S^0 + d_i S^i - (\partial_i\phi)(\partial^i\phi)$. Now introduce the quantity $\mathbf{S} = \int d^3x S^0$ so that $[\partial S^0(\phi(x),x)/\partial\phi(x)] = [\delta\mathbf{S}([\phi(\mathbf{x},t)],t)/\delta\phi(\mathbf{x},t)]$ where $\delta/\delta\phi(\mathbf{x},t) \equiv [\delta/\delta\phi(x)]_{\phi(x)=\phi(\mathbf{x},t)}$ is the space functional derivative. Putting this together gives then

$$\int d^3x\left[\frac{1}{2}\left(\frac{\delta\mathbf{S}}{\delta\phi(\mathbf{x},t)}\right)^2 + \frac{1}{2}(\nabla\phi)^2 + V(\phi)\right] + \partial_t\mathbf{S} = 0 \tag{3.81}$$

which corresponds to the standard noncovariant HJ equation. The time evolution of $\phi(\mathbf{x},t)$ is given by $\partial_t\phi(\mathbf{x},t) = \delta\mathbf{S}/\delta\phi(\mathbf{x},t)$ which arises from the time component of (3.79). Note that in deriving (3.81) it was necessary to use the space part of the equations of motion (3.79) (this does not play an important role in classical physics but is important here).

Now for the Bohmian formulation look at the SE $\hat{H}\Psi = i\hbar\partial_t\Psi$ where we write

$$\hat{H} = \int d^3x\left[-\frac{\hbar^2}{2}\left(\frac{\delta}{\delta\phi(\mathbf{x})}\right)^2 + \frac{1}{2}(\nabla\phi)^2 + V(\phi)\right]; \tag{3.82}$$

$$\Psi([\phi(\mathbf{x})],t) = \mathfrak{R}([\phi(\mathbf{x})],t)exp[i\mathfrak{S}([\phi(\mathbf{x})],t)/\hbar]$$

Then the complex SE equation is equivalent to two real equations

$$\int d^3x\left[\frac{1}{2}\left(\frac{\delta\mathfrak{S}}{\delta\phi(\mathbf{x})}\right)^2 + \frac{1}{2}(\nabla\phi)^2 + V(\phi) + Q\right] + \partial_t\mathfrak{S} = 0; \tag{3.83}$$

$$\int d^3x\left[\frac{\delta\mathfrak{R}}{\delta\phi(\mathbf{x})}\frac{\delta\mathfrak{S}}{\delta\phi(\mathbf{x})} + J\right] + \partial_t\mathfrak{R} = 0;\ Q = -\frac{\hbar^2}{2\mathfrak{R}}\frac{\delta^2\mathfrak{R}}{\delta\phi^2(\mathbf{x})};\ J = \frac{\mathfrak{R}}{2}\frac{\delta^2\mathfrak{S}}{\delta\phi^2(\mathbf{x})}$$

The second equation is also equivalent to

$$\partial_t\mathfrak{R}^2 + \int d^3x\frac{\delta}{\delta\phi(\mathbf{x})}\left(\mathfrak{R}^2\frac{\delta\mathfrak{S}}{\delta\phi(\mathbf{x})}\right) = 0 \tag{3.84}$$

and this exhibits the unitarity of the theory because it provides that the norm $\int[d\phi(\mathbf{x})]^2\Psi^*\Psi = \int[d\phi(\mathbf{x})]\mathfrak{R}^2$ does not depend on time. The quantity $\mathfrak{R}^2([\phi(\mathbf{x})],t)$ represents the probability density for fields to have the configuration $\phi(\mathbf{x})$ at time t. One can take (3.83) as the starting point for quantization of fields (note $exp(i\mathfrak{S}/\hbar)$ should be single valued). Equations (3.83) and (3.84) suggest a Bohmian interpretation with deterministic time evolution given via $\partial_t\phi$. Remarkably the statistical predictions of this deterministic interpretation are equivalent to those of the conventional interpretation. All quantum uncertainties are a consequence of the ignorance of the actual initial field configuration $\phi(\mathbf{x},t_0)$. The main reason for the consistency of this interpretation is the fact that (3.84) with $\partial_t\phi$ as above represents the continuity equation which provides that the statistical distribution $\rho([\phi(\mathbf{x})],t)$ of field configurations $\phi(\mathbf{x})$ is given by the quantum distribution $\rho = \mathfrak{R}^2$ at any time t, provided that ρ is given by $\mathfrak{R}^2$ at some initial time. The initial distribution is arbitrary in principle but a quantum H theorem explains why the quantum distribution is the most probable (cf. [195]). Comparing (3.83) with (3.81) we see that the quantum field satisfies an equation similar to the classical one, with the addition of a term resulting from the nonlocal quantum potential Q. The quantum equation of motion then turns out to be

$$\partial^\mu\partial_\mu\phi + \frac{\partial V(\phi)}{\partial\phi} + \frac{\delta\mathfrak{Q}}{\delta\phi(\mathbf{x};t)} = 0 \tag{3.85}$$

where $\mathfrak{Q} = \int d^3xQ$. A priori perhaps the main unattractive feature of the Bohmian formulation appears to be the lack of covariance, i.e. a preferred Lorentz frame is needed and this can be remedied with the DDW presentation to follow.

Thus one wants a quantum substitute for the classical covariant DDW HJ equation $(1/2)\partial_\phi S_\mu \partial_\phi S^\mu + V + \partial_\mu S^\mu = 0$. Define then the derivative

$$\frac{dA([\phi],x)}{d\phi(x)} = \int d^4x' \frac{\delta A([\phi],x')}{\delta\phi(x)} \tag{3.86}$$

where $\delta/\delta\phi(x)$ is the spacetime functional derivative (not the space functional derivative used before in (3.81)). In particular if $A([\phi],x)$ is a local functional, i.e. if $A([\phi],x) = A(\phi(x),x)$ then

$$\frac{dA(\phi(x),x)}{d\phi(x)} = \int d^4x' \frac{\delta A(\phi(x'),x')}{\delta\phi(x)} = \frac{\partial A(\phi(x),x)}{\partial\phi(x)} \tag{3.87}$$

Thus $d/d\phi$ is a generalization of $\partial/\partial\phi$ such that its action on nonlocal functionals is also well defined. An example of interest is a functional nonlocal in space but local in time so that

$$\frac{\delta A([\phi],x')}{\delta\phi(x)} = \frac{\delta A([\phi],x')}{\delta\phi(\mathbf{x},x^0)}\delta((x')^0 - x^0) \Rightarrow \tag{3.88}$$

$$\Rightarrow \frac{dA([\phi],x)}{d\phi(x)} = \frac{\delta}{\delta\phi(\mathbf{x},x^0)}\int d^3x' A([\phi],\mathbf{x}',x^0)$$

Now the first equation in (3.79) and the equations of motion become

$$\frac{1}{2}\frac{dS_\mu}{d\phi}\frac{dS^\mu}{d\phi} + V + \partial_\mu S^\mu = 0;\ \partial^\mu\phi = \frac{dS^\mu}{d\phi} \tag{3.89}$$

which is appropriate for the quantum modification. Next one proposes a method of quantization that combines the classical covariant canonical DDW formalism with the standard specetime asymmetric canonical quantization of fields. The starting point is the relation between the noncovariant classical HJ equation (3.81) and its quantum analogue (3.83). Suppressing the time dependence of the field in (3.81) we see that they differ only in the existence of the Q term in the quantum case. This suggests the following quantum analogue of the classical covariant equation (3.89)

$$\frac{1}{2}\frac{dS_\mu}{d\phi}\frac{dS^\mu}{d\phi} + V + Q + \partial_\mu S^\mu = 0 \tag{3.90}$$

Here $S^\mu = S^\mu([\phi],x)$ is a functional of $\phi(x)$ so S^μ at x may depend on the field $\phi(x')$ at all points x'. One can also allow for time nonlocalities (cf. [138]). Thus (3.91) is manifestly covariant provided that Q given by (3.83) can be written in a covariant form. The quantum equation (3.90) must be consistent with the conventional quantum equation (3.83); indeed by using a similar procedure to that used in showing that (3.79) implies (3.81) one can show that (3.90) implies (3.83) provided that some additional conditions are fulfilled.

First S^0 must be local in time so that (3.88) can be used. Second S^i must be completely local so that $dS^i/d\phi = \partial S^i/\partial\phi$, which implies

$$d_i S^i = \partial_i S^i + (\partial_i \phi)\frac{dS^i}{d\phi} \tag{3.91}$$

However just as in the classical case in this procedure it is necessary to use the space part of the equations of motion (3.79). Therefore these classical equations of motion must be valid even in the quantum case. Since we want a covariant theory in which space and time play equal roles the validity of the space part of the (3.79) implies that its time part should also be valid. Consequently in the covariant quantum theory based on the DDW formalism one must require the validity of the second equation in (3.89). This requirement is nothing but a covariant version of the Bohmian equation of motion written for an arbitrarily nonlocal S^μ (this clarifies and generalizes results in [118]). The next step is to find a covariant substitute for the second equation in (3.83). One introduces a vector $R^\mu([\phi], x)$ which will generate a preferred foliation of spacetime such that the vector R^μ is normal to the leaves of the foliation. Then define

$$\mathfrak{R}([\phi], \Sigma) = \int_\Sigma d\Sigma_\mu R^\mu;\ \mathfrak{S}([\phi], x) = \int_\Sigma d\Sigma_\mu S^\mu \tag{3.92}$$

where Σ is a leaf (a 3-dimensional hypersurface) generated by R^μ. Hence the covariant version of $\Psi = \mathfrak{R}exp(i\mathfrak{S})$ is $\Psi([\phi], \Sigma) = \mathfrak{R}([\phi], \Sigma)exp(i\mathfrak{S}([\phi], \Sigma)/\hbar)$. For R^μ one postulates the equation

$$\frac{dR^\mu}{d\phi}\frac{dS^\mu}{d\phi} + J + \partial_\mu R^\mu = 0 \tag{3.93}$$

In this way a preferred foliation emerges dynamically as a foliation generated by the solution R^μ of the equaitons (3.93) and (3.90). Note that R^μ does not play any role in classical physics so the existence of a preferred foliation is a purely quantum effect. Now the relation between (3.93) and (3.83) is obtained by assuming that nature has chosen a solution of the form $R^\mu = (R^0, 0, 0, 0)$ where R^0 is local in time. Then integrating (3.93) over d^3x and assuming again that S^0 is local in time one obtains (3.83). Thus (3.93) is a covariant substitute for the second equation in (3.83). It remains to write covariant versions for Q and J and these are

$$Q = -\frac{\hbar^2}{2\mathfrak{R}}\frac{\delta^2\mathfrak{R}}{\delta_\Sigma\phi^2(x)};\ J = \frac{\mathfrak{R}}{2}\frac{\delta^2\mathfrak{S}}{\delta_\Sigma\phi^2(x)} \tag{3.94}$$

where $\delta/\delta_\Sigma\phi(x)$ is a version of the space functional derivative in which Σ is generated by R^μ. Thus (3.93) and (3.90) with (3.94) represent a covariant substitute for the functional SE equivalent to (3.84). The covariant Bohmian equations (3.89) imply a covariant version of (3.85), namely

$$\partial^\mu\partial_\mu\phi + \frac{\partial V}{\partial\phi} + \frac{dQ}{d\phi} = 0 \tag{3.95}$$

Since the last term can also be written as $\delta(\int d^4xQ)/\delta\phi(x)$ the equation of motion (3.95) can be obtained by varying the quantum action

$$\mathfrak{A}_Q = \int d^4x\mathfrak{L}_Q = \int d^4x(\mathfrak{L} - Q) \tag{3.96}$$

Thus in summary the covariant canonical quantization of fields is given by equations (3.89), (3.90), (3.93), and (3.94). The conventional functional SE corresponds to a special class of solutions for which $R^i = 0,\ S^i$ are local, while R^0 and S^0 are local in time. In [137] a multifield generalization is also spelled out, a toy model is considered, and applications to quantum gravity are treated. The main result is that a manifestly covariant method of field quantization based on the DDW formalism is developed which treats space and time on an equal footing. Unlike the conventional canonical quantization it is not formulated in terms of a single complex SE but in terms of two coupled real equations. The need for a Bohmian formulation emerges from the requirement that the covariant method should be consistent with the conventional noncovariant method. This suggests that Bohmian mechanics (BM) might be a part of the formalism without which the covariant quantum theory cannot be formulated consistently.

4 Dirac Weyl Geometry

A sketch of Dirac Weyl geometry following [71] was given in [42] in connection with deBroglie-Bohm theory in the spirit of the Tehran school (cf. [28, 29, 129, 130, 163, 164, 165, 166, 167, 168, 169, 170, 173, 174, 175, 176, 177, 178, 179, 180, 181, 182, 183, 184, 185, 186, 187, 188]). We go now to [110, 111, 112, 113, 114, 115, 116, 117, 162] for a very brief discussion of versions of the Dirac Weyl theory involved in discussing magnetic monopoles, dark matter, quintessence, matter creation, etc. (see [54] for more in this direction). Thus go to [111] where in particular an integrable Weyl-Dirac theory is developed (the book [110] is a lovely exposition but the work in [111] is somwhat newer). Note, as remarked in [126] (where twistors are used), the integrable Weyl-Dirac geometry is desirable in order that the natural frequency of an atom at a point should not depend on the whole world line of the atom. The first paper in [111] is designed to investigate the integrable Weyl-Dirac (Int-W-D) geometry and its ability to create massive matter. For example in this theory a spherically symmetric static geometric formation can be spatially confined and an exterior observer will recognize it as a massive entity. This may be either a fundamental particle or a cosmic black hole both confined by a Schwarzschild surface. Here we only summarize some basic features in order to establish notation, etc. and sketch the preliminary theory (referring to [54] and the work of Israelit and Rosen for many examples). Thus in the Weyl geometry one has a metric $g_{\mu\nu} = g_{\nu\mu}$ and a length connection vector w_μ along with an idea of Weyl gauge transformation (WGT)

$$g_{\mu\nu} \to \tilde{g}_{\mu\nu} = e^{2\lambda}g_{\mu\nu};\ g^{\mu\nu} \to \tilde{g}^{\mu\nu} = e^{-2\lambda}g^{\mu\nu} \tag{4.1}$$

where $\lambda(x^\mu)$ is an arbitrary differerentiable function. One is interested in covariant quantities satisfying $\psi \to \tilde{\psi} = exp(n\lambda)\psi$ where the Weyl power n is described via

$\pi(\psi) = n$, $\pi(g_{\mu\nu}) = 2$, and $\pi(g^{\mu\nu}) = -2$. If $n = 0$ the quantity ψ is said to be gauge invariant (in-invariant). Under parallel displacement one has length changes and for a vector

$$(i)\ dB^{\mu} = -B^{\sigma}\Gamma^{\mu}_{\sigma\nu}dx^{\nu};\ (ii)\ B = (B^{\mu}B^{\nu}g_{\mu\nu})^{1/2};\ (iii)\ dB = Bw_{\nu}dx^{\nu} \tag{4.2}$$

(note $\pi(B) = 1$). In order to have agreement between (i) and (iii) one requires

$$\Gamma^{\lambda}_{\mu\nu} = \left\{ \begin{array}{c} \lambda \\ \mu\ \nu \end{array} \right\} + g_{\mu\nu}w^{\lambda} - \delta^{\lambda}_{\nu}w_{\mu} - \delta^{\lambda}_{\mu}w_{\nu} \tag{4.3}$$

where $\left\{ \begin{array}{c} \lambda \\ \mu\ \nu \end{array} \right\}$ is the Christoffel symbol based on $g_{\mu\nu}$. In order for (iii) to hold in any gauge one must have the WGT $w_{\mu} \to \tilde{w}_{\mu} = w_{\mu} + \partial_{\mu}\lambda$ and if the vector B^{μ} is transported by parallel displacement around an infinitesimal closed parallelogram one finds

$$\Delta B^{\lambda} = B^{\sigma}K^{\lambda}_{\sigma\mu\nu}dx^{\mu}\delta x^{\nu};\ \Delta B = BW_{\mu\nu}dx^{\mu}\delta x^{\nu}; \tag{4.4}$$

$$K^{\lambda}_{\sigma\mu\nu} = -\Gamma^{\lambda}_{\sigma\mu,\nu} + \Gamma^{\lambda}_{\sigma\nu,\mu} - \Gamma^{\alpha}_{\sigma\mu}\Gamma^{\lambda}_{\alpha\nu} + \Gamma^{\alpha}_{\sigma\nu}\Gamma^{\lambda}_{\alpha\mu}$$

is the curvature tensor formed from (4.3) and $W_{\mu\nu} = w_{\mu,\nu} - w_{\nu,\mu}$. Equations for the WGT $w_{mu} \to \tilde{w}_{\mu}$ and the definition of $W_{\mu\nu}$ led Weyl to identify w_{μ} with the potential vector and $W_{\mu\nu}$ with the EM field strength; he used a variational principle $\delta I = 0$ with $I = \int L\sqrt{-g}d^4x$ with L built up from $K^{\lambda}_{\sigma\mu\nu}$ and $W_{\mu\nu}$. In order to have an action invariant under both coordinate transformations and WGT he was forced to use R^2 (R the Riemannian curvature scalar) and this led to the gravitational field.

Dirac revised this with a scalar field $\beta(x^{\nu})$ which under WGT changes via $\beta \to \tilde{\beta} = e^{-\lambda}\beta$ (i.e. $\pi(\beta) = -1$). His in-invariant action integral is then ($f_{,\mu} \equiv \partial_{\mu}f$)

$$I = \int [W^{\lambda\sigma}W_{\lambda\sigma} - \beta^2 R + \beta^2(k-6)w^{\sigma}w_{\sigma} + 2(k-6)\beta w^{\sigma}\beta_{,\sigma} + \tag{4.5}$$

$$+k\beta_{,\underline{\sigma}}\beta_{,\sigma} + 2\Lambda\beta^4 + L_M]\sqrt{-g}d^4x$$

Here k is a parameter, Λ is the cosmological constant, L_M is the Lagrangian density of matter, and an underlined index is to be raised with $g^{\mu\nu}$. Now according to (4.4) this is a nonintegrable geometry but there may be situations when geometric vector fields are ruled out by physical constraints (e.g. the FRW universe). In this case one can preserve the WD character of the spacetime by assuming that w_{ν} is the gradient of a scalar function w so that $w_{\nu} = w_{,\nu} = \partial_{\nu}w$. One has then $W_{\mu\nu} = 0$ and from (4.4) results $\Delta B = 0$ yielding an integrable spacetime (Int-W-D spacetime). To develop this begin with (4.5) but with w_{ν} given by $w_{\nu} = \partial_{\nu}w$ so the first term in (4.5) vanishes. The parameter k is not fixed and the dynamical variables are $g_{\mu\nu}$, w, and β. Further it is assumed that L_M depends on $(g_{\mu\nu}, w, \beta)$. For convenience write

$$b_{\mu} = (log(\beta))_{,\mu} = \beta_{,\mu}/\beta \tag{4.6}$$

and use a modified Weyl connection vector $W_\mu = w_\mu + b_\mu$ which is a gauge invariant gradient vector. Write also $k - 6 = 16\pi\kappa$ and varying w in (4.5) one gets a field equation

$$2(\kappa\beta^2 W^\nu)_{;\nu} = S \tag{4.7}$$

where the semicolon denotes covariant differentiation with the Christoffel symbols and S is the Weylian scalar charge given by $16\pi S = \delta L_M/\delta w$. Varying $g_{\mu\nu}$ one gets also

$$G^\nu_\mu = -8\pi\frac{T^\nu_\mu}{\beta^2} + 16\pi\kappa\left(W^\nu W_\mu - \frac{1}{2}\delta^\nu_\mu W^\sigma W_\sigma\right) + \tag{4.8}$$

$$+2(\delta^\nu_\mu b^\sigma_{;\sigma} - b^\nu_{;\mu}) + 2b^\nu b_\mu + \delta^\nu_\mu b^\sigma_\sigma - \delta^\nu_\mu \beta^2 \Lambda$$

where G^ν_μ represents the Einstein tensor and the EM density tensor of ordinary matter is

$$8\pi\sqrt{-g}T^{\mu\nu} = \delta(\sqrt{-g}L_M)/\delta g_{\mu\nu} \tag{4.9}$$

Finally the variation with respect to β gives an equation for the β field

$$R + k(b^\sigma_{;\sigma} + b^\sigma b_\sigma) = 16\pi\kappa(w^\sigma w_\sigma - w^\sigma_{;\sigma}) + 4\beta^2\Lambda + 8\pi\beta^{-1}B \tag{4.10}$$

Note in (4.10) R is the Riemannian curvature scalar and the Dirac charge B is a conjugate of the Dirac gauge function β, namely $16\pi B = \delta L_M/\delta\beta$.

By a simple procedure (cf. [71]) one can derive conservation laws; consider e.g. $I_M = \int L_M\sqrt{-g}d^4x$. This is an in-invariant so its variation due to coordinate transformation or WGT vanishes. Making use of $16\pi S = \delta L_M/\delta w$, (4.9), and $16\pi B = \delta L_M/\delta\beta$ one can write

$$\delta I_M = 8\pi\int (T^{\mu\nu}\delta g_{\mu\nu} + 2S\delta w + 2B\delta\beta)\sqrt{-g}d^4x \tag{4.11}$$

Via $x^\mu \to \tilde{x}^\mu = x^\mu + \eta^\mu$ for an arbitrary infinitesimal vector η^μ one can write

$$\delta g_{\mu\nu} = g_{\lambda\nu}\eta^\lambda_{;\mu} + g_{\mu\lambda}\eta^\lambda_{;\nu};\ \delta w = w_{,\nu}\eta^\nu;\ \delta\beta = \beta_{,\nu}\eta^\nu \tag{4.12}$$

Taking into account $x^\mu \to \tilde{x}^\mu$ we have $\delta I_M = 0$ and making use of (4.12) one gets from (4.11) the energy momentum relations

$$T^\lambda_{\mu;\lambda} - Sw_\mu - \beta B b_\mu = 0 \tag{4.13}$$

Further considering a WGT with infinitesimal $\lambda(x^\mu)$ one has from (4.11) the equation $S + T - \beta B = 0$ with $T = T^\sigma_\sigma$. One can contract (4.8) and make use of (4.7) and $S + T = \beta B$ giving again (4.10), so that (4.10) is a corollary rather than an independent equation and one is free to choose the gauge function β in accordance with the gauge covariant nature of the theory. Going back to the energy-momentum relations one inserts $S + T = \beta B$ into (4.13) to get $T^\lambda_{\mu;\lambda} - Tb_\mu = SW_\mu$. Now go back to the field equation (4.8) and introduce the EM density tensor of the W_μ field

$$8\pi\Theta^{\mu\nu} = 16\pi\kappa\beta^2[(1/2)g^{\mu\nu}W^\lambda W_\lambda - W^\mu W^\nu] \tag{4.14}$$

Making use of (4.7) one can prove $\Theta^{\lambda}_{\mu;\nu} - \Theta b_{\mu} = -SW_{\mu}$ and using $T^{\lambda}_{\mu;\lambda} - TB_{\mu} = SW_{\mu}$ one has an equation for the joint energy momentum density

$$(T^{\lambda}_{\mu} + \Theta^{\lambda}_{\mu})_{;\lambda} - (T + \Theta)b_{\mu} = 0 \tag{4.15}$$

One can derive now the equation of motion of a test particle (following [162]). Consider matter consisting of identical particles with rest mass m and Weyl scalar charge q_s, being in the stage of a pressureless gas so that the EM density tensor can be written $T^{\mu\nu} = \rho U^{\mu}U^{\nu}$ where U^{μ} is the 4-velocity and the scalar mass density ρ is given by $\rho = m\rho_n$ with ρ_n the particle density. Taking into account the conservation of particle number one obtains from $T^{\lambda}_{\mu;\lambda} - Tb_{\mu} = SW_{\mu}$ the equation of motion

$$\frac{dU^{\mu}}{ds} + \left\{ \begin{matrix} \mu \\ \lambda\ \sigma \end{matrix} \right\} U^{\lambda}U^{\sigma} = \left(b_{\lambda} + \frac{q_s}{m}W_{\lambda}\right)(g^{\mu\lambda} - U^{\mu}U^{\lambda}) \tag{4.16}$$

In the Einstein gauge ($\beta = 1$) we are then left with

$$\frac{dU^{\mu}}{ds} + \left\{ \begin{matrix} \mu \\ \lambda\ \sigma \end{matrix} \right\} U^{\lambda}U^{\sigma} = \frac{q_s}{m}w_{\lambda}(g^{\mu\lambda} - U^{\mu}U^{\lambda}) \tag{4.17}$$

This gives a sketch of a powerful framework capable of treating many problems involving "mattter" and geometry. Connections to Section 2 are obvious and we have supplied earlier additional relations to fluctuations via Fisher information and quantum geometry (cf. also [40, 41, 42, 43, 54]). Many cosmological questions of great interest including dark matter, quintessence, etc. are also treated in [110, 111, 112, 113, 114, 115, 116, 117] and one can speculate about the original universe from many points of view. The inroads into cosmology here seem to be an inevitable consequence of the presence of Weyl-Dirac theory in dealing with quantum fluctuations via the quantum potential.

5 Remarks on Quantum Geometry

We gave a "hands on" sketch of quantum geometry in [43] and refer to [10, 11, 12, 16, 17, 35, 36, 59, 60, 61, 62, 63, 88, 96, 97, 102, 109, 123, 127, 128, 153, 154, 191, 192, 193, 201] for background and extensive theory. Here we follow [43, 59, 60, 61, 62, 63] and briefly extract from [43]. Roughly the idea is that for H the Hilbert space of a quantum system there is a natural quantum geometry on the projective space $P(H)$ with inner product $<\phi|\psi>= (1/2\hbar)g(\phi,\psi) + (i/2\hbar)\omega(\phi,\psi)$ where $g(\phi,\psi) = 2\hbar\Re(\phi|\psi)$ is the natural Fubini-Study (FS) metric and $g(\phi,\psi) = \omega(\phi, J\psi)$ $(J^2 = -1)$. On the other hand the FS metric is proportional to the Fisher information metric of the form $Cos^{-1}| <\phi|\psi> |$. Moreover (in 1-D for simplicity) $\mathfrak{F} \propto \int \rho Q dx$ is a functional form of Fisher information where Q is the quantum potential and $\rho = |\psi|^2$. Finally one recalls that in a Riemannian flat spacetime (with quantum matter and Weyl geometry) the Weyl-Ricci scalar curvature is proportional to Q. Thus assume H is separable with a complete orthonormal system $\{u_n\}$ and for any $\psi \in H$ denote by $[\psi]$ the ray generated by ψ while $\eta_n = (u_n|\psi)$. Define for $k \in \mathbf{N}$

$$U_k = \{[\psi] \in P(H);\ \eta_k \neq 0\};\ \phi_k : U_k \to \ell^2(\mathbf{C}) :\ \phi_k([\psi]) = \left(\frac{\eta_1}{\eta_k}, \cdots, \frac{\eta_{k-1}}{\eta_k}, \frac{\eta_{k+1}}{\eta_k}, \cdots\right) \tag{5.1}$$

where $\ell^2(\mathbf{C})$ denotes square summable functions. Evidently $P(H) = \cup_k U_k$ and $\phi_k \circ \phi_j^{-1}$ is biholomorphic. It is easily shown that the structure is independent of the choice of complete orthonormal system. The coordinaes for $[\psi]$ relative to the chart (U_k, ϕ_k) are $\{z_n^k\}$ given via $z_n^k = (\eta_n/\eta_k)$ for $n < k$ and $z_n^k = (\eta_{n+1}/\eta_k)$ for $n \geq k$. To convert this to a real manifold one can use $z_n^k = (1/\sqrt{2})(x_n^k + iy_n^k)$ with

$$\frac{\partial}{\partial z_n^k} = \frac{1}{\sqrt{2}}\left(\frac{\partial}{\partial x_n^k} + i\frac{\partial}{\partial y_n^k}\right);\ \frac{\partial}{\partial \bar{z}_n^k} = \frac{1}{\sqrt{2}}\left(\frac{\partial}{\partial x_n^k} - i\frac{\partial}{\partial y_n^k}\right) \tag{5.2}$$

etc. Instead of nondegeneracy as a criterion for a symplectic form inducing a bundle isomorphism between TM and T^*M one assumes here that a symplectic form on M is a closed 2-form which induces at each point $p \in M$ a toplinear isomorphism between the tangent and cotangent spaces at p. For $P(H)$ one can do more than simply exhibit such a natural symplectic form; in fact one shows that $P(H)$ is a Kähler manifold (meaning that the fundamental 2-form is closed). Thus one can choose a Hermitian metric $\mathfrak{G} = \sum g_{mn}^k dz_m^k \otimes d\bar{z}_n^k$ with

$$g_{mn}^k = (1 + \sum_i z_i^k \bar{z}_i^k)^{-1}\delta_{mn} - (1 + \sum_1 z_i^k \bar{z}_i^k)^{-2}\bar{z}_m^k z_n^k \tag{5.3}$$

relative to the chart $U_k, \phi_k)$. The fundamental 2-form of the metric $\mathfrak{G}$ is $\omega = i\sum_{m,n} g_{mn}^k dz_m^k \wedge d\bar{z}_n^k$ and to show that this is closed note that $\omega = i\partial\bar{\partial}f$ where locally $f = log(1 + \sum z_i^k \bar{z}_i^k)$ (the local Kähler function). Note here that $\partial + \bar{\partial} = d$ and $d^2 = 0$ implies $\partial^2 = \bar{\partial}^2 = 0$ so $d\omega = 0$ and thus $P(H)$ is a K manifold (cf. [128] for K geometry).

Now $P(H)$ is the set of one dimensional subspaces or rays of H; for every $x \in H/\{0\}$, $[x]$ is the ray through x. If H is the Hilbert space of a Schrödinger quantum system then H represents the pure states of the system and $P(H)$ can be regarded as the state manifold (when provided with the differentiable structure below). One defines the K structure as follows. On $P(H)$ one has an atlas $\{(V_h, b_h, C_h)\}$ where $h \in H$ with $\|h\| = 1$. Here (V_h, b_h, C_h) is the chart with domain V_h and local model the complex Hilbert space C_h where

$$V_h = \{[x] \in P(H);\ (h|x) \neq 0\};\ C_h = [h]^{\perp};\ b_h : V_h \to C_h;\ [x] \to b_h([x]) = \frac{x}{(h|x)} - h \tag{5.4}$$

This produces a analytic manifold structure on $P(H)$. As a real manifold one uses an atlas $\{(V_h, R \circ b_h, RC_h)\}$ where e.g. RC_h is the realification of C_h (the real Hilbert space with $\mathbf{R}$ instead of $\mathbf{C}$ as scalar field) and $R : C_h \to RC_h;\ v \to Rv$ is the canonical bijection (note $Rv \neq \Re v$). Now consider the form of the K metric relative to a chart

$(V_h, R\circ b_h, RC_h)$ where the metric g is a smooth section of $L_2(TP(H),\mathbf{R})$ with local expression $g^h : RC_h \to L_2(RC_h,\mathbf{R})$; $Rz \mapsto g^h_{Rz}$ where

$$g^h_{Rz}(Rv, Rw) = 2\nu\Re\left(\frac{(v|w)}{1+\|z\|^2} - \frac{(v|z)(z|w)}{(1+\|z\|^2)^2}\right) \tag{5.5}$$

The fundamental form ω is a section of $L_2(TP(H),\mathbf{R})$, i.e. $\omega^h : RC_h \to L_2(RC_h,\mathbf{R})$;
$Rz \to \omega^h_{Rz}$, given via

$$\omega^h_{Rz}(Rv, Rw) = 2\nu\Im\left(\frac{(v|w)}{1+\|z\|^2} - \frac{(v|z)(z|w)}{(1+\|z\|^2)^2}\right) \tag{5.6}$$

Then using e.g. (5.5) for the FS metric in $P(H)$ consider a Schrödinger Hilbert space with dynamics determined via $\mathbf{R}\times P(H) \to P(H) : (t,[x]) \mapsto [exp(-(i/\hbar)tH)x]$ where H is a (typically unbounded) self adjoint operator in H. One thinks then of Kähler isomorphisms of $P(H)$ (i.e. smooth diffeomorphisms $\Phi : P(H) \to P(H)$ with the properties $\Phi^*J = J$ and $\Phi^*g = g$). If U is any unitary operator on H the map $[x] \mapsto [Ux]$ is a K isomorphism of $P(H)$. Conversely (cf. [42]) any K isomorphism of $P(H)$ is induced by a unitary operator U (unique up to phase factor). Further for every self adjoint operator A in H (possibly unbounded) the family of maps $(\Phi_t)_{t\in\mathbf{R}}$ given via $\Phi_t : [x] \to [exp(-itA)x]$ is a continuous one parameter group of K isomorphisms of $P(H)$ and vice versa (every K isomorphism of $P(H)$ is induced by a self adjoint operator where boundedness of A corresponds to smoothness of the Φ_t). Thus in the present framework the dynamics of QM is described by a continuous one parameter group of K isomorphisms, which automatically are symplectic isomorphisms (for the structure defined by the fundamental form) and one has a Hamiltonian system. Next ideally one can suppose that every self adjoint operator represents an observable and these will be shown to be in $1-1$ correspondence with the real K functions.

One defines a (Riemann) metric (statistical distance) on the space of probability distributions $\mathcal{P}$ of the form

$$ds^2_{PD} = \sum(dp_j^2/p_j) = \sum p_j(dlog(p_j))^2 \tag{5.7}$$

Here one thinks of the central limit theorem and a distance between probability distributions distinguished via a Gaussian $exp[-(N/2)(\tilde{p}_j - p_j)^2/p_j]$ for two nearby distributions (involving N samples with probabilities p_j, $\tilde{p}_j$). This can be generalized to quantum mechanical pure states via (note $\psi \sim \sqrt{p}exp(i\phi)$ in a generic manner)

$$|\psi\rangle = \sum\sqrt{p_j}e^{i\phi_j}|j\rangle;\ |\tilde{\psi}\rangle = |\psi\rangle + |d\psi\rangle = \sum\sqrt{p_j+dp_j}e^{i(\phi_j+d\phi_j)}|j\rangle \tag{5.8}$$

Normalization requires $\Re(<\psi|d\psi>) = -1/2 < d\psi|d\psi>$ and measurements described by the one dimensional projectors $|j><j|$ can distinguish $|\psi>$ and $|\tilde{\psi}>$ according to the metric (5.7). The maximum (for optimal disatinguishability) is given by the

Hilbert space angle $cos^{-1}(|<\tilde{\psi}|\psi>|)$ and the corresponding line element ($PS \sim$ pure state)

$$\frac{1}{4}ds^2_{PS} = [cos^{-1}(|<\tilde{\psi}|\psi>|)]^2 \sim 1 - |<\tilde{\psi}|\psi>|^2 = <d\psi_\perp|d\psi_\perp> \sim \tag{5.9}$$

$$\sim \frac{1}{4}\sum \frac{dp_j^2}{p_j} + \left[\sum p_j d\phi_j^2 - (\sum p_j d\phi_j)^2\right]$$

(called the Fubini-Study (FS) metric) is the natural metric on the manifold of Hilbert space rays. Here

$$|d\psi_\perp> = |d\psi> - |\psi><\psi|d\psi> \tag{5.10}$$

is the projection of $|d\psi>$ orthogonal to $|\psi>$. Note that if $cos^{-1}(|<\tilde{\psi}|\psi>| = \theta$ then $cos(\theta) = |<\tilde{\psi}|\psi>|$ and $cos^2(\theta) = |<\tilde{\psi}|\psi>|^2 = 1 - Sin^2(\theta) \sim 1 - \theta^2$ for small θ. Hence $\theta^2 \sim 1 - cos^2(\theta) = 1 - |<\tilde{\psi}|\psi>|^2$. The term in square brackets (the variance of phase changes) is nonnegative and an appropriate choice of basis makes it zero. In [35] one then goes on to discuss distance formulas in terms of density operators and Fisher information but we omit this here. Generally as in [201] one observes that the angle in Hilbert space is the only Riemannian metric on the set of rays which is invariant uder unitary transformations. In any event $ds^2 = \sum(dp_i^2/p_i)$, $\sum p_i = 1$ is referred to as the Fisher metric (cf. [128]). Note in terms of $dp_i = \tilde{p}_i - p_i$ one can write $d\sqrt{p} = (1/2)dp/\sqrt{p}$ with $(d\sqrt{p})^2 = (1/4)(dp^2/p)$ and think of $\sum(d\sqrt{p_i})$ as a metric. Alternatively from $cos^{-1}(|<\tilde{\psi}|\psi>|$ one obtains $ds_{12} = cos^{-1}(\sum\sqrt{p_{1i}}\sqrt{p_{2i}})$ as a distance in $\mathcal{P}$. Note from (5.10) that $ds^2_{12} = 4cos^{-1}|<\psi_1|\psi_2>| \sim 4(1 - |(\psi_1|\psi_2)|^2 \equiv 4(<d\psi|d\psi> - <d\psi|\psi><\psi|d\psi>)$ begins to look like a FS metric before passing to projective coordinates. In this direction we observe from [128] that the FS metric can be expressed also via

$$\partial\bar{\partial}log(|z|^2) = \phi = \frac{1}{|z|^2}\sum dz_i \wedge d\bar{z}_i - \frac{1}{|z|^4}\left(\sum \bar{z}_i dz_i\right) \wedge \left(\sum z_i d\bar{z}_i\right) \tag{5.11}$$

so for $v \sim \sum v_i\partial_i + \bar{v}_i\bar{\partial}_i$ and $w \sim \sum w_i\partial_i + \bar{w}_i\bar{\partial}_i$ and $|z|^2 = 1$ one has $\phi(v, w) = (v|w) - (v|z)(z|w)$.

Now recall the material on fisher information in Section 1.2 and the results on the SE in Weyl space in Section 1.1 to confirm the connection of quantum geometry as above to Fisher information, Weyl curvature, and the quantum potential. Several features arise which deserve emphasis (cf. also [55])

- Philosophically the wave function seems to be inevitably associated to a cloud or ensemble (cf. Remarks 2.1 and 3.2). This provides meaning for $psi = Rexp(iS/\hbar)$ with $R = \sqrt{\rho}$ and $\rho = \psi^*\psi$ representing a probability density. Connections to hydrodynamics, diffusion, and kinetic theory are then natural and meaningful.

- From the ensemble point of view or by statistical derivations as in Section 1.1 one sees that spacetime geometry should also be conceived of in statistical terms at the quantum level. This is also connected with the relativistic theory and the quantum potential (in various forms) is exhibited as a fundamental ingredient of both QM and spacetime geometry.

- Bohmian type mechanics plays a fundamental role in providing unification of all these ideas. Similarly fractal considerations as in Nottale's scale relativity lead to important formulas consistent with the pictures obtained via Bohmian mechanics and the quantum potential.
- Quantum geometry in a projective Hilbert space is connected to all these matters as indicated in this section.

References

[1] L. Abbot and M. Wise, *Amer. Jour. Phys.*, **49** (1981), 37-39

[2] M. Abolhasani and M. Golshani, gr-qc 9709005; quant-ph 9808015

[3] F. Ben Adda and J. Cresson, *Quantum derivatives and the Schrödinger equation, Chaos, Solitons, and Fractals*, **19** (2004), 1323-1334

[4] R. Adler, M. Bazin, and M. Schiffer, *Introduction to general relativity,* McGraw-Hill, 1965

[5] M. Agop, P. Ioannou, C. Buzea, and P. Nica, *Chaos, Solitons, and Fractals*, **16** (2003), 321-338

[6] M. Agop, P. Ioannou, and C. Buzea, *Chaos, Solitons, and Fractals*, **13** (2002), 1137

[7] Y. Aharonov and D. Bohm, *Phys. Rev.* **115** (1959), 485

[8] V. Allori, D. Dürr, S. Goldstein, and N. Zanghi, quant-ph 0112005

[9] V. Allori and N. Zanghi, quant-ph 0112008 and 0112009

[10] J. Anandan, quant-ph 0012011; gr-qc 9505011, 9506011, 9712015; *Found. Phys.*, **21** (1991), 1265-1284; *Phys. Lett. A*, **147** (1990), 3-8; 19 (1994), 284-292; *Ann. Inst. H. Poincaré*, **49** (1988), 271-286

[11] J. Anandan and Y. Aharonov, *Phys. Rev. Lett.*, **65** (1990), 1697-1700; *Phys. Rev. D,* **38** (1988), 1863-1870

[12] L. Anderson and J. Wheeler, hep-th 0406159

[13] O. Arias, T. Gonzalez, Y. Leyva, and I. Quiros, gr-qc 0307016

[14] O. Arias and I. Quiros, gr-qc 0212006

[15] O. Arias, T. Gonzalez, and I. Quiros, gr-qc 0210097

[16] A. Ashtekar and T. Schilling, gr-qc 9706069

[17] A. Ashtekar, gr-qc 0112038 and 9901023

[18] J. Audretsch, *Phys. Rev. D,* **27** (1083), 2872-2884; *Comm. Math. Phys.*, **95** (1984), 41-51

[19] A. Barut and N. Zanghi, *Phys. Rev. Lett.*, **52** (1984), 2009-2012

[20] J. Bell, *Speakable and unspeakable in quantum mechanics*, Cambridge Univ. Press, 1987

[21] J. Bell, *Phys. Repts.*, **137** (1986), 49-54; Physics (1964), 195-200; *Rev. Mod. Phys.*, **38** (1966), 447-452

[22] K. Berndl, quant-ph 9509009

[23] K. Berndl, M. Daumer, and D. Dürr, quant-ph 9504010

[24] K. Berndl, D. Dürr, S. Goldstein, G. Peruzzi, and N. Zanghi, *Comm. Math. Phys.*, **173** 91995), 647-673 (quant-ph 9503013)

[25] K. Berndl, D. Dürr, S. Goldstein, and N. Zanghi, *Phys. Rev. A*, **53** (1996), 2062 (quant-ph 9510027)

[26] K. Berndl, M. Daumer, D. Dürr, S. Goldstein, and N. Zanghi, *Nuovo Cimento*, **110B** (1995), 737-750

[27] G. Bertoldi, A. Faraggi, and M. Matone, hep-th 9909201

[28] Y. Bisabr and H. Salehi, gr-qc 0109087

[29] Y. Bisabr, gr-qc 0302102; hep-th 0306092

[30] D. Bohm, B. Hiley, and P. Kaloyerou, *Phys. Rept.* **144** (1987), 323-375

[31] D. Bohm and B. Hiley, The undivided universe, Routledge, Chapman and Hall, 1993

[32] D. Bohm and B. Hiley, *Phys. Repts.*, **144** (1987), 323-348

[33] D. Bohm, *Phys. Rev.* **95** (1952), 166-179, 180-193

[34] L. deBroglie, *Electrons et photons*, Solvay Conf., Paris, pp. 105-141

[35] S. Braunstein and C. Caves, *Phys. Rev. Lett.*, **72** (1994), 3439-3443

[36] D. Brody and L. Hughston, quant-ph 9701051, 9706030, 9706037,9906085, and 9906086; gr-qc 9708032

[37] H. Brown and P. Holland, quant-ph 0302062

[38] R. Cardenas, T. Gonzalez, O. Martin, and I. Quiros, astro-ph 0210108

[39] R. Cardenas, T. Gonzalez, Y. Leiva, O. Martin, and I. Quiros, astro-ph 0206315

[40] R. Carroll, quant-ph 0401082, *Inter. Jour. Evol. Eqs.*, to appear

[41] R. Carroll, quant-ph 0403156, *Applicable Anal.*, to appear

[42] R. Carroll, gr-qc 0406004

[43] R. Carroll, quant-ph 0406203, *Foundations of Physics*, to appear

[44] R. Carroll, Quantum theory, deformation, and integrability, North-Holland, 2000

[45] R. Carroll, *Calculus revisited*, Kluwer, 2002

[46] R. Carroll, *Proc. Conf. Symmetry*, Kiev, 2003, Part I, pp. 356-367

[47] R. Carroll, *Canadian Jour. Phys.*, **77** (1999), 319-325

[48] R. Carroll, *Direct and inverse problems of mathematical physics*, Kluwer, 2000, pp. 39-52

[49] R. Carroll, *Generalized analytic functions*, Kluwer, 1998, pp. 299-311

[50] R. Carroll, quant-ph 0309023 and 0309159

[51] R. Carroll, *Nucl. Phys. B*, **502** (1997), 561-593; Springer *Lect. Notes Physics*, **502**, 1998, pp. 33-56

[52] R. Carroll and Y. Kodama, *Jour. Phys. A*, **28** (1995), 6373-6387

[53] R. Carroll and B. Konopelchenko, *Inter. Jour. Mod. Phys. A*, **11** (1996), 1183-1216

[54] R. Carroll, Quantum mechanics, general relativity, and information, book in preparation

[55] C. Castro, *Found. Phys.*, **22** (1992), 569-615; *Found. Phys. Lett.*, **4** (1991), 81

[56] C. Castro, J. Mahecha, and B. Rodriguez, quant-ph 0202026

[57] C. Castro, *Chaos, Solitons, and Fractals*, **11** (2000), 1663-1670; 12 (2001), 101-104, 1585-1606

[58] M. Célérier and L. Nottale, hep-th 0112213 and 0210027

[59] R. Cirelli, M. Gatti, and A. Mania, quant-ph 0202076

[60] R. Cirelli, A. Mania, and L. Pizzocchero, *Jour. Math. Phys.*, **31** (1990), 2891-2897 and 2898-2903

[61] R. Cirelli, P. Lanzavecchia, and A. Mania, *Jour. Phys. A*, **16** (1983), 3829-3835

[62] R. Cirelli and L. Pizzocchero, *Nonlinearity*, **3** (1990), 1057-1080

[63] R. Cirelli and P. Lanzavecchia, *Nuovo Cimento B*, **79** (1984), 271-283

[64] J. Cresson, math.GM 0211071

[65] J. Cresson, Scale calculus and the Schrödinger equation; *Scale geometry, I*, preprints 2003

[66] J. Cresson, *Nondifferentiable variational principles*, preprint 2003

[67] R. Czopnik and P. Garbaczewski, quant-ph 0203018; cond-mat 0202463

[68] F. Darabi and P. Wesson, gr-qc 0003045

[69] M. Daumer, S. Goldstein, and N. Zanghi, *Jour. Stat. Phys.*, **88** (1997), 967-977 (quant-ph 9601013)

[70] M. Davidson, quant-ph 0110050, 0112157, and 0106124

[71] P. Dirac, *Proc. Royal Soc. London A*, **209** (1951), 291-296, 212 (1952), 330-339, 333 (1973), 403-418

[72] T. Djama, quant-ph 0111121, 0201003, 0311057, and 0311059

[73] D. Dürr, S. Goldstein, and N. Zanghi, quant-ph 9511016 and 0308039

[74] D. Dürr, S. Goldstein, and N. Zanghi, quant-ph 0308038

[75] D. Dürr, S. Goldstein, R. Tumulka, and N. Zanghi, quant-ph 0208072; 0303156; 0303056; 0311127

[76] D. Dürr, S. Goldstein, and S. Zanghi, *Jour. Stat. Phys.*, **67** (1992), 843-907

[77] D. Dürr, S. Goldstein, K. Münch-Berndl, and N. Zanghi, *Phys. Rev. A*, **60** (1999), 2729 (quant-ph 9801070)

[78] D. Dürr, S. Goldstein, and N. Zanghi, quant-ph 9512031

[79] D. Dürr, S. Goldstein, R. Tumulka, and N. Zanghi, quant-ph 0407116

[80] K. Falconer, Fractal geometry, Wiley, 1990; *The geometry of fractal sets*, 1988; *Techniques in fractal geometry*, Wiley, 1997

[81] A. Faraggi and M. Matone, *Phys. Rev. Lett.*, **78** (1997), 163-166

[82] A. Faraggi and M. Matone, *Inter. Jour. Mod. Phys. A*, **15** (2000), 1869-2017

[83] E. Floyd, *Inter. Jour. Mod. Phys. A*, **14** (1999), 1111-1124; 15 (2000), 1363-1378; *Found. Phys. Lett.*, **13** (2000), 235-251; quant-ph 0009070, 0302128 and 0307090

[84] E. Floyd, *Phys. Rev. D*, **29** (1984) 1842-1844; 26 (1982), 1339-1347; 34 (1986), 3246-3249; 25 (1982), 1547-1551; *Jour. Math. Phys.*, **20** (1979), 83-85; 17 (1976), 880-884; *Phys. Lett. A*, **214** (1996), 259-265; *Inter. Jour. Theor. Phys.*, **27** (1998), 273-281

[85] B. Frieden, Physics from Fisher information, Cambridge Univ. Press, 1998

[86] P. Garbaczewski, cond-mat 0211362 and 0301044; quant-ph 0408192

[87] H. Georgii and R. Tumulka, math.PR 0312294

[88] G. Gibbons, *Jour. Geom. and Phys.*, **8** (1992), 147-162

[89] S. Goldstein, J. Taylor, R. Tumulka, and N. Zanghi, quant-ph 0405039 and 0407134

[90] S. Goldstein, *Bohmian mechanics*, Stanford Encycl. Philos., http://plato.stanford.edu/archives/win2002/

[91] S. Goldstein and R. Tumulka, quant-ph 0105040

[92] S. Goldstein and R. Tumulka, quant-ph 9902018

[93] S. Goldstein, quant-ph 9512027 and 9901005

[94] S. Goldstein and J. Lebowitz, quant-ph 9512028

[95] R. Hakim, *Jour. Math. Phys.*, **8** (1967), 1315-1344

[96] M. Hall, quant-ph 9806013, 9903045, 9912055, 0103072, 0107149, 0302007

[97] M. Hall and M. Reginatto, quant-ph 0102069 and 0201084

[98] M. Hall, K. Kumar, and M. Reginatto, hep-th 0206235; quant-ph 0103041

[99] M. Hall, K. Kumar, and M. Reginatto, he-th 0307259

[100] H. Halvorson and R. Clifton, quant-ph 0103041

[101] B. Hatfield, *Quantum field theory of point particles and strings*, Addison-Wesley, 1992

[102] A. Heslot, *Phys. Rev. D,* **31** (1985), 1341-1348

[103] P. Holland, *The quantum theory of motion*, Cambridge Univ. Press, 1997

[104] P. Holland, *Foundations of Physics*, **38** (1998), 881-911; *Nuovo Cimento B,* **116** (2001), 1043 and 1143; *Phys. Lett. A,* **128** (1988), 9; quant-ph 0305175

[105] P. Holland, quant-ph 0405145

[106] P. Holland, *Phys. Repts.*, **224** (1993), 95; quant-ph 0302076

[107] G. Horton, C. Dewdney, and H. Nesteruk, quant-ph 0103114; *Jour. Phys. A*, **33** (2000), 7337-7352

[108] M. Inaba, *Inter. Jour. Mod. Phys. A*, **16** (2001), 2965-2973

[109] J. Isidro, hep-th 0110151

[110] M. Israelit, *The Weyl-Dirac theory and our universe*, Nova Science Pub., 1999

[111] M. Israelit, *Found. Phys.*, **29** (1999), 1303-1322; 32 (2002), 295-321 and 945-961

[112] M. Israelit, gr-qc 9608035, 9608047, and 9611060

[113] M. Israelit and N. Rosen, *Found. Phys.*, **22** (1992), 555; 24 (1994), 901

[114] M. Israelit and N. Rosen, *Astrophys. Jour.*, **342** (1989), 627; *Astrophys. Space Sci.*, **204** (1993), 317

[115] M. Israelit and N. Rosen, *Found. Phys.*, **25** (1995), 763; 26 (1996), 585

[116] M. Israelit, *Found. Phys.*, **19** (1989), 33-55

[117] M. Israelit and N. Rosen, *Found. Phys.*, **13** (1983), 1023-1045

[118] I. Kanatchikov, *Phys. Lett. A,* **25** (2001), 25-36 (hep-th 0012084); quant-ph 9712058; gr-qc 9810076 and 9912094; hep-th 9810165 and 0301001

[119] G. Kaniadakis and A. Scarfone, *Rep. Math. Phys.*, **51** (2003), 225 (cond-mat 0303334); *Jour. Phys. A*, **35** (2002), 1943 (quant-ph 0202032)

[120] G. Kaniadakis, *Physica A*, **307** (2002), 172 (quant-ph 0112049)

[121] G. Kaniadakis, *Phys. Lett. A*, **310** (2003), 377 (quant-ph 0303159); *Found. Phys. Lett.*, *16* (2003),99 (quant-ph 0209033)

[122] H. Kastrup, *Phys. Repts.*, **101** (1983), 1

[123] T. Kibble, *Comm. Math. Phys.*, **65** (1979), 189-201

[124] F. London, *Zeit. f. Phys.*, **42** (1927), 375

[125] B. Mandelbrot, *Gaussian self affinities and fractals*, Springer, 2002

[126] S. Merkulov, *Class. Quant. Grav.*, **1** (1984), 349-354

[127] D. Minic and C. Tze, *Phys. Rev. D,* **68** (2003), 061501(R); hep-th 0305193, 0309239, and 0401028

[128] A. Moroianu, *Lectures on Kähler geometry*, math.DG 0402223

[129] H. Motavali, H. Salehi, and M. Golshani, hep-th 0011062

[130] H. Motavali and M. Golshani, hep-th 0011064

[131] M. Nagasawa, *Schrödinger equations and diffusion theory*, Birkäuser, 1993

[132] M. Nagasawa, *Stochastic processes in quantum physics*, Birkhäuser, 2000

[133] E. Nelson, *Quantum fluctuations*, Princeton Univ. Press, 1985; Dynamical theory of Brownian motion, Princeton Univ. Press, 1967

[134] H. Nikolic, quant-ph 0302152 and 0307179

[135] H. Nikolić, *Found. Phys. Lett.*, **17** (2004), 363-380 (quant-ph 0208185)

[136] H. Nikolić, *Int. Jour. Mod. Phys.* **D12** (2003), 407-444 (hep-th 0202204); *Phys. Lett. B*, **527** (2002), 119-124 (gr-qc 0111024) and 529 (2002), 265

[137] H. Nikolić, hep-th 0407228

[138] H. Nikolić, quant-ph 0406173; gr-qc 04031121

[139] L. Nottale, *Fractal space-time and microphysics: Toward a theory of scale relativity*, World Scientific, 1993

[140] L. Nottale, *Chaos, Solitons, and Fractals*, **7** (1996), 877-938; 12 (2000), 1577-1583

[141] L. Nottale, M. Célérier, and T. Lehner, hep-th 0307093

[142] L. Nottale, *Chaos, solitons, and Fractals*, **10** (1999), 459-468

[143] L. Nottale, *La relativité dans tous ses états*, Hachette, 2000

[144] L. Nottale, *Chaos, Solitons, and Fractals*, **16** (2003), 539-564

[145] L. Nottale and J. Schneider, *Jour. Math. Phys.*, **25** (1984), 1296-1300

[146] H. Ohanian and R. Ruffing, Gravity and spacetime, Norton, 1994

[147] L. O'Raifeartaigh, The dawning of gauge theory, Princeton Univ. Press, 1997

[148] G. Ord, *Chaos, Solitons, and Fractals*,**8** (1997), 727-741; 9 (1998), 1011-1029; *Jour. Phys. A*, **16** (1983), 1869-1884

[149] G. Ord, *Chaos, Solitons, and Fractals*, **11** (2000), 383-391; 17 (2003), 609-620

[150] G. Ord and J. Gualtieri, *Chaos, Solitons, and Fractals*, **14** (2002), 929-935

[151] G. Ord and R. Mann, quant-ph 0206095; 0208004

[152] G. Ord and A. Deakin, *Phys. Rev. A*, **54** (1996), 3772-3778

[153] D. Petz, *Jour. Math. Phys.*, **35** (1994), 780-795; quant-ph 0106125

[154] D. Petz and C. Sudar, *Jour. Math. Phys.*, **37** (1996), 2662-2673; quant-ph 0102132

[155] J. Pissondes, *Chaos, Solitons, and Fractals*, **9** (1998), 1115-1142; *Jour. Phys. A*, **32** (1999), 2871

[156] I. Quiros, gr-qc 0004014, 0011056, and 9904004; hep-th 0009169 and 0010146

[157] I. Quiros, R. Bonal, an R. Cardenas, gr-qc 9908075

[158] I. Quiros, R. Cardenas, and R. Bonal, gr-qc 0002071

[159] I. Quiros, R. Bonal, and R. Cardenas, gr-qc 9905071

[160] M. Reginatto, quant-ph 9909065

[161] M. Reginatto, *Phys. Rev. A*, **58** (1998), 1775

[162] N. Rosen, *Found. Phys.*, **12** (1982), 213-248; 13 (1983), 363-372

[163] H. Salehi, H. Motavali, and M. Golshani, hep-th 0011063

[164] H. Salehi, hep-th 9909157, 9912115, and 0302178

[165] H. Salehi, H. Sepangi, and F. Darabi, gr-qc 0002058

[166] H. Salehi and S. Mirabotalebi, gr-qc 0402074

[167] H. Salehi and H. Sepangi, hep-th 9810207

[168] H. Salehi, H. Motavali, and M. Golshani, hep-th 0011063

[169] H. Salehi, H. Sepangi, and F. Darabi, gr-qc 0002058

[170] H. Salehi and Y. Bisabr, hep-th 0301208 and 0001095

[171] E. Santamato, *Phys. Rev. D*, **29** (1984), 216-222

[172] E. Santamoto, *Phys. Rev. D* **32** (1985), 2615-2621; *Jour. Math. Phys.*, **25** (1984), 2477-2480

[173] A. Shojai and M. Golshani, quant-ph 9612022 and 9612019

[174] F. Shojai and M. Golshani, *Inter. Jour. Mod. Phys. A*, **13** (1998), 2135-2144 (gr-qc 9903047)

[175] F. Shojai and M. Golshani, *Int. Jour. Mod. Phys. A*, **13** (1998), 677-693

[176] F. Shojai, A. Shojai, and M. Golshani, *Mod. Phys. Lett. A*, **13** (1998), 2725-2729 and 2915-2922

[177] A. Shojai, F. Shojai, and M. Golshani, *Mod. Phys. Lett. A*, **13** (1998), 2965-2969 (gr-qc 9903048)

[178] A. Shojai, *Inter. Jour. Mod. Phys. A*, **15** (2000), 1757-1771 (gr-qc 0010013)

[179] F. Shojai and A. Shojai, gr-qc 0105102 and 0109052

[180] F. Shojai and A. Shojai, *Physica Scripta*, **64** (2001), 413 (quant-ph 0109025)

[181] F. Shojai and A. Shojai, *Gravitation and Cosmology*, **9** (2003), 163 (gr-qc 0306099)

[182] F. Shojai and A. Shojai, gr-qc 0404102

[183] F. Shojai and A. Shojai, *Inter. Jour. Mod. Phys. A*, **15** (2000), 1859-1868 (gr-qc 0010012)

[184] A. Shojai and M. Golshani, quant-ph 9812019, 9612023, 9612020, and 9612021

[185] F. and A. Shojai, *Class. Quant. Grav.*, **21** (2004), 1-9 (gr-qc 0311076), gr-qc 0409020 and 0409035

[186] A. and F. Shojai, quant-ph 0109025; gr-qc 0306100

[187] F. Shojai, gr-qc 9907093

[188] A. and F. Shojai, astro-ph 0211272

[189] S. Teufel, K. Berndl, and D. Dürr, quant-ph 9609005

[190] S. Teufel and R. Tumulka, math-ph 0406030

[191] S. Tiwari, quant-ph 0109048

[192] S. Tiwari, quant-ph 0109048; *Phys. Rev. A*, **56** (1997), 157-161

[193] S. Tiwari, *Superluminal phenomena in modern perspective*, Rinton Press, 2003

[194] R. Tumulka, math.PR 0312326; quant-ph 0408113 and 0210207

[195] A. Valentini, quant-ph 0104067, 0106098, 0309107, 0112151, and 0403034; *Phys. Lett. A*, **156** (1991), 5

[196] R. Wald, General relativity, Univ. Chicago Press, 1984

[197] D. Wallace, quant-ph 0112148 and 0112149

[198] H. Weyl, *Space, time, and matter*, Dover, 1952

[199] J. Wheeler, *Phys. Rev. D*, **41** (1990), 431

[200] J. Wheeler, hep-th 9706214, 0002068, and 0305017; gr-qc 9411030

[201] W. Wootters, *Phys. Rev. D*, **23** (1981), 357-362

[202] J. Zambrini, *Phys. Rev. A*, **38** (1987), 3631-3649; 33 (1986), 1532-1548; *Jour. Math. Phys.*, **27** (1986), 2307-2330

In: Trends in Quantum Gravity Research
Editor: David C. Moore, pp. 65-107
ISBN: 1-59454-670-3

Chapter 2

Quantum States of Neutrons in the Earth's Gravitational Field: State of the Art Applications and Perspectives

***V.V. Nesvizhevsky*[1] *and K.V. Protasov*[2]**
[1] Institut Laue-Langevin, Grenoble, France
[2] Laboratoire de Physique Subatomique et de Cosmologie, IN3P3-CNRS-UJF, Grenoble, France

Abstract

Gravitationally bound quantum states of matter were observed for the first time thanks to the unique properties of ultra-cold neutrons (UCN). The neutrons were allowed to fall towards a horizontal mirror which, together with the Earth's gravitational field, provided the necessary confining potential well. In this paper we discuss the current status of the experiment, as well as possible improvements: the integral and differential measuring modes; the flow-through and storage measuring modes; resonance transitions between the quantum states in the gravitational field or between magnetically split sub-levels of a gravitational quantum state.

This phenomenon and the related experimental techniques could be applied to various domains ranging from the physics of elementary particles and fields (for instance, spin-independent or spin-dependent short-range fundamental forces or the search for a non-zero neutron electric charge) to surface studies (for instance, the distribution of hydrogen in/above the surface of solids or liquids, or thin films on the surface) and the foundations of quantum mechanics (for instance, loss of quantum coherence, quantum-mechanical localization or experiments using the very long path of UCN matter waves in medium and in wave-guides).

In the present article we focus on transitions between the quantum states of neutrons in the gravitational field, consider the characteristic parameters of the problem and examine various methods for producing such transitions. We also analyze the feasibility of experiments with these quantum transitions and their optimization with respect to particular physical goals.

1 Introduction

The quantum motion of a particle with mass m in the terrestrial gravitational field and the acceleration g above an ideal horizontal mirror is a well-known problem in quantum mechanics which allows an analytic solution involving special functions known as Airy functions. The solutions of the corresponding Schrödinger equation with linear potential were discovered in 1920th [1] and can be found in major textbooks on quantum mechanics [2–7]. For a long time, this problem was considered only as a good theoretical exercise in quantum mechanics. The main obstacle for observing these quantum states experimentally was the extreme weakness of the gravitational interaction with respect to electromagnetic one, which meant that the latter could produce considerable false effects. In order to overcome this difficulty, an electrically neutral long-life particle (or quantum system) must be used for which an interaction with a mirror can be considered as an ideal total reflection. Ultracold neutrons (UCN) were discussed in this respect in refs. [8, 9]. UCN [10, 11] represent an extremely small initial part of total neutron flux. A reactor with very high neutron flux is therefore required. These quantum states were observed and investigated for the first time in a series of experiments [12–15] performed at the high-flux reactor at the Institut Laue-Langevin in Grenoble. Other quantum optics phenomena investigated with neutrons are presented in ref. [16].

To observe the gravitationally bound states, two experimental techniques were used. The first one, the so-called "integral" flow-through mode, is a measurement of the neutron flux through a narrow horizontal slit between a mirror below and an absorber/scatterer above it, which is used to scan the neutron density distribution above the mirror. This experimental technique allowed us to observe, for the first time, the non-continuous (discrete) behavior of the neutron flux. This observation was interpreted as being due to quantum states of neutrons corresponding to their vertical motion in the slit. Another, more sophisticated, so-called "differential" mode is based on specially developed position-sensitive neutron detectors with a very high spatial resolution, which made it possible to begin more detailed studies of this system and, in particular, to measure the spatial distributions of neutrons as a function of their height above a mirror (the square of the neutron wave function).

The present article does not claim to give an exhaustive overview of the different, rapidly developing applications of this beautiful phenomenon; it simply focuses on areas of particular interest to our research at present. In section 2, we start by giving a brief presentation of the phenomenon itself and in section 3 we describe the first experiment in which the ground quantum state was observed. Section 4 is devoted to a discussion of the "differential" measuring mode. Some of the interesting consequences of this experiment in different domains of physics (such as the search for exotic particles and spin-independent or spin-dependent short-range fundamental interactions; foundations of quantum mechanics) are discussed in section 5. Particular attention is paid to further developments of this experiment. In section 6, we present for the first time a feasibility analysis and theoretical description of the observation of resonance transitions between the quantum states. Such transitions could be induced by various interactions: by strong forces (if the mechanical oscillations of a bottom mirror are applied with a frequency corresponding to the energy difference between the quantum states), by electromagnetic forces (oscillating magnetic field), or probably even, at the limit of experimental feasibility, by gravitational forces (oscillating mass in the vicinity of the experimental setup). Some other methodological applications are also discussed.

2 The Properties of the Quantum States of Neutron in the Earth's Gravitational Field

The wave function $\psi(z)$ of the neutron in the Earth's gravitational field satisfies the Schrödinger equation:

$$\frac{\hbar^2}{2m}\frac{d^2\psi(z)}{d^2z}+(E-mgz)\psi(z)=0. \quad (2.1)$$

An ideal mirror at $z=0$ could be approximated as an infinitely high and sharp potential step (infinite potential well). Note that the neutron energy in the lowest quantum state, as will be seen a little later, is of the order of 10^{-12} eV and is much lower than the effective Fermi potential of a mirror, which is close to 10^{-7} eV. The range of increase of this effective potential does not exceed a few nm, which is much shorter than the neutron wavelength in the lowest quantum state ~10 μm. This effective infinite potential gives a zero boundary condition for the wave function:

$$\psi(z=0)=0. \quad (2.2)$$

The exact analytical solution of equation (1) which is regular at $z=0$, is the so-called Airy-function

$$\psi(z)=C\,\mathrm{Ai}\left(\frac{z}{z_0}\right). \quad (2.3)$$

Here

$$z_0=\sqrt[3]{\frac{\hbar^2}{2m^2g}} \quad (2.4)$$

represents a characteristic scale of the problem, C being the normalization constant. For neutrons at the Earth's surface the value of z_0 is equal to 5.87 μm. The equation (2.2) imposes the quantization condition:

$$z_n=z_0\lambda_n \quad (2.5)$$

where λ_n are zeros of the Airy function. They define the quantum energies:

$$E_n = mgz_0\lambda_n . \tag{2.6}$$

For the 4 lowest quantum states they are equal to:

$$\lambda_n = \{2.34,\ 4.09,\ 5.52,\ 6.79,\ \ldots\} \tag{2.7}$$

and for the corresponding energies, we obtain:

$$E_n = \{1.4,\ 2.5,\ 3.3,\ 4.1,\ \ldots\}\ \text{peV}. \tag{2.8}$$

It is useful to obtain an approximate quasi-classical solution of this problem [2–4,7]. This approximation is known to be valid, for this problem, with a very high accuracy, which is of the order of 1% even for the lowest quantum state. In accordance with the Bohr-Sommerfeld formula, the neutron energy in quantum states E_n^{qc} ($n = 1, 2, 3, \ldots$) is equal to:

$$E_n^{qc} = \sqrt[3]{\left(\frac{9m}{8}\right)\left(\pi\hbar g\left(n - \frac{1}{4}\right)\right)^2} \tag{2.9}$$

The exact energies E_n as well as the approximate quasi-classical values E_n^{qc} have the same property: they depend only on m, g and on the Planck constant $\hbar$, and do not depend on the properties of the mirror.

The simple analytical expression (2.9) shows that the energy of n-th state increases as $E_n^{qc} \sim n^{2/3}$ with increasing n. In other words, the distance between the neighbor levels decreases with increasing n.

In classical mechanics, a neutron with energy E_n in a gravitational field could rise to the maximum height of:

$$z_n = E_n / mg . \tag{2.10}$$

In quantum mechanics, the probability of observing a neutron in n-th quantum state with energy E_n at a height z is equal to the square of the modulus of its wave function $|\psi_n|^2$ in this quantum state. For the 4 lowest quantum states, neutron residence probability $|\psi_n|^2$ as a function of height above a mirror z is presented in Fig. 1 (see [2–6,12,13]). Formally, these functions do not equal zero at any height $z > 0$. However, as soon as a height z is greater than some critical value z_n, specific for every n-th quantum state and approximately equal to the height of the neutron classical turning point, then the probability of observing a neutron approaches zero exponentially fast. Such a pure quantum effect of the penetration of neutrons to a classically forbidden region is the tunneling effect. For the 4 lowest quantum states, the values of the classical turning points are equal to:

$$z_n = \{13.7, 24.0, 32.4, 39.9, \ldots\}\ \mu\text{m}. \tag{2.11}$$

An asymptotic expression for the neutron wave functions $\psi_n(z)$ at large heights $z > z_n$ [3, 4, 7] in the classically forbidden region is:

$$\psi_n(\xi_n(z)) \to C_n \xi_n^{-1/4} \exp\left(-\frac{2}{3}\xi_n^{3/2}\right), \tag{2.12}$$

for $\xi_n \to \infty$. Here C_n are known normalization constants and

$$\xi_n = \frac{z_0}{z_n} - \lambda_n . \tag{2.13}$$

As soon as such a height z_n is reached, the neutron wave function $\psi_n(z)$ starts approaching zero exponentially fast.

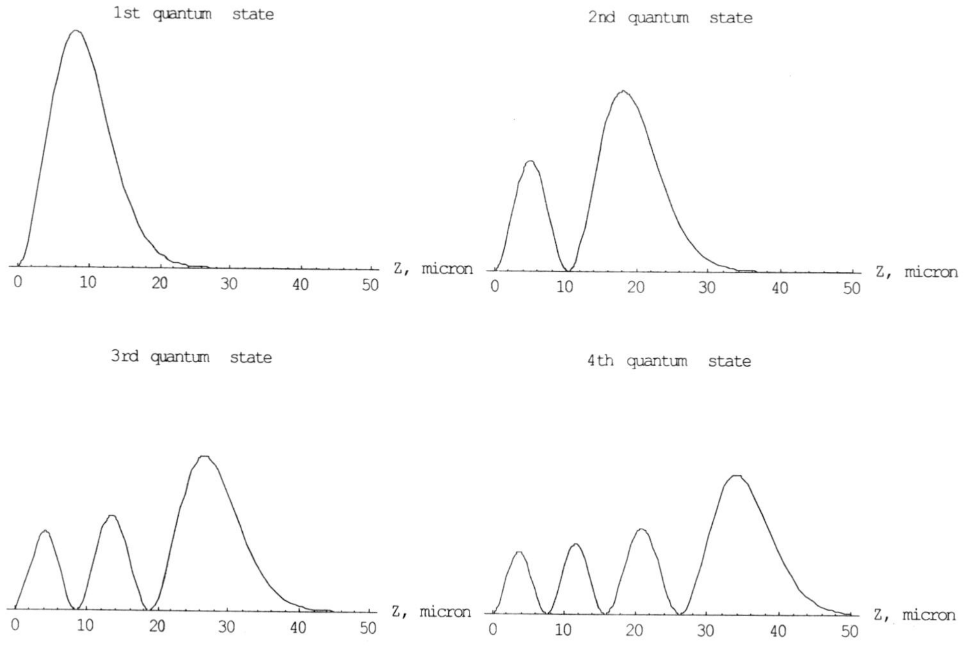

Fig. 1. Neutron presence probability as a function of height above the mirror z for the 1st, 2nd, 3rd and 4th quantum states.

3 Discovery of the Ground Quantum State in the "Integral" Flow-Through Mode

Such a wave-function shape allowed us to propose a method for observing the neutron quantum states. The idea is to measure the neutron transmission through a narrow slit Δz between a horizontal mirror on the bottom and a scatterer/absorber on top (which we shall refer to simply as a scatterer if not explicitly called otherwise). If the scatterer is much higher than the turning point for the corresponding quantum state $\Delta z \gg z_n$, then neutrons pass such a slit without significant losses. When the slit decreases, the neutron wave function $\psi_n(z)$ starts penetrating up to the scatterer and the probability of neutron losses increases. If the slit size is smaller than the characteristic size of the neutron wave function in the lowest quantum state z_1, then such a slit is not transparent for neutrons. Precisely this phenomenon was measured in a series of our recent experiments [12–15].

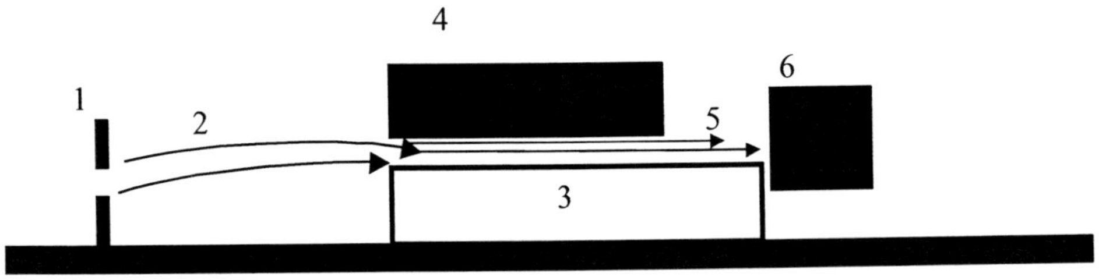

Fig. 2. A basic scheme of the first experiment. From left to right: the vertical bold lines indicate the upper and lower plates of the input collimator (1); the solid arrows correspond to classical neutron trajectories (2) between the input collimator and the entrance slit between the mirror (3, the empty rectangle below) and the scatterer (4, the black rectangle above). The dotted horizontal arrows illustrate the quantum motion of neutrons above the mirror (5), and the black box represents a neutron detector (6). The size of the slit between the mirror and the scatterer could be changed and measured.

A basic scheme of this experiment is presented in Fig. 2. The experiment (also described in ref.[17]) consists of measuring of the neutron flux (with an average velocity of 5–10 m/s) through a slit between a mirror and a scatterer as a function of the slit size. The size of the slit between the mirror and the scatterer can be finely adjusted and precisely measured. The scatterer's surface, while macroscopically smooth and flat, is microscopically rough, with roughness elements measuring in microns. In the classical approximation, one can imagine that this scatterer eliminates those neutrons whose vertical velocity component is sufficient for them to reach its surface. Roughness elements on the scatterer's surface lead to the diffusive (non-specular) reflection of neutrons and, as a result, to the mixing of the vertical and horizontal velocity components. Because the horizontal component of the neutron velocity in our experiment greatly exceeds its vertical component, such mixing leads to multiple successive impacts of neutrons on the scatterer/absorber and, as a result, to the rapid loss of the scattered neutrons. The choice of the absorbing material on the surface of the scatterer/absorber does not play a role, as has been verified experimentally in ref. [15]. Therefore the main mechanism causing the disappearance of neutrons is their scattering on the rough surface of the scatterer/absorber. This is why it is simply called a scatterer hereafter.

The neutron flux at the front of the experimental setup (in Fig. 2 on the left) is uniform over height and isotropic over angle in the ranges which exceed the slit size and the angular acceptance of the spectrometer respectively by more than one order of magnitude. The spectrum of the horizontal neutron velocity component is shaped by the input collimator with two plates, which can be adjusted independently to a required height. The background caused by external thermal neutrons is suppressed by "4π shielding" of the detector. A low-background detector measures the neutron flux at the spectrometer exit. Two discrimination windows in the pulse height spectrum of the ^{4}He detector are set as follows: 1) a "peak" discrimination window corresponds to the narrow peak of the reaction $n+^3\text{He}\rightarrow t+p$ and provides low background; 2) a much broader range of amplitudes allowes the "counting of all events". This method make it possible to suppress the background efficiently: when the scatterer height is zero and the neutron reactor is "on" then the count rate corresponds, within statistical accuracy, to the detector background measured with the neutron reactor "off".

Ideally, the vertical and horizontal neutron motions are independent. This is valid if the neutrons are reflected specularly from the horizontal mirror and if the influence of the scatterer, or that of any other force, is negligible to those neutrons which penetrate through the slit. If so, the horizontal motion of the neutrons (with an average velocity of 5–10 m/s) is ruled by the classical laws, while in the vertical direction we observe the quantum motion with an effective velocity of a few centimeters per second and with a corresponding energy (2.9) of a few peV (10^{-12} eV). The degree of validity of each condition is not obviously a priori and was therefore verified in related experiments.

The length of the reflecting mirror below the moving neutrons is determined from the energy-time uncertainty relation $\Delta E\Delta t\sim\hbar$, which may seem surprising given the macroscopic scale of the experimental setup. The explanation is that the observation of quantum states is only possible if the energy separation between neighboring levels ($\Delta E_n = E_{n+1}-E_n\sim 1/n^{1/3}$, see (2.8)) is greater (preferably, much greater) than the level width δE. As the quantum number n increases, the energy separation ΔE_n between the neighboring levels decreases until the levels ultimately merge into a classical continuum. Clearly, the lower quantum states are simpler and more convenient to measure in methodological terms. As to the width of a quantum state, it is determined by its lifetime or (in our case) by the observation time, i.e. by the neutron's flight time above the mirror. Thus, the length of the mirror is determined by the minimum time of observation of the neutron in a quantum state and should fulfill the condition $\Delta\tau\geq 0.5$ ms. In our experiments, the average value of the horizontal neutron velocity was chosen to be close to 10 m/s or to 5 m/s, implying that a mirror 10 cm in length was long enough.

The vertical scale of the problem, on the other hand, is determined by the momentum-coordinate uncertainty relation $\Delta v_z\cdot\Delta z\sim\hbar/m$. The reason is that the smaller the vertical component of the neutron velocity, the larger the neutron wavelength corresponding to this motion component. However, the classical height to which a neutron can rise in the gravitational field cannot be less than the quantum-mechanical uncertainty in its position, i.e. less than the neutron wavelength. In fact, it is this condition which specifies the lowest bound state of a neutron in a terrestrial gravitational field. The uncertainty in height is then ~15 μm, whereas the uncertainty in the vertical velocity component is ~1.5 cm/s.

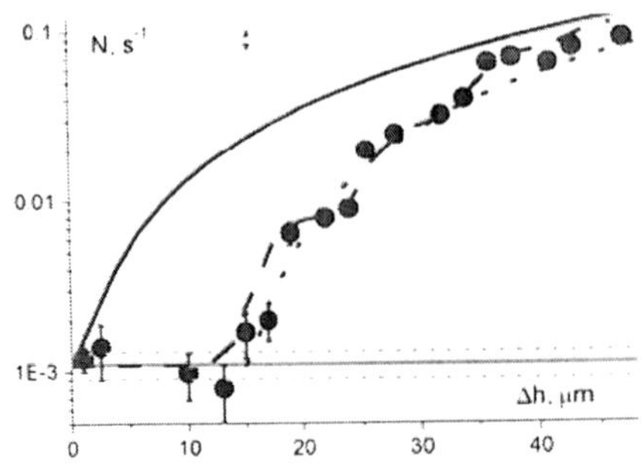

Fig. 3. Neutron flux through a slit between a horizontal mirror and a scatterer above it is given as a function of the distance between them obtained in the first experiment [12,13]. Experimental data are averaged over 2-μm intervals. The dashed line represents quantum-mechanical calculations in which both the level populations and the energy resolution of the experiment are treated as free parameters being determined by the best fit to the experimental data. The solid line corresponds to classical calculations. The dotted line is for a simplified model involving only the lowest quantum state.

The results of the first measurement presented in Fig. 3 (see refs. [12, 13]) differ considerably from the classical dependence and agree well with the quantum-mechanical prediction. In particular, it is firmly established that the slit between the mirror and the scatterer is opaque if the slit is narrower than the spatial extent of the lowest quantum state, which is approximately 15 μm. The dashed line in Fig. 3 shows the results of a quantum-mechanical calculation, in which the level populations and the height (energy) resolution were treated as free parameters. The solid line shows the classical dependence normalized so that, at sufficiently large heights (above 50–100 μm), the experimental results are described well by the line. The dotted line given for illustrative purposes describes a simplified situation with the lowest quantum state alone, i.e. in drawing this line only the uncertainty relation was taken into account. As can be seen from Fig. 3, the statistics and energy resolution of the measurements are still not good enough to detect quantum levels at a wide slit, but the presence of the lowest quantum state is clearly revealed.

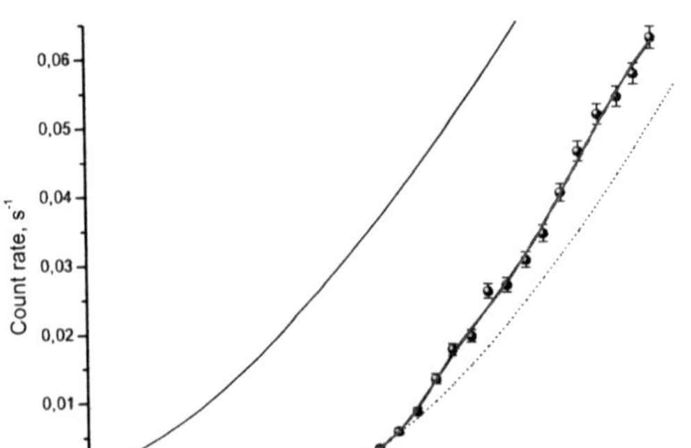

Fig. 4. Neutron flux through a slit between a horizontal mirror and a scatterer above it is given as a function of the distance between them obtained in the second experiment [15].

However, as was shown experimentally (Fig. 4) and explained theoretically in ref. [15], even when the height (energy) resolution and statistics are improved considerably compared to those in refs. [12, 13], further significant improvement of resolution in the "integral" measuring mode presented is scarcely achievable due to one fundamental constraint: the finite sharpness of the dependence on height of the probability of neutron tunneling through the gravitational barrier between the allowed heights for neutrons and the height of the scatterer [15]. As is demonstrated in this article, the neutron flux $F(\Delta z)$ as a function of the scaterrer position Δz above the turning point z_n ($\Delta z > z_n$) can be written within the quasi-classical approximation, for a given level, as:

$$F(\Delta z) \sim \mathrm{Exp}\left(-\alpha \mathrm{Exp}\left(-\frac{4}{3}\left(\frac{\Delta z - z_n}{z_0}\right)^{\frac{3}{2}}\right)\right), \tag{3.1}$$

where z_0 is given in (2.4) and α is a constant. The exponent factor after this constant represents here the probability for the neutron to pass from the classically allowed region to the scatterer/absorber, i.e. the probability of tunneling through the gravitation barrier. This dependence describes the experimental data reasonably well (see Fig. 4) and gives a simple explanation for the existence of intrinsic resolution related to the tunneling effect. Roughly speaking, to resolve experimentally the nearest states $n+1$ and n, the distance $z_{n+1} - z_n$ should be smaller than a characteristic scale of the function (3.1), which is approximately equal to $z_0 = 5.87\ \mu\mathrm{m}$. This condition can be satisfied only for the ground state because even for the first excited state the difference $z_2 - z_1 \approx 8\ \mu\mathrm{m}$ is comparable with z_0.

Nevertheless, the theoretical description of the measured experimental data within the model of the tunneling of neutrons through this gravitational barrier shows reasonable agreement between the extracted parameters of the quantum states and their theoretical prediction. In order to increase the accuracy of this experiment further in the mode which involves scanning the neutron density using a scatterer at various heights, we are working in two directions: First of all, further development [18] of the theoretical description of this experiment could allow us to reduce the theoretical uncertainties in the determination of quantum states parameters to the velev of a few percent. On the other hand, experimental efforts related to improving the accuracy of the absolute positioning of the scatterer [19, 20] would produce a comparable level of accuracy.

To summarize this section, it can be said that the lowest quantum state of neutrons in the gravitational field was clearly identified using the "flow-through" mode, which measures the neutron flux as a function of an absorber/scatterer height. This observation itself already makes many interesting applications possible. Higher quantum states could also be resolved. However, such a measurement is much more complicated because the energy (or height) resolution of the present method is limited by one main factor: the finite sharpness of the dependence on height of neutron tunneling through the gravitational barrier between the classically allowed height and the scatterer height.

4 Studies of the Neutron Quantum States in "Differential" Flow-Through Mode

In order to resolve higher quantum states clearly and measure their parameters accurately, we must adopt other methods, such as for example, the "differential" method, which uses position-sensitive neutron detectors with a very high spatial resolution, which were developed specifically for this particular task [21].

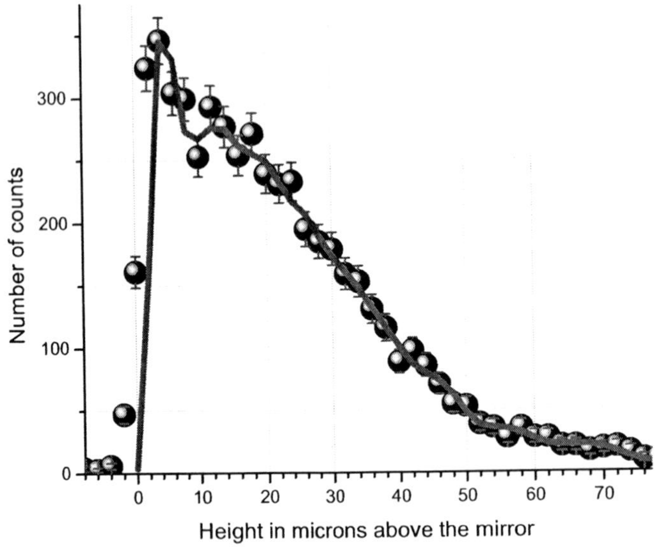

Fig. 5. The results of the measurement of the neutron density above a mirror in the Earth's gravitational field are obtained using a high-resolution plastic nuclear-track detector with uranium coating. The horizontal axis corresponds to a height above the mirror in microns. The vertical axis gives the number of events in an interval of heights. The solid line shows the theoretical expectation under the assumption that the spatial resolution is infinitely high. Calculated populations of the quantum states correspond to those measured by means of two scatterers using the method shown in Fig. 6.

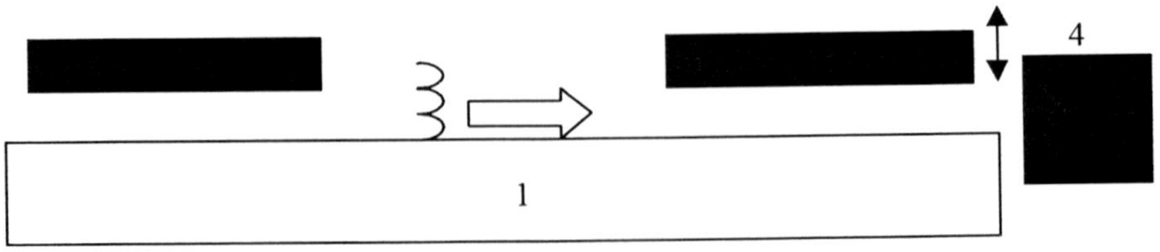

Fig. 6. A scheme of the experiment with a long bottom mirror (1, shown as the open box) and with two scatterers (2, 3, shown as the black boxes). The first scatterer (2, on the left) shapes the neutron spectrum. It is installed at the constant height of 42 μm. The second scatterer (3, on the right) analyses the resulting neutron spectrum. Its height is varied. The detector (4), shown as the black box, measures the total neutron flux at the exit of the slit between the mirror and the analyzing scatterer. The distance between the scatterers is equal to 9 cm.

The direct measurement of the spatial density distribution in a standing neutron wave is preferable to its investigation with the aid of a scatterer whose height can be adjusted. The former technique is differential, since it permits the simultaneous measurement of the probability that neutrons reside at all heights of interest. The latter technique is integral, since the information on the probability that neutrons reside at a given height is in fact obtained by the subtraction of the values of neutron fluxes measured for two close values of the scatterer height. Clearly, the differential technique is much more sensitive than the integral one and makes it possible to gain the desired statistical accuracy much faster. This is of prime importance considering the extremely low counting rate in this experiment, even with the use of the highest UCN flux available today. Furthermore, the scatterer employed in the integral technique inevitably distorts the measured quantum states by deforming their eigen-functions and shifting their energy values. The finite accuracy of taking these distortions into account results in systematic errors and ultimately limits the attainable accuracy of the measurement of the quantum state parameters. For these and other reasons, the use of a position-sensitive detector to directly measure the probability of neutron residence above the mirror is highly attractive. However, until now there were no neutron detectors with the spatial resolution of ~1 μm needed for this experiment. We therefore had to develop such a detector and measuring technique. The result was a plastic track nuclear detector (CR39) with a thin uranium coating ($^{235}UF_4$), described in ref. [21]. The tracks created by the entry into the plastic detector of a daughter nucleus produced by the neutron-induced fission of a ^{235}U nucleus were increased to ~1 μm in diameter by means of chemical development in an alkaline solution. The developed detector was scanned with an optical microscope over a length of several centimeters with an accuracy of ~1 μm. The sensitive ^{235}U layer is thin enough (<1 μm) for the coordinates of neutron entry into the uranium layer to almost coincide with the coordinates of daughter nucleus entry into the plastic. On the other hand, the sensitive layer is thick enough to ensure high UCN detection efficiency (~30 %). The measuring technique and the preliminary analysis of the results are described in ref. [15].

The feasibility of this technique was demonstrated in the second experiment and the results are presented in Fig. 5 [15]. This is the first direct measurement of the neutron density above the mirror with a spatial resolution of 1-2 μm. The theoretical curve presented in Fig. 5 is calculated with known neutron wave functions and with the quantum level populations and the zero height above the mirror as free parameters. The spatial detector resolution is assumed to be perfect. A comparison of the experimental data with the theoretical prediction suggests that: firstly, the measured presence probability for neutrons above the mirror on the whole domain of Δz corresponds closely to the theoretical prediction; secondly, the spatial detector resolution can be estimated, for instance, using the steepest portion of the dependence near the zero height, which is equal to ~1.5 μm; finally, even a relatively small neutron density variation of ~10%, which is to be expected for the mixture of several quantum states employed in this experiment, can be measured using this technique. It should be noted that this measurement was performed in the special geometry of the mirror and the scaterrers shown in Fig.6. A long bottom mirror (1) was used with two scatterers (2) and (3). The first scatterer gives the neutron spectrum the desirable shape and is installed at the constant height of 42 μm. The second one analyses the resulting neutron spectrum; its height is varied. The detector (4), shown as the black box, measures the total neutron flux at the exit of the slit between the mirror and the analyzing scatterer. The distance between the scatterers is equal to 9 cm.

However, the measurement presented in Fig. 5 is merely a test of the detector for spatial resolution and is not optimized for studying the neutron quantum states in this system. In ref. [20], the measurement with the position-sensitive detector was analyzed from the standpoint of its optimization for the identification of neutron quantum states. Fig. 1 depicts the probability $\psi_n^2(z)$ of neutron detection at a height z above the mirror surface for 4 pure quantum states. Clearly, every dependence $\psi_n^2(z)$ has n maxima and $n-1$ minima between them with zero values at the minima, which is characteristic of any standing wave. An ideal experiment would consist of the extraction of one or several pure quantum states higher than the first one ($n > 1$) and the direct measurement of neutron detection probability against the height above the mirror with the aid of a position-sensitive detector with a spatial resolution of ~1 μm.

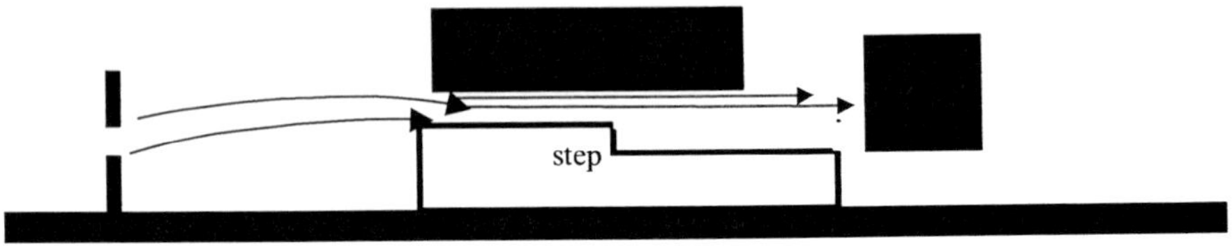

Fig. 7. A scheme of the experiment with a small negative step on the lower mirror, which allows the transition of neutrons to higher quantum states (to the region to the right of the step).

Let us consider a possible method for carrying out such an experiment. One or two lower quantum states can be selected with a scatterer by the conventional method adopted in all our previous experiments, which showed that the spectrometer resolution is sufficient for this. The method for transferring neutrons from the lower quantum states to the higher quantum states was considered in ref. [22]. It involves the fabrication of a small negative step on the lower mirror, as shown in Fig. 7. Neutrons are in quantum states both to the left of the step and to the right of the step. However, the corresponding wave functions have shifted relative to each other by the step height Δz_{step}. By passing through the step, neutrons are redistributed from the nth quantum state prior to the step $\psi_{\text{before}}(z) = \psi_n(z + \Delta z_{\text{step}})$ over the quantum state $\psi_{\text{after}}(z) = \psi_n(z)$ after the step with some probabilities $\beta_{nk}^2(\Delta z_{\text{step}})$. In this case, the step can be treated as an infinitely fast perturbation and therefore the transition matrix element $\beta_{nk}(\Delta z_{\text{step}})$ is:

$$\beta_{nk}(\Delta z_{\text{step}}) = \int_0^\infty \psi_n(z + \Delta z_{\text{step}})\psi_k(z)dz\,. \tag{4.1}$$

Fig. 8 shows the calculated probability $\beta_{1k}^2(\Delta z_{\text{step}})$ of transition from the 1st quantum state, prior to passing through the step, to the 1st, 2nd, 3rd and 4th quantum states after passing through the step.

When the negative step is large enough, for instance is equal to (–15 μm), the probability β_{11}^2 to detect neutrons in the lowest quantum state after passing through the step is extremely small. The similar probability β_{n1}^2 for neutron transitions from higher initial quantum states is also low. Any overlap integral β_{n1}^2 for $\Delta z_{\text{step}} = -15\ \mu\text{m}$ is small, since the spatial dimension of the neutron wave function in the lowest quantum state $\psi_1(z)$ is smaller than 15 μm.

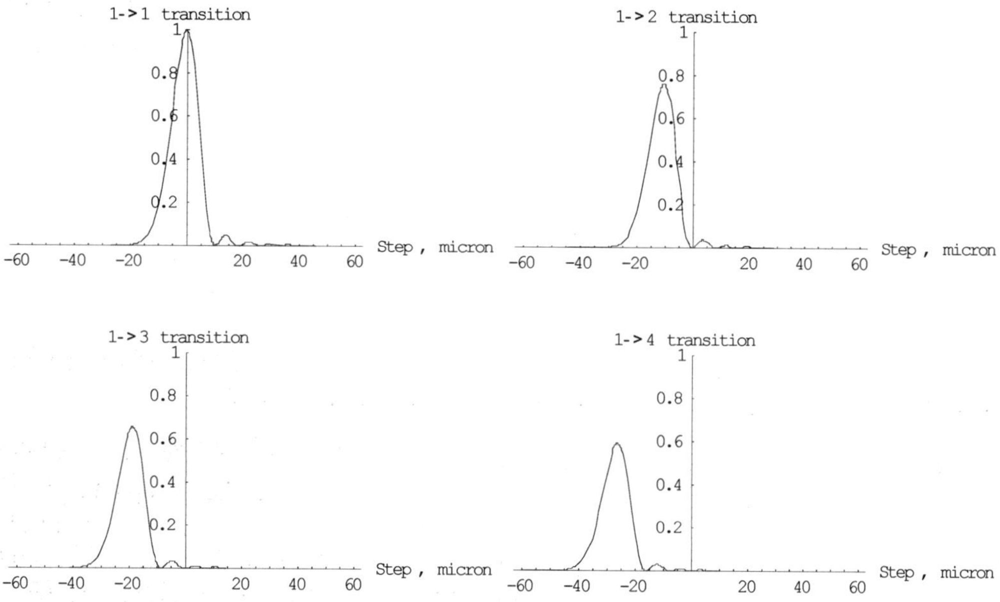

Fig. 8. Probability of neutron transition from the 1st quantum state, prior to transit through the step, to the 1st, 2nd, 3rd and 4th quantum states on transit through the step as a function of the step height Δz_{step}.

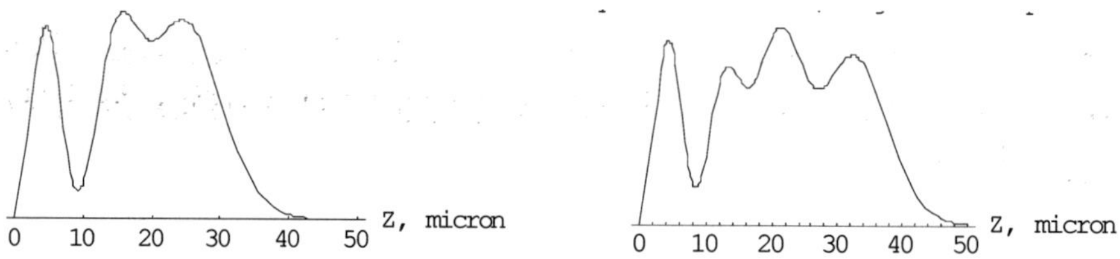

Fig. 9. Probability of neutron residence versus height above the mirror on neutron transit through a negative 15-μm step for two cases: one and two lowest quantum states prior to the passage through the step.

Fig. 9 shows the probability of neutron detection above the mirror depending on the height after the neutron passes through the negative 15-μm step. The probability is plotted in two cases: for one and two quantum states prior to passing through the step. It is evident that the expected spatial variation of neutron density is clearly defined and can be measured. The reason for such a strong neutron density variation in the case of the elimination of the lowest quantum state is simple: we can see from Fig. 1 that only the lowest quantum state has a peak near 10 μm. The remaining low-lying quantum states possess a minimum at this height. Therefore, several lower quantum states ($n > 1$) are "coherently" combined: the probability of neutron detection at a height of ~10 μm is systematically much lower than for neighboring heights.

This idea was demonstrated in the last experiment performed in the summer of 2004 [23]. A neutron beam with a horizontal velocity component of ~5 m/sec and a vertical velocity component of 1–2 cm/sec, which corresponds to the energy of the lowest neutron quantum state in the gravitational field above a mirror, is selected using a bottom mirror (1) and a scatterer/absorber (3) positioned above it at a height of ~20 μm. A second mirror (2) is installed 21 μm lower than the first mirror (1). The precision of the optical components' adjustment and the neutron detection resolution are equal to ~1 μm.

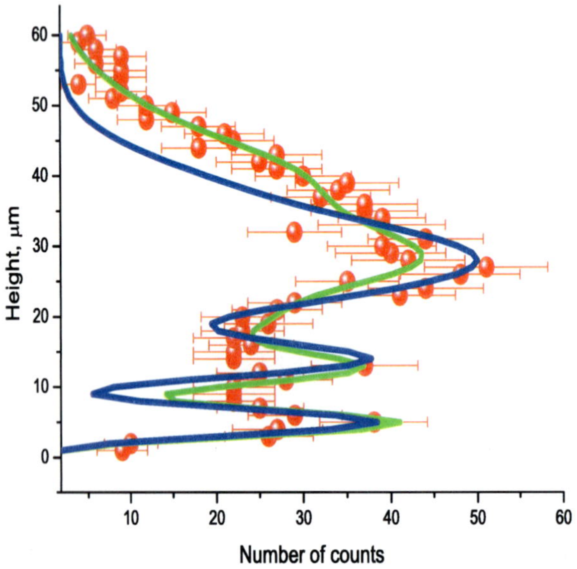

Fig. 10. The neutron density distribution in the gravitational field is measured using position-sensitive detectors of extra-high spatial resolution. The circles indicate experimental results. The solid curve corresponds to the theoretical expectation under the assumption of an ideally efficient scatterer able to select a single quantum state above the mirror (1) and no parasitic transitions between the quantum states above the mirror (2). The dotted curve corresponds to the more realistic fit using precise wave-functions and free values for the quantum states populations (for simplicity, the intereference terms between different levels are neglected). The detector background is constant in the range from –3 mm to +3 mm below and above the presented part of the detector.

Typical results of a few days' detector exposure in such an experiment are presented in Fig. 10. Even if the analysis of these data has not yet been completed and the fine details of the quantum states can not be extracted, we can see clearly that the experimental approach developed here allows us to obtain a very pronounced variation of the wave function and can thus be considered as very promising.

The characteristic behavior of the neutron wave functions in the quantum states in the gravitational field above the mirror, as well as the successful initial testing of the position-sensitive detector with a uranium coating, suggest that it will be possible to identify neutron quantum states by directly measuring the neutron detection probability above a mirror using the position-sensitive detector. It should be noted that this detector could be also used to measure the velocity distribution in quantum states. To do so, we need simply to shift the detector a few centimeters downstream to the bottom mirror edge: the spatial spread of the picture thus obtained will not be sensitive to the initial position of the neutron above the mirror but to its velocity.

Thus, the two techniques considered and the available fluxes of UCN are already sufficient for a broad range of applications. Let us analyze them briefly, before considering further developments of this experiment, related to resonance transitions between different quantum states and thus to a much more precise measurement of the parameters of these quantum states.

5 Use of Neutron Quantum States in Different Domains of Physics

As we have already mentioned in section 3, further development [18] of the theoretical description of this experiment and experimental efforts related to improving the accuracy of the absolute positioning of the scatterer [19, 20] could allow us to achieve close to a few percent accuracy in the determination of quantum state parameters. It should also be noted that the direct measurement of the spectral variation of neutron density above mirror in the quantum states seem to be quite promising. For this reason we are rather confident that, even at this early stage we can already obtain some interesting physical results using this method.

For instance, as shown in ref. [24] and presented here in section 5.1, a competitive upper limit for short-range fundamental forces was obtained simply from the very fact that the gravitationally bound quantum states exist. Moreover, if any additional short-range interaction were to exist (of whatever nature: new hypothetical particles, supplementary spatial dimensions, etc.), this would change the parameters of the neutron quantum states. Therefore, the precise measurement of these parameters gives an upper limit for unknown interactions.

This experiment can also be used to search for the axion – a hypothetical particle which strongly violates CP invariance; the characteristic distance for this interaction is comparable to the characteristic length of our problem z_0. This is discussed in section 5.2 and can be considered within the more general context of studies of spin-gravity interaction.

This method could be used for studies related to the foundations of quantum mechanics, such as for instance, the quantum-mechanical localization (also known as quantum revivals, see section 5.3) [25], or various extensions of quantum mechanics [26, 27] (see section 5.4). One should note here that the present method provides two unique opportunities: on the one hand, it provides a rare combination of quantum states and gravitation that is favorable for testing possible extensions of quantum mechanics; on the other hand, UCN can be reflected from the surface up to $\sim 10^5$ times without loss, i.e. much more than for optical phenomena, which means that any kind of localization can be better studied using UCN. Finally, as presented in section 5.5, this method could be useful for such problems of high long-term interest as the loss of quantum coherence in the systems with gravitational interaction (see, for instance, refs. [28, 29]).

5.1 Search for Non-newtonian Gravity

According to the predictions of unified gauge theories, super-symmetry, super-gravity and string theory, there exist a number of light and massless particles [30]. An exchange of such particles between two bodies gives rise to an additional force. Additional fundamental forces at short distances have been intensively studied, in particular over the past few years in the light of the hypothesis about "large" supplementary spatial dimensions proposed by Antoniadis, Arkami-Hamed, Dimopoulos and Dvali [31] and based on earlier ideas presented in [32–35]. A review of theoretical works and recent experimental results can be found in [36–40]. This hypothesis could be verified using neutrons because the absence of an electric charge makes it possible to strongly suppress the false electromagnetic effects [41]. It was noticed in [42] that the measurement of the neutron quantum states in the earth's gravitational field is sensitive to such extra forces in the sub-micrometer range. In the case of $n = 3$ extra dimensions, the characteristic range lies just within the nanometre domain [31, 41] which is accessible in this experiment. The first attempt to establish a model-dependent boundary in the range 1–10 μm was presented in [40].

An effective gravitational interaction in the presence of an additional Yukawa-type force is conventionally parameterized as:

$$V_{\text{eff}}(r) = G\frac{m_1 m_2}{r}\left(1+\alpha_G e^{-r/\lambda}\right) \tag{5.1}$$

Here, G is the Newtonian gravitational constant, m_1 and m_2 are interacting masses, r their relative distance, α_G and λ are the strength and characteristic range of this hypothetical interaction.

The dependence of neutron flux on the slit size is sensitive to the presence of quantum states of neutrons in the potential well formed by the earth's gravitational field and the mirror. In particular, the neutron flux was found to be equal to zero within the experimental accuracy if the slit size Δz was smaller than the characteristic spatial size (a quasi-classical turning point height) of the lowest quantum state of $\sim 15\ \mu m$ in this potential well. The neutron flux at the slit size $\Delta z < 10\ \mu m$ in the second experiment [15] was lower by at least a factor of 200 than that for the lowest quantum state ($\Delta z \approx 20\ \mu m$).

If an additional short-range force of sufficiently high strength were to act between the neutrons and the mirror then it would modify the quantum states parameters: an attractive force would "compress" the wave functions towards the mirror, while a repulsive force would shift them up. In this experiment, no deviation from the expected values was observed within the experimental accuracy. This accuracy is defined by the uncertainty in the slit size, which can be conservatively estimated as ~30% for the lowest quantum state [15].

As we mentioned in section 2, the motion of neutrons in this system over the vertical axis z could be considered, in a first, relatively good approximation, as a one-dimensional problem for which the mirror provides an infinitely high potential. The interaction between neutrons and the Earth is described by the first term in eq. (5.1) and can be approximated by the usual linear potential $(r = R + z)$:

$$V(z) = mgz \tag{5.2}$$

with $g = GMm / R^2$, R being the Earth's radius, M its mass.

The second term in eq. (5.1) introduces an additional interaction. Due to the short range of this interaction, its main contribution is provided by the interaction of neutrons with a thin surface layer of the mirror and the scatterer.

Let us first estimate the interaction of neutrons with the mirror due to this additional term if this interaction is attractive. If the mirror's density is constant and equal to ρ_{m}, then an additional potential of the interaction between the neutrons and the mirror, in the limit of small λ, is given by [24]:

$$V'(z) = -U_0 e^{-z/\lambda} \tag{5.3}$$

with $U_0 = 2\pi G \alpha_G m \rho_{\mathrm{m}} \lambda^2$.

The simplest upper limit on the strength of an additional interaction follows from the condition that this additional interaction does not itself create any bound state. It is known [7] that for an exponential attractive ($U_0 > 0$) potential (5.3) this means that

$$\frac{U_0 m \lambda^2}{\hbar^2} < 0.72 \cdot \tag{5.4}$$

This condition gives a boundary for an additional potential strength:

$$\alpha_G = 0.72 \frac{2}{\pi} \frac{\rho}{\rho_{\mathrm{m}}} \frac{\hbar}{mg\lambda^2} \frac{\hbar}{m\lambda} \frac{R}{\lambda} \tag{5.5}$$

ρ being the Earth's average density. In this experiment, both densities are close to each other $\rho \approx \rho_{\mathrm{m}}$, therefore their ratio ρ / ρ_{m} is close to 1. However, a suitable choice of mirror material (coating) would easily allow us to gain a factor of 3–5 in the sensitivity in future experiments. We obtain the following numerical boundary:

$$\alpha_G = 1 \times 10^{15} \left(\frac{1\ \mu\mathrm{m}}{\lambda} \right)^4 . \tag{5.6}$$

Here, 1 μm is chosen as a natural scale for this experiment. This limit is presented in Fig. 11 in comparison with the limits from the Casimir-like and van der Waals force measurement experiments [38], as well as from experiments on protonium atoms. An additional force between a nucleus and an antiproton would change the spectrum of such an atom. The most precise measurement of the energy spectrum of antiprotonic atoms was done for $^3He^+$ and $^4He^+$ atoms by the ASAKUSA collaboration at the antiproton decelerator at CERN [44]. No deviation was found from the values expected within the QED calculations [43]. An 3σ upper limit on $|\alpha_G|$ from this experiment was established in [24]:

$$|\alpha_G| = 3.3 \times 10^{28}. \tag{5.7}$$

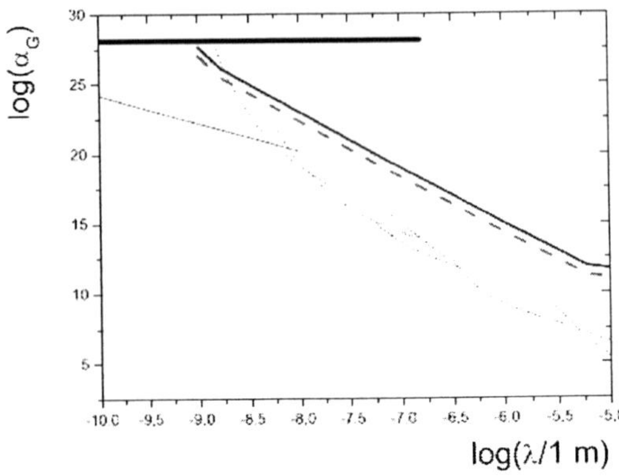

Fig. 11. The constraints on α_G following from this experiment [12, 13] (the solid line) in comparison with that from the measurement of the Casimir and the van derWaals forces [35] (the short dashed lines). The long dashed line shows a limit which can be easily obtained by an improvement of this experiment. The solid horizontal line represents the limit established from the atomic experiment [41]. Dash-dotted line shows the limit which would be obtained if one equals the strength of this additional hypothetical interaction to the value of effective Fermi potential for Pb [43].

It is necessary to note that, in the realistic case, one has to establish a condition of non-existence of an additional bound state for the sum of (5.2) and (5.3) but not for the interaction (5.3) alone. The presence of the linear potential modifies slightly the critical value in (5.4). For instance, for $\lambda = 1\ \mu m$ it is approximately equal to 1.0 and for $\lambda = 0.1\ \mu m$ it is equal to 0.74. For smaller λ, this value tends to 0.72. It is possible to explain qualitatively why the strength of an additional interaction should be higher in the presence of the *mgz*-potential than without it. When a bound state has just appeared, then its wave function is extremely spread. If a supplementary "external" confining potential is added, it does not allow the wave function to be spread and thus a stronger potential is needed to create a bound state.

The range of presented λ is 1 nm–10 μm. A deviation from a straight line in the solid curve at 1 nm is due to the finite range of increase of the mirror effective nuclear potential (impurities on the surface and its roughness). The same effect at $\lambda \approx 10\ \mu m$ is due to "interference" between the potentials (5.2) and (5.3).

Unfortunately, this experiment does not allow us to establish a competitive limit for a repulsive interaction. In this case, there could be no "additional" bound state. Here, instead of the condition of "non-existence" of a bound state, one could consider the critical slit size for which the first bound state appears in this system. Such an approach would be model-dependent due to uncertainties in the description of the interaction of neutrons with the scatterer. Nevertheless, it is possible to obtain a simple analytical expression for small λ and to show explicitly a difference in the sensitivity of this experiment to an attractive and to a repulsive additional interaction.

$$\alpha_G = \frac{1}{\pi}\frac{\rho}{\rho_m}\frac{\hbar}{mg\lambda^2}\frac{\hbar}{m\lambda}\frac{R}{\lambda}\exp(\lambda_0/\lambda) \quad (5.8)$$

with $\lambda_0 = \delta E_n / mg$, δE_n being the precision of determination of the n-th quantum state energy.

A direct comparison of relation (5.8) to (5.5) shows that the limit (5.8) at small λ is sufficiently less restrictive than the limit for an attractive one (5.5) due to the exponential factor. On the other hand, it would be possible to achieve as strict a limit for a repulsive interaction as for an attractive one, if the mirror was coated with a material with negative Fermi potential.

As a conclusion, let us emphasize that even though this experiment was never designed to search for additional short-range forces it provides the competitive limit (5.5) in the nanometer range. However, it could be easily improved in the same kind of experiment by making some obvious modifications. For instance, one could choose a mirror material (coating) with a higher density. A significant improvement to such a limit would only seem possible by using the "storage" method, which would allow a gain in accuracy of a few orders of magnitude.

A more significant gain in the sensitivity could be achieved in dedicated neutron experiments. Simply as a qualitative illustration of the potential capacities of experiments with neutrons, it can be said that if the strength of this additional hypothetical interaction were equal to the value of effective Fermi potential for Pb [46] this equality would produce the limit presented by the dash-dotted line in Fig. 11.

5.2 Search for the Axion and Spin-Gravity Interaction

Axions are well-known as a possible solution to the strong CP problem as well as an interesting darkmatter candidate [47]. One of the most remarkable predictions associated with the axion is that it yields a parity and time-reversal violating, monopole-dipole coupling between spin and matter [48]. Experimental and astrophysical observations imply that the mass of the axion must lie between 1 μeV and 1 meV, corresponding to a range between 20 cm and 0.2 mm [49]. This range is commonly referred to as the "axion window." An exhastive review of theoretical and experimental activities to search for the axion can be found in [30].

Axions mediate a CP violating monopole-dipole Yukawa-type gravitational interaction potential [48]

$$V(\vec{r}) = \hbar g_p g_s \frac{\vec{\sigma}\cdot\vec{n}}{8\pi mc}\left(\frac{1}{\lambda r}+\frac{1}{r^2}\right)e^{-r/\lambda} \quad (5.9)$$

between spin and matter where $g_p g_s$ is the product of couplings at the scalar and polarized vertices and λ is the range of the force. Here r is the distance between the neutron and the nucleus and $\vec{n} = \vec{r}/r$ a unitary vector.

Untill now, only a few experiments placed upper limits on the product coupling $g_p g_s$ in a system of magnetized media and test masses. Of the experiments covering the axion window, one of them [50] had peak sensitivity near 100 mm (2 μeV axion mass) and another [51] had peak sensitivity near 10 mm (20 μeV axion mass).

Let us make an initial qualitative estimation of the limit of the axion coupling constant which can be established from the existing experiment. The upper limit for which the peak of sensitivity is clearly close 10 μm.

By analogy with the demonstration presented in the previous section where an additional interaction between (5.1) the neutron and the mirror's nuclei created an additional neutron-mirror interaction potential (5.3), in the case of the interaction (5.9), a neutron with a given projection of spin on the vertical (*g*) axis will see an additional potential with the following shape created by the whole mirror:

$$U(z_0) = \frac{g_p g_s}{4\pi} \frac{\pi \hbar \rho_m \lambda}{2m^2 c} e^{-z_0/\lambda} . \tag{5.10}$$

This potential, considered as a perturbation, will produce a positive energy shift ε_0 (in the first order of the perturbation theory) for one of two possible spin projections and a negative energy shift $-\varepsilon_0$. Thus obtained, the energy splitting can be constrained from the experimental data. For instance, we can propose a very rough and robust upper limit if we says that this splitting is smaller than at least half of the energy difference between two gravitational levels:

$$2\varepsilon_0 \leq \frac{1}{2}(E_2 - E_1) \equiv \frac{1}{2}\Delta E \tag{5.11}$$

Therefore the limit of the axion coupling constant will be given by

$$\frac{g_p g_s}{\hbar c} = \frac{2\Delta E m^2}{\hbar^2 \rho_m \lambda} \tag{5.12}$$

(here the exponential function is replaced by 1, because the size of the wave function is of the order of ten micrometers whereas the range of the interaction, for the axion window, is higher than 100 microns).

To obtain a naive estimation for $\lambda = 1$ mm, we can suppose that $\Delta E = 1$ peV (i.e. the energy difference between two gravitational levels), $\rho_m = 4000$ kg/m^3:

$$\frac{g_p g_s}{\hbar c} = 2 \cdot 10^{-17} . \tag{5.13}$$

This limit is at least a few orders of magnitude better than the limit obtained in the experiments [50, 51].

In principle, a very competitive constraint could be obtained using the present flow-through method for spin-dependent short-range forces in a dedicated experiment with polarized neutrons. By alternating the neutron spin in such an experiment an accuracy of $\sim 10^{-3} - 10^{-4}$ could easily be achieve (instead of 1 considered in the estimation given here). The main simplification in the case of spin-dependent forces is the relative nature of the measurement, because the neutron spin can be easily flipped with a high accuracy. In contrast to that, spin-independent forces can not be "switched off". We would therefore need an absolute measurement in this case.

Let us emphasize that this discussion can be seen as a part of the wider search for spin-gravity interaction. The idea that a nuclear particle may possess a gravitoelectric dipole moment was proposed about forty years ago by Kobsarev and Okun [52] and by Leitner and Okubo [53]. A brief review of experimental and theoretical activity on this question can be found in [54]. Here we would like to emphasize that this problem has been discussed at length in a number of recent articles, with arguments for [52] and against [56] this kind of term (5.10) in the interaction of fermions with an external gravitational field, and that the inrtoduction of polarized neutrons into our experiment does not represent a difficult experimental challenge.

5.3 Quantum Revivals

The application of this experiment to quantum mechanical localization (also known as quantum revivals) was considered in detail in a recent review article by Robinett [25]. Let us remind the reader of the main ideas presented there and the feasibility of such a measurement in our experimental setup.

Quantum revivals are characterized by initially localized quantum states which have a short-term, quasi-classical time evolution, which then can spread significantly over several orbits, only to reform later in the form of a quantum revival in which the spreading reverses itself, the wave packet relocalizes, and the semi-classical periodicity is once again evident.

The study of the time-development of wave packet solutions of the Schrödinger equation often makes use of the concept of the overlap $\langle \psi_t | \psi_0 \rangle$ of the time-dependent quantum state $| \psi_t \rangle$ with the initial state $| \psi_0 \rangle$. This overlap is most often referred to as the autocorrelation function.

For one-dimensional bound state systems, where a wave packet is expanded in terms of energy eigenfunctions $\psi_n(x)$ with quantized energy eigenvalues E_n in the form

$$\psi(x,t) = \sum_{n=1}^{\infty} a_n \psi_n(x) e^{-iE_n t/\hbar} \tag{5.14}$$

with

$$a_n = \int_{-\infty}^{\infty} \psi_n^*(x)\psi(x,t)dx \tag{5.15}$$

the autocorrelation function can be written as:

$$A(t) = \sum_{n=1}^{\infty} |a_n|^2 e^{iE_n t/\hbar} \tag{5.16}$$

and the evaluation of $A(t)$ in this form for initially highly localized wave packets will be investigated experimentally.

If a localized wave packet is excited with an energy spectrum which is tightly spread around a large central value of the quantum number n_0 so that $n_0 \gg \Delta n \gg 1$, it is possible to expand the individual energy eigenvalues, $E_n \equiv E(n)$, about this value, giving

$$E(n) \approx E(n_0) + E'(n_0)(n-n_0) + \frac{E''(n_0)}{2!}(n-n_0)^2 + \frac{E'''(n_0)}{3!}(n-n_0)^2 + \ldots \tag{5.17}$$

This gives the time-dependence of each individual quantum eigenstate through the factors:

$$e^{iE_n t/\hbar} = e^{i\omega_0 t} \cdot e^{i2\pi(n-n_0)/T_{cl}} \cdot e^{i2\pi(n-n_0)^2/T_{rev}} \cdot e^{i2\pi(n-n_0)^3/T_{super}}, \tag{5.18}$$

where each term in the expansion (after the first which is an unimportant overall phase not observable experimentally) defines an important characteristic time scale, via:

$$T_{cl} = \frac{2\pi\hbar}{|E'(n_0)|},\ T_{rev} = \frac{4\pi\hbar}{|E''(n_0)|} \text{ and } T_{super} = \frac{12\pi\hbar}{|E'''(n_0)|}. \tag{5.19}$$

The second term in the expansion is associated with the classical period of motion in the bound state. It can also be shown that the wave packet near the revival time T_{rev} returns to something like its initial form, exhibiting the classical periodicity. In the special case when T_{rev}/T_{cl} is an integer, the revival occurs exactly in phase with the original time-development, and is exact (in that $|A(t)|$ returns to exactly unity). For some realistic systems, with higher order terms in the expansion in Eq. (5.17), the superrevival time, T_{super} also becomes very important.

To obtain the order of magnitude of the different characteristic times introduced previously, one can consider a neutron in the second state. For this state, the value of the classical turning point (2.11) is equal to $z_2 = 24\ \mu\text{m}$. The classical periodicity of the system is given by

$$T_{cl} = 2\sqrt{\frac{2z_2}{g}} \approx 4.4 \text{ ms}. \qquad (5.20)$$

The revival time appears to be equal to

$$T_{rev} = \frac{16mz_0^2}{\pi\hbar} \approx 46 \text{ ms}. \qquad (5.21)$$

With the neutrons of 5 m/s velocity, a 25 cm long mirror is needed to observe this revival phenomenon.

All the methodical developements for this kind of experiments are already available: the position-sensitive detector discussed in section 3 can provide the spatial resolution of 1 μm, the absorber/scatterer and a suitable mirror geometry (see sections 2 and 3) make it possible to chose the necessary number of quantum states, and the phase of the wave function can be fixed by a special collimator at the entry to the system.

5.4 Search for a Logarithmic Term in the Schrödinger Equation

As discussed in refs. [26, 27], an extension of quantum mechanics with an additional logarithmic term in the Schrödinger equation assumes quasi-elastic scattering of UCN at the surface, with extremely small, but nevertheless measurable, energy changes. Such spectral measurements of high resolution with UCN were themselves methodologically challenging. They were also motivated by a long-standing anomaly in the storage of UCN in traps [57]. These experiments [58, 59] allowed the authors to constrain such quasi-elasticity at ~10^{-11} eV per collision, under the assumption of a "random walk" in phase space at each neutron collision with the wall: a non-zero result at this level was reported in ref. [58] at the limit of experimental sensitivity, but was not confirmed later in ref. [59], measured in the same setup with slightly better statistical sensitivity but with worse energy resolution.

A significant increase in the accuracy of neutron gravitational spectrometry using the high-resolution position-sensitive neutron detectors presented here allows us to improve many times over the upper limit for the probability and for the minimum energy transfer values for the quasi-elastic scattering of UCN at the surface [60]. Moreover, we can now consider energy changes at a single reflection, rather then having to follow the integral effects of many collisions, as in refs. [58, 59]. In addition to this, the present limit concerns one specific component of the neutron velocity along the vertical axis before reflection and after it. Also any deviation from conventional quantum mechanics can be verified in a more direct way with the quantum limit used here for the minimum possible initial energy, or velocity.

Such constraints, however, present a broader interest and could be considered in a more general model-independent way: how precisely do we know that UCN conserve their energy at wall reflections or during UCN storage in material traps?

Let us remind the reader of the details of the experimental set up used in the last run. A neutron beam with a horizontal velocity component of ~5 m/sec and a vertical velocity component of 1-2 cm/sec, which corresponds to the energy of the lowest neutron quantum state in the gravitational field above a mirror, is selected using a bottom mirror (1) and a scatterer/absorber (3) positioned above it at a height of ~20 μm. A second mirror (2) is installed 21 μm lower than the first mirror (1). If the UCN bounce elastically on the mirror (2) surface in the zone between the scatterer's (3) exit edge and the position-sensitive detector (4), the measured spatial variation of the neutron density as a function of height would correspond to that shaped by the mirrors (1,2) and the scatterer (3) in the zone upstream of the scatterer's (3) exit edge. If they do not, then the excess number of neutrons observed in the higher position would be attributed to their quasi-elastic reflection from the mirror (2) surface. The experimental setup is designed in such a way that any known parasitic effects (vibration of the mirrors and the scatterer, residual magnetic field gradients, quasi-specular reflections of UCN from mirrors or at residual dust particles) should be small enough not to cause a significant change in the spectrum of vertical neutron velocities (see refs. [8-9, 19-22]).

We will not discuss the possible microscopic mechanisms of quasi-elastic reflections of UCN at surfaces; we shall simply consider this problem in phenomenological terms. A simple conservative upper limit for the quasi-elastic scattering/heating probability (versus average energy transfer) following UCN reflection from the lower polished glass mirror could be calculated, assuming an ideal scatterer able to select a single quantum state above the mirror (1) in Fig. 7. Populations of all quantum states above the mirror (2) can be precisely calculated in this case [22]. They provide the neutron density distribution, presented by the solid curve in Fig. 10. We know in fact that a few neutrons at higher quantum states should survive [15] producing a density distribution close to one presented by the dotted curve in Fig. 10. However, we do not attempt to take such neutrons into account and intentionally sacrifice the sensitivity of the present limit in favor of maximum reliability and transparency. Such an estimation could be further improved with the present experimental data using a more sophisticated theoretical analysis based on ref. [15]. It would however be slightly model-dependent in such a case. For the simplified approach chosen, the solid line in Fig. 10 is considered as "background" for the measurement of quasi-elasticity and any additional events above this line would be supposed to be due to quasi-elastic scattering. Fig. 12 illustrates the results of the treatment of the experimental data presented in Fig. 10.

The straightforward calculation of such a constraint provides the solid curve in Fig. 12 under the following assumptions: 1) all additional events higher than the solid curve in Fig. 10 are attributable to quasi-elastic scattering/heating; 2) the energy is assumed to change in one step (due to the low probability of such an event); 3) we take the number of quasi-classical collisions in such a system [15].

The rather sharp decrease with height of the neutron density on a characteristic scale of a few microns simplifies considerably the present calculation. For large enough ΔE values, any excess counts above the constant background level $\Delta N_{\text{bg}} / \Delta h$ in the height range $h > 60\ \mu\text{m}$ are attributed to quasi-elastic scattering/heating. Quasi-elastically scattered neutrons could be observed at any height between zero and $(E_0 + \Delta E)/mg$, where E_0 is the initial energy of vertical motion and ΔE is the energy gain. If $\Delta E \gg E_0$, the total number of background events is approximately equal to $\frac{\Delta N_{\text{bg}}}{\Delta h}\frac{\Delta E}{mg}$, neglecting the initial spectral line width $h < 60\ \mu m$. At 3σ confidence level, we would observe an excess N_{qel} of events at $h > 60\ \mu m$, if it is equal to:

$$N_{qel} = 3\sqrt{\frac{\Delta N_{bg}}{\Delta h}\frac{\Delta E}{mg}}. \tag{5.22}$$

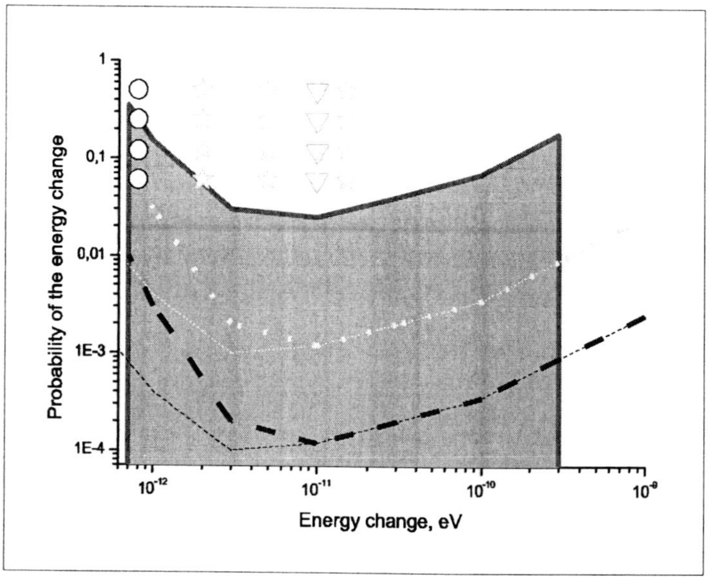

Fig. 12. The solid curve corresponds to constraints for quasi-elastic scattering of UCN at a flat glass surface: the total probability of such a scattering per one quasi-classical bounce versus average energy transfer at "3σ" confidence level. The dotted curve shows the possible improvement of such constraints in the flow-through measuring mode. The dashed curve indicates a further increase in sensitivity in the storage measuring mode. The circles correspond to theoretical predictions for the present experiment in accordance with refs. [14-15, 17]. The stars indicate analogous predictions for measurements with the experimental setup [8-9, 19-22] inclined to various angles. The triangles show the value of the energy change expected in refs. [14-15, 17] (for a higher initial neutron velocity than that in the present experiment). The thin dotted and dashed curves indicate schematically the constraints if the initial spectral shape line were to be taken into account.

With the horizontal velocity component v_{hor} and the mirror length L between the scatterer's exit edge and the detector (see Fig. 7), the total number N_{qel} of quasi-classical bounces is:

$$N_{bounces} = \frac{L}{\frac{2}{g}\sqrt{\frac{2E_0}{m}}v_{hor}}. \tag{5.23}$$

Thus, with the total number N_0 of neutrons in the initial spectral line, we would be able to observe quasi-elastic scattering at 3σ confidence level if its probability $P_{\text{qel}}(\Delta E)$ is equal to:

$$P_{\text{qel}}(\Delta E) = \frac{N_{\text{qel}}}{N_0 \cdot N_{\text{bounces}}} = \frac{3}{N_0 L}\sqrt{\frac{\Delta N_{\text{bg}}}{\Delta h}\frac{\Delta E}{mg}\frac{2}{g}\sqrt{\frac{2E_0}{m}}v_{\text{hor}}} \tag{5.24}$$

As is evident from eq. (5.24), $P_{\text{qel}}(\Delta E)$ increases as $\sqrt{\Delta E}$, thus decreasing the sensitivity of the present constraint at large energy changes. The sensitivity is also lower at energy changes smaller than the initial spectral line width of ~60 μm (here the constraint is estimated numerically). Therefore the best sensitivity is achieved at the energy change comparable to one or few initial spectral line widths, as shown in Fig. 12.

The constraint presented shows the high degree of elasticity of neutron reflections in the range $\Delta E \sim 10^{-12} - 3 \cdot 10^{-10}$ eV ΔE; this is important for the further development of precision neutron spectrometry experiments. Further improvements in the sensitivity of such constraints by an order of magnitude are feasible in the flow-through measuring mode, by improved shielding of the neutron detectors (a factor $\sqrt{\frac{\Delta N_{\text{bg}}}{\Delta h}}$ in eq. (5.24)), by increasing the length of the bottom mirror (a factor $1/L$ in eq. (5.24)), by further increasing the scatterer efficiency, and by using a narrower initial neutron spectrum (a factor $\sqrt{E_0}$ in eq. (5.24)). On the other hand, a broader initial spectrum could allow us to increase the factor N in eq. (5.24) and therefore to improve the sensitivity at higher ΔE values (sacrificing the sensitivity at lower ΔE values).

An almost order-of-magnitude gain in the minimum measurable energy change could be achieved by providing a proper theoretical account (in accordance with ref. [15], for instance) of the spectrum-shaping properties of the scatterer, or by a differential measurement of the vertical spectrum evolution using bottom mirrors of different lengths. Possible improvements in the flow-through mode are illustrated by the dotted curve in Fig. 12. One should note that any jumps in energy by a value significantly lower than 1 peV would clearly contradict to the observation of quantum states of neutrons in the gravitational field [12–15, 21–23] and therefore they are not analyzed in the present article. The minimum energy increase considered corresponds to the energy difference between neighboring quantum states in the gravitational field.

A much higher increase in sensitivity could be achieved in the storage measuring mode with the long storage of UCN at specular trajectories in a closed trap (the dashed curve in Fig. 12 or better).

As an example of a possible application of the present constraint, let us compare it to the theoretical prediction in accordance with refs. [58,59]. This model assumes the replacement of "continuous interaction" of UCN with a gravitational field by a sequence of "collisions with the field". The time interval $\delta\tau$ between the "collisions" is defined as the time during which the mass "does not know that there is an interaction" since the kinetic energy change δE (by falling) is too small to be resolved. From the uncertainty principle:

$$\delta\tau \cdot \delta E \approx \frac{\hbar}{2}, \text{ or } \delta E \approx \sqrt{\frac{\hbar m g v_{\text{vert}}}{2}} = \sqrt{33 v_{\text{vert}}} \text{ (peV)} \tag{5.25}$$

where v_{vert} is in m/s.

For the vertical velocity component $v_{\text{vert}} \approx 2.5$ cm/s in our present experiment, the expected energy change is $\delta E \approx 8 \cdot 10^{-13}$ eV (shown as the circle in Fig. 12). The "100%" probability of quasi-elastic scattering is slightly higher than the 3σ experimental constraint (the solid line in Fig. 12). However, considering the expected probability value of ~10% and low experimental sensitivity at small ΔE values, one needs to further improve the sensitivity of the present constraint.

On the other hand, a slight modification of the experimental setup would allow us to verify clearly the considered hypothesis. Namely, the whole apparatus should be turned by a significant angle relative to the direction of the gravitational field. In this case, the vertical velocity component is comparable to the longitudinal velocity of 5–10 m/s. The transversal velocity component (relative to the bottom mirror) is very small, just equal to the one in the experiment [12–15, 21–23]. All sensitivity estimations for quasi-elastic scattering/heating are analogous to those given above (see Fig. 12). However, the theoretically predicted effect could be as high as ~10^{-11} eV (depending on the inclination angle) – just in the range of the best sensitivity of the present constraint: the stars in Fig. 12. In order to measure a hypothetical cooling of UCN at their quasi-elastic reflections, we must first of all select a higher quantum state ($n > 1$) and then follow the evolution of the corresponding neutron spectrum. The sensitivity estimations in the energy range $0 < \Delta E < E_0$ would be about as strong as those for the quasi-elastic heating if the experiment was optimized for this purpose. Such measurements would be significantly easier to perform than the measurement of the gravitationally bound quantum states because they do not require such record levels of energy and spatial resolution.

5.5 Search for the Loss of Quantum-Mechanical Coherence

The fundamental loss of quantum coherence because of gravitational interaction is an issue of high long-term scientific interest. As it was pointed out even in the first publication [28], neutron interference experiments could be sensitive to this phenomenon. The quantity defining the sensitivity of such an experiment is the characteristic time of observation of an interference pattern. In the experiment [61] with thermal neutrons this value was about 300 μs (which corresponds to the energy $2 \cdot 10^{-21}$ GeV). In our experimental setup, in the flow-through measuring mode the observation time could be as high as ~60 ms (10^{-23} GeV). A measurement of the localization phenomenon, described in this article, could give a direct estimation of the effect of the fundamental loss of quantum coherence. A much longer observation time would be possible in the storage measuring mode in our experiment. On the other hand, even better constraints for the loss of quantum coherence would be obtained by measuring neutron oscillations between two quantum states due to a small mixing interaction (for instance, a magnetic one) in some analogy to the experiment mentioned in ref. [29].

6 Transitions between the Quantum States

The observation of transitions between the quantum states would allow a qualitatively new step in this research. These transitions can be initiated in various ways and by different forces (strong, electromagnetic, gravitational). In this section we will study, for the first time, different options, giving estimations of probabilities of these transitions.

The mechanical vibration of a mirror would be the simplest way of inducing such transitions. This vibration means a periodical variation of the boundary condition created due to the effective Fermi potential of the bottom mirror (i.e. due to strong forces). In fact, we already observed this kind of transitions induced by nuclear forces, in our last experiment. To suppress neutrons in the ground state, the mirror was assembled in a special way so as to produce a negative step (Fig. 7). This trick can be considered as an infinitely fast change of the Hamiltonian which produces a change in the occupation numbers, i.e. the transitions between the levels.

Another way to produce transitions between the levels is to introduce a varying gradient of magnetic field (i.e. by electromagnetic forces). Until now, all magnetic effects have been considered as parasitic and able to blur the gravitational levels. Considerable efforts were therefore needed to avoid undesirable interaction between the neutron magnetic moment and an external magnetic field. Now that once the existence of the gravitational levels is well-established, a controlled magnetic field can be introduced to manage transitions between the levels. Experimentally, it is easy to produce such a gradient with any form of time-dependence, in particular, perfectly harmonic oscillations.

However, the most interesting way to produce the transitions is by variation of the gravitational field. This could be done, for instance, by the rotation of a massive body close to the experimental set up. This kind of transition is, of course, very difficult to observe. The aim of this study is therefore to evaluate the feasibility of performing this kind of experiment with current and future neutron facilities.

A measurement of transitions between the gravitational levels can be used to study the properties of neutrons. For instance, if we look for transitions induced by a variable electric field and we establish an upper limit on such transitions, we can establish a limit for the electric charge of the neutron.

6.1 General Expressions for Transition Amplitudes

Let us remind the reader of the main formulas [3] which will be used hereafter concerning the transitions between the quantum levels (two states of the discrete spectrum). We are interested in the transitions induced by a periodical external interaction considered as a perturbation and which, in order to obtain simple analytic expressions, is considered to be harmonic.

The Hamiltonian $\widehat{H}$ of the problem can be written in the form

$$\widehat{H} = \widehat{H_0} + \widehat{V}(t) \tag{6.1}$$

where $\widehat{H_0}$ corresponds to the unperturbed gravitational problem $(z > 0)$

$$\widehat{H_0} = \frac{\hat{p}^2}{2m} + mgz \tag{6.2}$$

and

$$\hat{V}(t) = V_0(z)e^{i\omega t} + V_0(z)e^{-i\omega t} \tag{6.3}$$

is a harmonic perturbation with $V_0(z)$ which depends on z. This particular harmonic form of excitation is chosen to obtain analytic results and can in some cases be achieved in an experiment.

A solution Ψ of the Schrödinger equation

$$i\hbar\frac{\partial\Psi}{\partial t} = \left(\widehat{H_0} + \hat{V}(t)\right)\Psi \tag{6.4}$$

can thus be written as a sum

$$\Psi = \sum_k a_k(t)\Psi_k^{(0)} \tag{6.5}$$

over solutions $\Psi_k^{(0)} = \psi_k(z)e^{-iE_k^{(0)}t/\hbar}$ of the unperturbed Schrödinger equation

$$i\hbar\frac{\partial\Psi_k^{(0)}}{\partial t} = \widehat{H_0}\Psi_k^{(0)}. \tag{6.6}$$

If we put (6.5) in (6.4) and taking into account (6.6) we obtain:

$$i\hbar\sum_k \Psi_k^{(0)}\frac{da_k}{dt} = \sum_k a_k\hat{V}(t)\Psi_k^{(0)}. \tag{6.7}$$

By multiplying the last equation by $\Psi_m^{*(0)}$ and by integrating over z, we obtain:

$$i\hbar\frac{da_m}{dt} = \sum_k V_{mk}(t)a_k \tag{6.8}$$

with the matrix element:

$$V_{mk}(t) = \int \Psi_m^{*(0)}\hat{V}(t)\Psi_k^{(0)}dz. \tag{6.9}$$

The last differential equation describes an evolution of the quantum system. If we suppose that at a moment $t = 0$, the system was, for instance, in a ground state ($k = 1$)

$$a_1(0) = 1 \text{ and } a_m(0) = 0 \text{ for any } m \neq 1,$$

then we can calculate, at least numerically, a probability to find the system in the state n for any moment t as

$$P_n(t) = |a_n(t)|^2 . \qquad (6.10)$$

As we said previously, the choice of a perturbation interaction $\widehat{V}(t)$ in a harmonic form (6.3) allows us to obtain an analytic expression for the probability (6.10) if we consider the problem of only two coupled states. Physically, this situation is produced when the frequency ω of the excitation is close to the difference $\omega_{n0} = (E_n^{(0)} - E_1^{(0)})/\hbar$ (resonance regime). Let us suppose therefore that the difference $\varepsilon = \omega_{n1} - \omega$ is very small and that in the matrix element (6.9) of the perturbation (6.3), we can leave only the dominant term with this small frequency:

$$V_{n1}(t) = \int \Psi_n^{*(0)} \widehat{V}(t) \Psi_1^{(0)} dz \approx e^{i\varepsilon t} \int \psi_n^* V_0(z) \psi_1 dz \equiv F_{n1} e^{i\varepsilon t} . \qquad (6.11)$$

By omitting all other terms, we obtain a system of two coupled equations relating the amplitudes of presence in the ground and in the n-th state:

$$\begin{aligned} i\hbar \frac{da_n}{dt} &= F_{n1} e^{i\varepsilon t} a_1, \\ i\hbar \frac{da_1}{dt} &= F_{n1}^* e^{-i\varepsilon t} a_n. \end{aligned} \qquad (6.12)$$

This system can be easily solved, for instance, by the standard Laplace transformation. If we suppose that at $t = 0$ the system is in the ground state, the probability of finding it in the n-th excited state appears to be equal to:

$$P_n(t) = |a_n(t)|^2 = \frac{\Omega_0^2}{\Omega^2} \sin^2 \Omega t \qquad (6.13)$$

with

$$\Omega_0 = \frac{F_{n1}}{\hbar} \text{ and } \Omega^2 = \Omega_0^2 + \frac{\varepsilon^2}{4}.$$

This is a well-known Raby formula describing an oscillation of the system between the two coupled states with the frequency 2Ω. The probability of presence in an excited state oscillates between 0 and Ω_0^2/Ω^2.

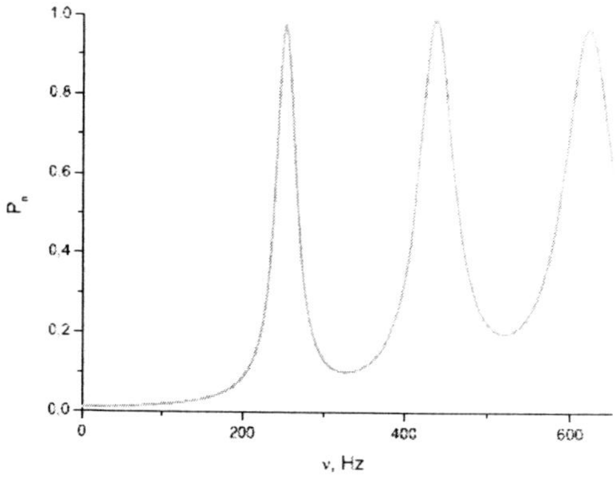

Fig. 13. The transition probability from the ground to the excited states as a function of frequency of a perturbation interaction $\nu = \omega/2\pi$.

The maximum probability as a function of frequency has a resonance-like behavior

$$P_{\max}(\omega) = \frac{\Omega_0^2}{\Omega^2} = \frac{\dfrac{(2F_{n1})^2}{\hbar^2}}{(\omega-\omega_{n1})^2 + \dfrac{(2F_{n1})^2}{\hbar^2}} \tag{6.14}$$

and is presented in Fig. 13. The resonance frequencies given in this figure correspond to transitions from the ground state $n=1$ to the first three excited states. The energy spectrum of the system becomes denser with increasing n (the levels become closer to each other). The width of this resonance is defined by the matrix element of the perturbation F_{n1} and is equal to

$$\Gamma_n = 4F_{n1}. \tag{6.15}$$

To resolve the two nearest states with n and $n+1$, their energy difference $E_{n+1} - E_n = \hbar\omega_{nn+1}$ should be smaller than the corresponding width:

$$\hbar\omega_{nn+1} > \Gamma_n. \tag{6.16}$$

In other words, the matrix element F_{n1} should not be very big

$$F_{n1} < \frac{\hbar\omega_{nn+1}}{4} \tag{6.17}$$

to populate only one excited state.

For an exact resonance $\varepsilon = 0$, formula (6.13) becomes

$$P_n(t) = \sin^2 \Omega_0 t \,. \tag{6.18}$$

For a very small period of time (or a very small matrix element F_{n1}), this probability is seen to depend quadratically on time:

$$P_n(t) \approx \Omega_0^2 t^2 \tag{6.19}$$

(this formula is valid even when not in an exact resonance). We can say that this probability becomes close to 1 if:

$$\frac{F_{n1} t}{\hbar} \approx 1 \,. \tag{6.20}$$

We can say that to have a non-negligible transition probability, the observation time τ should be of the order of:

$$\tau \approx \frac{\hbar}{F_{n1}} \,. \tag{6.21}$$

By combining this condition with the condition of the resolution of two neighboring states (6.17), we conclude that, to observe a resonance transition, the neutron life time in the system should be higher than:

$$\tau > \frac{4}{\omega_{nn+1}} \,. \tag{6.22}$$

For instance, for a transition between the ground state $n = 1$ and first excited state $n = 2$, the corresponding frequency is approximately equal to 150 Hz and we obtain $\tau > 4$ ms. For neighbor higher excited states, this time should be even greater. We would remind the reader that in the last experiment the time of presence of neutron in the system was close to 25 ms.

If the condition (6.22) is not satisfied, the transitions may also occur but in several states simultaneously. This is true in particular, in the case mentioned in the introduction to this chapter: the transition due to the sudden change of the mirror height (negative step). The neutrons from the ground state (before the step) populate a few low excited states (after the step). The transition exists but it is not a resonance one.

6.2 Transitions Induced by a Magnetic Field

As we mentioned at the beginning of this section, the magnetic field $\boldsymbol{B}$ easily couples to the neutron magnetic moment $\hat{\boldsymbol{\mu}}$ by:

$$\widehat{H}_{\text{int}} \equiv \widehat{V}(t) = -\hat{\boldsymbol{\mu}}\boldsymbol{B} \tag{6.23}$$

and thus can be used to induce transitions between the gravitational levels. To obtain the desirable effect, one can introduce an oscillating magnetic field with a gradient along the z axis (which is also the direction of the magnetic field itself):

$$B_z = \beta z \sin \omega t \,. \tag{6.24}$$

For this interaction, the matrix element F_{n1} is equal to:

$$F_{n1} = \mu_{\text{n}} \beta z_{n1} \tag{6.25}$$

where μ_{n} is the neutron magnetic moment and

$$z_{n1} = \int \psi_n^* z \psi_1 dz \,.$$

This matrix element can be calculated numerically with the well-known Airy function and, for instance for $n = 2$, appears to be equal to

$$z_{21} = 0.653 z_0 \tag{6.26}$$

where $z_0 = 5.87\ \mu\text{m}$ is the characteristic length of the problem introduced previously.

This can be easily achieved, even in the current experimental setup. The gradient of the magnetic field β necessary to introduce a transition between the first two levels with a probability close to one (6.20) is equal to:

$$\beta = \frac{\hbar}{\mu_{\text{n}} z_{12} t} \,. \tag{6.26}$$

For the present experiment with $t = 25$ ms, we obtain $\beta \approx 10$ Gs/cm, which can be achieved without major difficulty. It is planned to conduct this experiment in the very near future.

Let us emphasize that the studies of transitions induced by a magnetic field would represent a very efficient tool for the search for the loss of quantum coherence induced by gravity. The time evolution (6.13) of the two-level system is modified in the presence of such effects and can be constrained experimentally without any major difficulty. This is another reason why the experiments on magnetic transitions between the gravitational levels are of high priority.

6.3 Transitions Induced by a Gravitational Field

The most interesting transition would be the one induced by gravitational interaction, for instance, by a massive body motion in the vicinity of the setup. Compared to the Coulomb interaction, this process is analogous to an excitation of the Coulomb atom by an electric charge moving near the atom. In a field theory picture, this excitation is induced by a virtual photon. In the case of a transition between the gravitational levels induced by a moving body, one would speak of a virtual graviton. Strictly speaking, the theoretical description of both processes does not require the explicit introduction of these virtual particles. We could not therefore say that the detection of the gravitational transition would be an unambiguous demonstration of the existence of the graviton. Nevertheless, this experiment would be a very important step towards this goal.

The main difficulty is obviously due to the very weak interaction constant. Let us therefore simply estimate the probability of such a transition in order to charge on the feasibility of its observation in the near future, let us say within a decade.

Let us suppose that a transition is induced by an oscillating body moving just above the neutron situated at distance z above the mirror. Thus the distance between the neutron and the body is equal to:

$$r = \frac{L}{2} + a(1 - \cos\omega t) + \Delta - z \tag{6.27}$$

where L is the linear size of the body, a is an amplitude of oscillations and Δ is the minimal distance between the body and the mirror. This oscillating body will introduce an additional gravitational interaction:

$$\widehat{H}_{\text{int}} = G\frac{mM}{r} \tag{6.28}$$

M being the mass of the body. z is small with respect to L and this interaction Hamiltonian can be developed in series on z. The linear term is equal to:

$$\widehat{H}_{\text{int}} \approx G\frac{mM}{L}\frac{z}{L}\frac{4}{\left(1+2\zeta(1-\cos\omega t)\right)^2} \tag{6.29}$$

Here Δ is neglected with respect to L and $x = a/L$ is introduced. The function:

$$f(t) = \frac{1}{\left(1+2\zeta(1-\cos\omega t)\right)^2} \tag{6.30}$$

is not harmonic but it is quite easy to calculate its development in Fourier series:

$$f(t) = \sum_{n=-\infty}^{\infty} c_n e^{in\omega t} \tag{6.31}$$

with the coefficients:

$$c_n = \frac{\omega}{2\pi} \int_0^{2\pi/\omega} f(t) e^{-in\omega t} dt = \frac{1}{\zeta\sqrt{1+4\zeta}} \eta^{n+1} \frac{(n+1)-(n-1)\eta^2}{\left(1-\eta^2\right)^2} \tag{6.32}$$

where

$$\eta = 1 + \frac{1}{2\zeta} - \frac{1}{2\zeta}\sqrt{1+4\zeta}\,. \tag{6.33}$$

As a function of ζ, η increases continuously from 0 to 1.

In particular, the coefficient corresponding to the first harmonic c_1 is equal to:

$$c_1 = \frac{2\zeta}{\left(1+4\zeta\right)^{3/2}}\,. \tag{6.34}$$

This coefficient as a function of ζ has a maximum for $\zeta = 1/2$ and its maximum value is equal to

$$c_1 = 0.192\,. \tag{6.35}$$

Let us note that the coefficient describing the difference between the neighbor harmonics is equal to:

$$\eta\left(\zeta = 1/2\right) = 2 - \sqrt{3} \approx 0.268\,. \tag{6.36}$$

This means that the higher terms decrease quite rapidly with n and, in first approximation, the interaction (6.30) can be considered as harmonic and can be represented, for the optimal choice $\zeta = 1/2$, as:

$$\widehat{H}_{\text{int}} \approx 3.08G\frac{mM}{L}\frac{z}{L}e^{i\omega t}\,. \tag{6.37}$$

This Hamiltonian is now considered as a perturbation $\widehat{V}(t)$ and its matrix element F_{21} between the first two states' wave functions is equal to:

$$F_{21} = 0.77G\frac{mM}{L}\frac{z_{21}}{L}, \tag{6.38}$$

where z_{21} is the same as in (6.26).

Obviously, for any realistic experiment, the condition (6.20) will be hardly fulfilled for the gravitational interaction. Even if the neutron life time is chosen as the time of observation (this hypothesis implies successfully storing neutrons at a given gravitational level over a very long period, which is actually an extremely challenging task) and the characteristic size of the oscillating body equals to $L = 20$ cm with high density $\rho = 20 \cdot 10^3$ kg/m^3 (i.e. $M = 160$ kg), the value of the product (6.20) appears to be small:

$$\frac{F_{21}t}{\hbar} \approx 0.01. \tag{6.39}$$

This means that the probability of a corresponding transition would be of the order of 10^{-4}. With existing sources of UCN, the detection of those transitions would scarcely seem possible.

However, the probability of transition increases if we choose other levels, for instance, highly excited neighboring levels. We can show that the matrix element z_{n+1n} behaves, for $n \gg 1$, as:

$$z_{n+1n} \approx 0.57\xi n^{2/3}, \tag{6.40}$$

i.e. increases quite rapidly with *n*, whereas the transition frequency will decrease:

$$\omega_{n+1n} \sim n^{-1/3}. \tag{6.41}$$

Hence the technical realization of the experiment would be even simpler.

Note also that increasing the size *L* of the oscillating body produces very limited gain because the amplitude of the transition depends linearly on *L*: $F_{21} \sim L$, whereas its mass will grow very rapidly $M \sim L^3$ and make the experiment much more complicated.

Taking into account these circumstances, we can conclude that an experimental observation of a transition between two gravitational levels, induced by the motion of a body seems relatively unlikely in the near future with the existing neutron sources.

6.4 Transitions Induced by an Electrical Field

By studying transitions between the levels induced by an oscillating electric field, we can establish an upper limit on (or find) the neutron charge.

As an example, let us consider a system where the mirror is one of the plates of a condenser. If we apply a varying electric field a perturbation Hamiltonian:

$$\widehat{H}_{\text{int}} \equiv \widehat{V}(t) = e_{\text{n}} E e^{i\omega t} z \tag{6.42}$$

e_{n} being the neutron charge, E the strength of the electric field.

For this interaction, the matrix element F_{n1} is equal to:

$$F_{n1} = e_{\text{n}} E z_{n1}. \tag{6.43}$$

Thus an upper limit for the probability of transition to the nth state P_{lim} will give an upper limit on the neutron charge:

$$e_{\text{n}} < \frac{\hbar}{E z_{n1} t} \sqrt{P_{\text{lim}}}\,. \tag{6.44}$$

In an experiment with the electric field $E \approx 10^7$ V/m, with t compatible with the neutron life time, $P_{\text{lim}} \approx 10^{-3}$ must be achieved in order to obtain an actual limit on the neutron electric charge equal to $e_{\text{n}} < 10^{-21} e$. The best limit is produced with the interferometer experiment using very cold neutrons [62]. It should be noted that ultracold neutrons were also used to establish the limit [63].

6.5 Transitions Induced by the Combined Effect of Different Excitations

Nevertheless a much tighter constraint for the neutron electric charge can be obtained and a transition induced by a moving massive body can be observed experimentally. The idea is to conduct an interference experiment where we measure a transition induced by two different causes, for instance, by a variable gradient of the magnetic field and a varying electric field or an oscillating body.

The matrix element F_{nk} of such a transition would be equal to the sum:

$$F_{nk} = F_{nk}^{\text{big}} + F_{nk}^{\text{small}} \tag{6.45}$$

of a big F_{nk}^{big} (for instance, magnetic) and a small F_{nk}^{small} (electric or gravitational) terms. The transition probability would be proportional to:

$$P_+(t) \sim \left|F_{nk}\right|^2 \approx \left|F_{nk}^{\text{big}}\right|^2 + 2F_{nk}^{\text{small}}F_{nk}^{\text{big}}. \tag{6.46}$$

By an adequate choice of the relative phase of these two perturbations, we can obtain another probability:

$$P_-(t) \sim \left|F_{nk}^{\text{big}}\right|^2 - 2F_{nk}^{\text{small}}F_{nk}^{\text{big}}. \tag{6.47}$$

Thus, an asymmetry is proportional to:

$$A = \frac{P_+(t) - P_-(t)}{P_+(t) + P_-(t)} \approx 2\left|\frac{F_{nk}^{\text{small}}}{F_{nk}^{\text{big}}}\right|. \tag{6.48}$$

With the estimation obtained previously (6.39), this kind of measurement seems to be conceivable in future experiments.

Of course, exactly the same idea of combined perturbations can be used to improve the limit on the neutron electric charge.

Conclusion

Gravitationally bound quantum states of neutrons were recently discovered in the measurement of neutron transmission through a narrow horizontal slit between a mirror below and an absorber/scatterer above it. The first experiment allowed us to identify clearly the ground quantum state in this system [12, 13]. Later, with improved height (energy) resolution and statistics, we were also able to measure also the first excited quantum state [15]. We showed that the process of the loss of neutrons in an absorber/scatterer could be very precisely described using a model of neutron tunneling through the gravitational barrier between the classically allowed height and the absorber/scatterer height [15]. Further progress with this experiment using the flow-through measuring mode is limited to a large degree by one fundamental factor: the finite sharpness of the dependence on height of neutron tunneling through this gravitational barrier. Nevertheless, with a more suitable and precise theoretical description [18] and improvements to the absolute distance calibration [19, 20], we can expect to achieve a few percent accuracy in the determination of quantum state parameters.

In order to resolve higher excited quantum states clearly and measure their parameters accurately, we investigate another method based on position-sensitive neutron detectors of very high spatial resolution [21, 22]. The direct measurement of the spatial density distribution in a standing wave is preferred to its investigation with the aid of an absorber/scatterer whose height can be adjusted. The former technique is differential, since it permits the simultaneous measurement of the probability that neutrons reside at all heights of interest. The latter technique is integral, since the information on the probability that neutrons reside at a given height is in fact obtained by the subtraction of the values of neutron fluxes measured for two close values of the scatterer height. Clearly, the differential technique is much more statistically sensitive. Furthermore, the scatterer employed in the integral technique inevitably distorts the measured quantum states; the finite accuracy of taking these distortions into account results in methodological errors and ultimately limits the attainable accuracy of the measurement of the quantum state parameters. The feasibility of the differential technique was demonstrated in refs. [15, 23].

The two techniques considered and the available fluxes of UCN are already sufficient for a broad range of applications. Thus, as this was shown in ref. [24], this experiment could be used to establish a competitive limit for short-range fundamental forces. However, it is from other specially designed neutron experiments that further progress in the nanometer range of distances can be expected. In order to be competitive in the micrometer range, we have to improve accuracy by many orders of magnitude, which can only be possible using the technique of resonance transitions between the quantum states. This experiment can also be used to search for the axion – a hypothetical particle which strongly violates CP-invariance; the characteristic distance for this interaction should be comparable to the characteristic length of our problem z_0. The very fact that the neutron quantum states exist provides the best constraint at this distance. An improvement by many orders of magnitude would seem to be easily achievable. This method could also be used for studies related to the foundations of quantum mechanics, such as for instance, quantum-mechanical localization (revivals phenomenon) or various extensions of quantum mechanics. For instance, it could be used to clearly rule out, or confirm, the presence of the logarithmic term in the Schrödinger equation in some models. It should be also noted that the present method provides two unique opportunities: on the one hand, it provides a rare combination of quantum states and gravitation that is favorable for testing possible extensions of quantum mechanics; on the other hand, UCN can be reflected from the surface $\sim 10^5$ times without loss, i.e. much more than for optical phenomena, which means that any kind of localization can be better studied with UCN. Finally, this method could be very useful for such problems of high interest as the fundamental loss of quantum coherence in systems with gravitational interaction.

Other methodological applications of the gravitationally bound quantum states and related techniques lie outside the subject of the present discussion of quantum gravity phenomena. We will therefore not discuss them in detail but simply mention a number of them. These experiments helped us to find an alternative approach to the problem of the neutron-tight valve for UCN traps able to operate in the broad range of temperatures needed for precision experiments with UCN storage. This is of crucial importance for precision neutron lifetime experiments. The existing solutions suffer from highly disturbing side effects: the so-called gravitational valve [64] modifies the spectrum of the stored UCN, whereas the so-called liquid valve [65,66] means the unavoidable use of fomblin oil with the accompanying effect of quasi-elastic scattering [67,68], producing large false effects also. Other methodological applications include the possibility of studying the distribution of hydrogen above/below solid or liquid surfaces, or the investigation of thin film on surfaces. These two subjects will be considered in more detail in separate publications.

A qualitatively new step in accuracy could be achieved even with the existing UCN density if the resonance transitions between the gravitationally bound quantum states were observed. These transitions could be initiated in various ways and by different forces (strong, electromagnetic, gravitational). In this article we presented for the first time a feasibility analysis and theoretical description of the observation of resonance transitions between the quantum states. All the above-mentioned applications of gravitationally bound quantum states for various physical problems would benefit considerably from the increase in accuracy which the technique of resonance transitions could bring. Moreover, a new class of highly competitive experiments could be considered, such as better constraints for the electric neutrality of neutrons, or the resonance transitions between the quantum states due to the gravitational interaction. It is clear that any increase in the energy resolution in measurements of the resonance transitions between the quantum states requires a high density of UCN. We therefore consider new approaches in order for significantly increasing the UCN densities, such as the thermalization of neutrons in gels of ultracold nanoparticles [69].

Acknowledgements

The experiments presented here were carried out by a dynamic and enthusiastic collaboration involving, at different stages, H. Abele, S. Baeßler, T.A. Baranova, H.G. Börner, G. Divkovic, A.M. Gagarski, T.M. Kuzmina, L. Lucovac, S. Nahrwold, G.A. Petrov, A.K. Petukhov, S. Raeder, F.J. Rueß, J. Schrauwen, S.M. Soloviev, T. Stöferle, A.V. Strelkov, V.G. Tishchenko, A.Yu. Voronin and A. Westphal, as well as the authors of the present article. We all benefited enormously from discussions with the advice help and support of K. Ben-Saidane, D. Berruyer, Th. Brenner, J. Butterworth, R. Corner, F. Descamps, D. Dubbers, E. Engel, N.P. Filatov, P. Geltenbort, B. Gerard, N. Havercamp, A. Hillairet, M. Jentschel, P. van Isaker, A.G. Kharitonov, C. Krantz, A.J. Leadbetter, A. Lensch, E.V. Lychagin, J.F. Marchand, D. Mund, A.Yu. Muzychka, R. Onofrio, B.G. Peskov, S.V. Pinaev, S. Roccia, V.A. Rubakov, R. Rusnyak, M.E. Shaposhnikov, I.A. Snigireva, E. Tomasi-Gustafson, S. Tinniswood, P.G. Tinyakov, J.-P. Varini, A.A. Zhdanov and M. Zimer.

References

[1] Breit, G. *Phys. Rev.*1928, 32, 273.
[2] Goldman, I.I.; Krivchenkov, V.D.; Kogan, V.I.; Galitsci, V.M. *Problems in Quantum Mechanics*; Academic Press: New York, 1960.
[3] Landau, L.D.; Lifshits, E.M. *Quantum Mechanics*; Pergamon Press: Oxford, 1977.
[4] Flügge, S. *Practical Quantum Mechanics*; Springer-Verlag: Berlin, 1974; Vol. 1.
[5] ter Haar, D. *Selected Problems in Quantum Mechanics*; Academic Press: New York, 1964.
[6] Sakurai, J.J. *Modern Quantum Mechanics*; Benjamin/Cummings: Menlo Park, 1985.
[7] Galitzky, V.M.; Karnakov, B.M.; Kogan, V.I. *La mécanique quantique. Problèmes résolus* EDP Sciences: Les Ulis, 2002; Vol. 1; 2003 ; Vol. 2.
[8] Luschikov, V.I. *Physics Today* 1977, 42, 51.
[9] Luschikov, V.I.; Frank, A.I.; *JETP Lett.* 1978, 28, 559.
[10] Golub, R.; Richardson, D.J.; Lamoreux, *S.K. Ultracold Neutrons*; Higler: Bristol, 1991.
[11] Ignatovich, V. K. *The Physics of Ultracold Neutrons*; Clarendon: Oxford, 1990.
[12] Nesvizhevsky, V.V.; Börner, H.G.; Petoukhov, A.K.; Abele, H.; Bäßler, S.; Rueß, F.J.; Stöferle, Th.; Westphal, A.; Gagarski, A.M.; Petrov, G.A.; Strelkov, A.V. *Nature* 2002, 415, 297.

[13] Nesvizhevsky, V.V.; Börner, H.G.; Gagarski, A.M.; Petoukhov, A.K.; Petrov, G.A.; Abele, H.; Bäßler, S.; Divkovic, G.; Rueß, F.J.; Stöferle, Th.; Westphal, A.; Strelkov, A.V.; Protasov, K.V.; Voronin, A.Yu. *Phys. Rev.* 2003, D67, 102002-1.
[14] Nesvizhevsky, V.V.; Petoukhov, A.K.; Börner, H.G.; Protasov, K.V.; Voronin, A.Yu.; Westphal, A.; Bäßler, S.; Abele, H.; Gagarski, A.M. *Phys. Rev.* 2003, D68, 108702.
[15] Nesvizhevsky, V.V.; Petoukhov, A.K.; Börner, H.G.; Baranova, T.A.; Gagarski, A.M.; Petrov, G.A.; Protasov, K.V.; Voronin, A.Yu.; Bäßler, S.; Abele, H.; Westphal, A.; Lucovac, L. *Eur.Phys.J.* 2005, C40, 479.
[16] Rauch H.; Werner, S. Neutron Interferometry; Oxford University: New York, 2000.
[17] Nesvizhevsky, V.V. *Phys. Usp.* 2004, 46, 93 [Uspekhi Fizicheskikh Nauk 2004, 146, 102].
[18] Voronin, A.Yu.; Abele, H.; Bäßler, S.; Nesvizhevsky, V.V.; Petoukhov, A.K.; Protasov, K.V.; Westphal, A. to appear in *Phys. Rev. D*, 2006.
[19] Schrauwen,V. Diploma thesis, UGent, Belgium, 2004.
[20] Nesvizhevsky, V.V.; et al., to be published.
[21] Nesvizhevsky, V.V.; Börner, H.G.; Gagarski, A.M.; Petrov, G.A.; Petoukhov, A.K.; Abele, H.; Bäßler, S.; Stöferle, Th.; Soloviev, S.M. *Nuclear Instruments and Methods in Physics Research* 2000, A440, 754.
[22] Nesvizhevsky, V.V. *Phys. Usp.* 2004, 47, 515 [Uspekhi Fizicheskikh Nauk 2003, 147, 569].
[23] Nesvizhevsky, V.V.; Börner, H.G.; Petoukhov, A.K.; Gagarski, A.M.; Petrov, G.A.; Lukovac, L.; Abele, H.; Wahrwold, S.; Baeßler, S.; Raeder, S.; Kuzmina, T.M.; Schrauwen, J.; Tischenko, V.M.; Protasov, K.V.; Voronin, A.Yu. *ILL Annual Report* 2004.
[24] Nesvizhevsky, V.V.; Protasov, K.V.; *Class. Quant. Grav.* 2004, 21, 4557.
[25] Robinett, R.W. *Phys. Rep.* 2004, 392, 1
[26] Steyerl, A.; Malik, S.S. *Ann. Phys.* 1992, 217, 222.
[27] Steyerl, A.; Malik, S.S. *Phys. Lett.* 1996, A217, 194.
[28] Ellis, J.; Hagelin, J.S.; Nanopoulos, D.V.; Srednicki, M. *Nucl. Phys.* 1984, B241, 381.
[29] Mavromatos, N. (2004). CPT Violation and decoherence in quantum gravity. gr-qc/0407005.
[30] Murayama, H.; Raffelt, G. Hagmann, C.; van Bibber, K.; Rosenberg, L.J. Review of Particle Physics, *Phys. Lett.* 2004, B592, 389.
[31] Arkani-Hamed, N.; Dimopoulos, S.; Dvali, G. *Phys. Lett.* 1998, B429, 263;
[32] Arkani-Hamed, N.; Dimopoulos, S.; Dvali, G. *Phys. Rev.* 1999, D59, 086004;
[33] Antoniadis, I.; Arkani-Hamed, N.; Dimopoulos, S.; Dvali, G. *Phys. Lett.* 1998, B436, 257
[34] Rubakov, V.A.; Shaposhnikov, M.E. *Phys. Lett.* 1983, B125, (1983) 136;
[35] Rubakov, V.A.; Shaposhnikov, M.E. *Phys. Lett.* 1983, B125, (1983) 139.
[36] Visser, M. *Phys. Lett.* 1985, B159, 22.
[37] Antoniadis, I. *Phys. Lett.* 1990, B246, 377.
[38] Lykken, J. *Phys. Rev.* 1996, D54, 3693.
[39] Hewett, J.; March-Russell, J. Review of Particle Physics, *Phys. Lett.* 2004, B592, 1056.
[40] Adelberger, E.G.; Heckel, B.R.; Nelson, A.E. *Ann. Rev. Nucl. Part. Sci.* 2003, 53, 77.

[41] Bordag, M.; Mohideen, U.; Mostepanenko, V.M. *Phys. Rep.* 2001, 353, 1;
[42] Mostepanenko, V.M. In The Gravitational Constant: Generalized Gravitational Theories and Experiments; de Sabbaata, V.; Gilles, G.T.; Melnikov, V.N.; Eds.; *NATO Sciences Series*, Kluwer: Dordrecht, 2004; p. 269.
[43] Carugno, G.; Fontana, Z.; Onofrio, R.; Rizzo, C. *Phys. Rev.* 1997, D55, 6591;
[44] Bressi, G.; Carugno, G.; Galavni, A.; Onofrio, R.; Ruoso, G. Class. *Quant. Grav.* 2000, 17, 2365;
[45] Bressi, G.; Carugno, G.; Galavni, A.; Onofrio, R.; Ruoso, G.; Veronese, F. *Class. Quant. Grav.* 2001, 18, 3943;
[46] Bressi, G.; Carugno, G.; Galavni, A.; Onofrio, R.; Ruoso, G. *Phys. Rev. Lett.* 2004, 88, 041804.
[47] Long, J.C.; Chan, H.W.; Price, J.C. *Nucl. Phys.* 1999, B539, 23.
[48] Frank, A.; van Isaker, P.; Gomes-Camacho, *J. Phys. Lett.* 2004, B582, 15.
[49] Bertolami, O.; Nunes, F.M.; Class. *Quant. Grav.* 2003, 20, L61.
[50] Abele, H.; Westphal, A. Quantum states of neutrons set limits for non-Newtonian interaction, *ILL Annual Report* 2002, 76;
[51] Abele, H.; Baessler, S.; Westphal, A. *Lect. Notes Phys.* 2003, 631, 355.
[52] Hory, M.; Eades, J.; Hayano, R.S.; Ishikawa, T.; Sakaguchi, J.; Widmann, E.; Yamaguchi, H.; Torii, H.A.; Juhász, B.; Horváth, D.; Yamazaki, T. *Phys. Rev. Lett.* 2003, 91, 123401
[53] Korobov, V.I. *Phys. Rev.* 2003, A67, 062501;
[54] Korobov, V.I. *Phys. Rev.* 2003, A68, 019902 (erratum).
[55] Watson, P. (2004). A limit on short-range modifications of gravity. hep-ph/0406308.
[56] Kolb, E. W.; Turner, M.S. The Early Universe; Addison-Wesley: Redwood, CA, 1990.
[57] Moody J. E.; Wilczek, F. Phys. Rev. 1984, D30, 130.
[58] Rosenberg, L. J. In Proceedings of the Workshop on Particle and Nuclear Astrophysics in the Next Millennium, Snowmass; World Scientific: Singapore, 1994.
[59] Youdin, A.N.; Krause, Jr., D.; Jagannathan, K.; Hunter, L.R.; Lamoreaux, S.K. *Phys. Rev. Lett.* 1996, 77, 2170.
[60] Wei-Tou Ni; Sheau-shi Pan; Hsien-Chi Yeh; Li-Shing Hou; Juling Wan. *Phys. Rev. Lett.* 1999, 82, 2439.
[61] Kobzarev, I.Yu.; Okun, L.B. Sov. Phys. JETP 1963, 16, 1343 [*Zh. Exp. Teor. Phys.* 1962, 43, 1904.]
[62] Leitner, J.; Okubo, S. *Phys. Rev.* 1964, 136, 1542.
[63] Mashhoon, B. Class. *Quant. Grav.* 2000, 17, 2399.
[64] Obukhov, Yu.N. Phys. *Rev. Lett.* 2001, 86, 192.
[65] Bini, D.; Cherubini, Ch.; Mashhoon, B. *Class. Quant. Grav.* 2004, 21, 3893.
[66] Alfimenkov, V.P.; Strelkov, A.V.; Shvetsov, V.N.; Nesvizhevsky, V.V.; Serebrov, A.P.; Tal'daev, R.R.; Kharitonov, A.G. *JETP Lett.* 1992, 55, *84 [Pis'ma Zh. Eksp. Teor. Fiz.* 1992, 55, 92].
[67] Steyerl, A.; Malik, S.S.; Geltenbort, P.; Neumaier, S.; Nesvizhevsky, V.V.; Utsuro, M.; Kawabata, Y. *Journal de Physique III* 1997, 7, 1941.
[68] Bestle, T.; Geltenbort, P.; Just, H.; Kawabata, Y.; Malik, S.S.; Nesvizhevsky, V.V.; Neumaier, S.; Okumura, K.; Steyerl, A.; Utsuro, M. *Phys. Lett.* 1998, A244, 217.
[69] Nesvizhevsky, V.V. To appear in *Int. J. Mod. Phys.* 2005, D14, 511.

[70] Collela, R.; Overhauser, A.W.; Werner, S.A. *Phys. Rev. Lett.* 1975, 34, 1472.
[71] Baumann, J.; Gähler, R.; Kalus, J.; Mampe, W. *Phys. Rev.* 1988, D37, 3107.
[72] Borisov, Yu.V.; Borovikova, N.V.; Vasilyev, A.V.; Grigorieva, L.A.; Ivanov, S.N.; Kashukeev, N.T.; Nesvizhevsky, V.V.; Serebrov, A.P.; Yaidjiev, P.S. *J. Tech. Phys.* 1988, 58 *[Zh. Tech. Fiz.* 1988, 58, 951].
[73] Nesvizhevsky, V.V.; Serebrov, A.P.; Tal'daev, R.R.; Kharitonov, A.G.; Alfimenkov, V.P.; Strelkov, A.V.; Shvetsov, V.N. Sov. *Phys. JETP* 1992, 75, 405. [*Zh. Eksp. Teor. Fiz.* 1992, 102, 740].
[74] Mampe, W.; Ageron, P.; Bates, C.; Pendlebury, J.M.; Steyerl, A. *Phys. Rev. Lett.* 1989, 63, 593.
[75] Arzumanov, S.; Bondarenko, L.; Chernyavsky, S.; Drexel, W.; Fomin, A.; Geltenbort, P.; Morozov, V.; Panin, Yu.; Pendlebury, J.; Schreckenbach, K. *Phys. Lett.* 2000, B483, 15.
[76] Nesvizhevsky, V.V.; Strelkov, A.V.; Geltenbort, P.; Iaydjiev, P.S. *Eur. J. Appl. Phys.* 1999, 6, 151;
[77] Bondarenko, L.N.; Geltenbort, P.; Korobkina, E.I.; Morozov, V.I.; Panin, Yu.N. *Phys. At. Nucl.* 2002, 65, 11 [*Yad. Fiz.* 2002, 65, 11]
[78] Nesvizhevsky, V.V. *Phys. At. Nucl.* 2002, 65, 400 [*Yad. Fiz.* 2002, 65, 426].

In: Trends in Quantum Gravity Research
Editor: David C. Moore, pp. 109-137
ISBN 1-59454-670-3

Chapter 3

QUANTUM MECHANICS, QUANTUM GRAVITY, AND APPROXIMATE LORENTZ INVARIANCE FROM A CLASSICAL PHASE-BOUNDARY UNIVERSE

Michael Grady
Department of Physics, SUNY Fredonia, Fredonia , NY 14063 , USA

Abstract

A classical dynamical system in a four-dimensional Euclidean space with universal time is considered. The space is hypothesized to be originally occupied by a uniform substance, pictured as a liquid, which at some time became supercooled. Our universe began as a nucleation event initiating a liquid to solid transition. The universe we inhabit and are directly aware of consists of only the three-dimensional expanding phase boundary - a crystalline surface. Random energy transfers to the boundary from thermal fluctuations in the adjacent bulk phases are interpreted by us as quantum fluctuations, and give a physical realization to the stochastic quantization technique. Fermionic matter is modeled as screw dislocations; gauge bosons as surface acoustic waves. Minkowski space emerges dynamically through redefining local time to be proportional to the spatial coordinate perpendicular to the boundary. Lorentz invariance is only approximate, and the photon spectrum (now a phonon spectrum) has a maximum energy. Other features include a geometrical quantum gravitational theory based on elasticity theory, and a simple explanation of the quantum measurement process as a spontaneous symmetry breaking. Present, past and future are physically distinct regions, the present being a unique surface where our universe is being continually constructed.

1 Introduction

In the following, a new picture of the big bang and the underlying structure of the universe is proposed, based on a classical field theory in a four-dimensional Euclidean space with a universal time (a 4+1 dimensional theory)[1, 2, 3, 4]. The big bang is treated as a nucleation event for a first-order phase transition (pictured as a liquid to solid transition) and our universe is the three-dimensional phase boundary between the expanding solid and preexisting liquid phases. This classical brane-theory appears to have the potential to explain a diverse set of phenomena – Lorentz invariance, quantum fluctuations and zero-point energy, quantum superposition and measurement, elementary fermions and bosons, gauge forces, gravity, the big bang and a non-decelerating expansion of the universe. It is possibly rich enough to give a "theory of everything" from a relatively simple base-theory consisting of a small number of elementary atoms or molecules and basic elastic forces holding them together. In this model, all of the forces and particles of standard particle theory are secondary effects, consisting of the collective excitations and dislocations of the base-theory, just as in condensed matter physics where such excitations play a pivotal role, reducing the elementary degrees of freedom to a mere background for the more interesting and important collective excitations.

We begin by assuming a four-dimensional Euclidean space, filled with a uniform fluid at some temperature, undergoing thermal fluctuations. In addition to the four spatial dimensions, there is also a universal time. Another possibility would be to start already with a five-dimensional Minkowski space, however this does not seem to be necessary. This liquid was cooling, became supercooled, and at some point a solid crystal nucleated. This was the big bang. The universe begins as a fluctuation, already at a finite size, because in order to grow rather than shrink, the initial crystal must be large enough that the positive surface energy is less than the negative volume energy relative to the liquid. In such a model there is no physical singularity at the beginning and there is no reason for the universe to be particularly hot or dense at this time either (more on this later). The *surface* of the solid, the phase boundary, is an expanding three-dimensional space, our universe. This differs from other "bubble universe" pictures, where the universe is the *interior* of a *3-d* bubble. In fact, it bears an uncanny resemblance to the simple "expanding balloon" model which is often used as an example of a uniformly expanding curved space. However the present model differs in that the interior and exterior of the balloon are real spaces, though not directly observed by us. We are directly aware only of the phase boundary separating the phases, which we refer to as the "present". As the crystal grows, this hypersurface, our universe, expands. Already there is a variance with the usual $\Lambda = 0$ Friedmann universes. Namely, our universe is closed, but will expand forever. The pressure on the surface caused by the energy difference of the two phases acts something like a repulsive cosmological constant. This universe actually expands faster as time goes on, not slower. If, as is likely, dissipation is present, it will eventually approach a constant rate. (This assumes a constant amount of supercooling – if the base liquid cools more, the expansion rate could continue to increase as the degree of supercooling increases. Without dissipation, the expansion rate increases exponentially). Recent astrophysical evidence shows that the expansion rate is not slowing, but may even be speeding up[5] which is consistent with this scenario.

In the following, the emergence of a quantum field theory on the surface and the origin of quantum fluctuations is discussed in section two. The relation between real and imaginary time path integrals is clarified as a difference between non-equilibrium and equilibrium statistical mechanics. Section three deals with the dynamical realization of Lorentz invariance and special relativity, including possible tests of the theory, and consequences for cosmic ray physics. In Section four, the description of photons as surface acoustic waves is explored. The Plank relation, $E = h\nu$, and zero point energy are derived, with Plank's constant being essentially the four-dimensional temperature. Section five describes the interpretation and realization of quantum superpositions and quantum measurements. Section six discusses four-dimensional dislocations as candidates for elementary fermions. The possibility of modeling quarks as partial dislocations, which, in ordinary crystals are naturally confined, is explored. Section seven outlines the likely gravitational theory that results from the relationship between the curvature of the surface and the presence of dislocations and interstitials, following previous analogies drawn by many authors between elasticity theory and general relativity. Modeling fermions as screw dislocations introduces a natural relation between spin and torsion, as in the Einstein-Cartan theory, which may be a good continuum approximation to the underlying lattice theory. Whatever gravitational theory that results is automatically a quantum theory of gravity since the 4-D thermal fluctuations are present in the surface. Section eight discusses the cosmology of the model, including possible difficulties in fitting observations. Section nine discusses the rather different nature of time that the model presents and relates it to Whitehead's conception of time. The different causality structure due to the model having a preferred frame is discussed (special relativity is only approximately realized).

2 Quantum Field Theory from a Classical Field Theory

The basic theory needed to describe this expanding phase boundary is non-equilibrium classical statistical mechanics. The boundary itself may be describable in terms of dynamical critical phenomena[6]. The solid, in some sense, lies in the past, since we have been there earlier, although it still exists in the present when observed from the higher dimension. The liquid represents the future, since that is where we are going, but it also exists now, as an undifferentiated, fluctuating medium. To distinguish the current states of the solid and the liquid from our own past and future, they may be called the "current past" and "current future". They differ from our past and future because changes may have occurred after the solid was formed, and the future certainly will be different when we arrive there. To the extent that the solid is frozen, however, our past may be accurately preserved within it. We may not be aware of the existence of the liquid due to its uniformity. However, the boundary which we inhabit is in thermal contact with both the liquid and solid phases, and can certainly exchange energy with them. Actually, since the surface is continuously colonizing new parts of the liquid, the mountain, in this case, is moving to Mohammed. Energy fluctuations that were present in the adjacent liquid will be incorporated into the "new surface" an instant later. These will interact with propagating surface modes which are passed from the "old surface" to the "new surface" as each layer is added. Thus waves riding the interface will experience random energy fluctuations from this thermal contact. These random 4-d thermal fluctuations could explain quantum fluctuations.

It is well known that in ordinary quantum theory, if Minkowski space is analytically continued to Euclidean space, quantum fluctuations behave as higher-dimensional thermal fluctuations, i.e. the Feynman path integral becomes an ordinary statistical mechanical partition function in 4 (+1) dimensions (in equilibrium statistical mechanics there is an implied time dimension). Plank's constant is proportional to the temperature of the four-dimensional Euclidean space. The existence of a 1-1 mapping between quantum field theory and statistical mechanics in one more dimension opens the possibility that the physical reality that quantum theories are describing actually corresponds to a higher dimensional classical theory, one for which, if all degrees of freedom were accounted for, would constitute a dynamical system of some kind. Aside from the important new feature of an extra dimension, this is essentially the point of view of Nelson[7], whose stochastic quantization technique attempts to explain the fluctuations of quantum mechanics through interaction with an otherwise unobservable fluctuating background field. Stochastic quantization was extended to field theory by Parisi and Wu[8], who showed the equivalence of the Euclidean path integral to a stochastic process controlled by a Langevin equation, which operated in a fictitious new time, completely unrelated to ordinary time. Whereas this can be seen as simply a mathematical tool, some have speculated that the reformulation could be closer to reality. Of course, to the extent that mathematical formulations are equivalent, it does not really matter to the physicist which is "more real", however if our current theories are only approximations, then such considerations make sense in trying to find a more correct and accurate theory. If a stochastic differential equation explains quantum fluctuations, then this would likely be the case, since in most cases one can picture such equations as approximations resulting from more detailed deterministic dynamical systems for which some degrees of freedom have been averaged over.

The main problem in making sense of this connection between quantum systems and classical systems in one higher dimension is the analytic continuation to imaginary time, and the lack of any apparent connection between the "Langevin time" of a Langevin simulation and real time. However, if one considers the behavior of fields that live on an expanding phase boundary in a 4-d Euclidean space, such a connection can be made. If one accepts the Langevin time itself as real time, then there will be a connection between it and the fourth spatial coordinate at the surface (the coordinate perpendicular to the surface), due to the motion of the surface. For the sake of simplicity it will be assumed to travel at constant speed. For observers riding the surface, the fourth spatial coordinate will be nearly indistinguishable from time, since they increase in lockstep. In a following section it will be argued that this identification leads to a "spatialization" of time from which all of the properties of special relativity arise - in particular it will be seen that clocks constructed from dislocations and surface modes do not keep universal time, but rather the local time of special relativity.

The remaining question is why quantum field theory is given in terms of a real-time path integral with an oscillating exponential rather than the imaginary-time version with a real exponential. It is perhaps not a question of real or imaginary time which is a mathematical transformation with no apparent physical basis, but the rather less exotic notion of real vs. imaginary frequency describing oscillatory vs. overdamped motions. This can also be seen as the difference between non-equilibrium and equilibrium statistical mechanics. If the universe were a single phase in equilibrium then it could be described by an equilibrium statistical mechanical ensemble. Correlation functions would be decaying real exponentials. The corresponding Langevin equation would be dominated by dissipative forces and the corresponding path integral would be Euclidean (i.e. the imaginary time version). However, an expanding phase boundary is a decidedly non-equilibrium object. It breaks time translation invariance and at least one spatial translational invariance. One may also have propagating modes present on the surface, due to conservation laws. Such propagating modes exhibit oscillatory rather than dissipative behavior, and occur in many 3-d systems[6]. They lead to various complications in the theory of dynamical critical phenomena, and are a crucial feature in the dynamical theory of phase transitions. In many cases these systems are still describable by a stochastic differential equation - a *complex* Langevin equation, where non-dissipative forces play a crucial role[6, 12]. Solutions are oscillating but contain random phase and amplitude fluctuations. The Fourier transforms of correlation functions contain real-axis poles.

A number of authors have shown that the Parisi-Wu stochastic quantization can be performed directly in Minkowski space, the result being a complex Langevin equation which will be exhibited shortly[9, 10]. This completes the logical connection. To sum, fields which represent dislocations or collective modes on a moving phase boundary in a 4-d Euclidean space are likely describable by a complex Langevin equation, which approximates the behavior of the larger deterministic dynamical system which fills the entire Euclidean space. This complex Langevin equation has an equivalent path-integral representation (meaning the two systems have the same correlation functions), which resembles the Minkowski space path integral of quantum field theory. Some details will likely be different, however. For instance, it does not seem likely that dissipation will be entirely absent from the surface. This could be countered by an energy input, resulting in a steady-state rather than an isolated system. Such a system lacks time-reversal invariance at some level, which could have observable consequences (and perhaps help to explain CP non-conservation in the $K^0 - \bar{K}^0$ system).

The Langevin equation is a first-order differential equation with a fluctuating random force. It was first applied to the case of Brownian motion of a small particle in a background of randomly moving molecules colliding with it. If v represents the velocity of the particle, then the Langevin equation is

$$\dot{v} = -\gamma v + F + \eta(t) \tag{1}$$

The γv term is the frictional force exerted by the fluid, F is an applied external force (if present) such as an electric field, and $\eta(t)$ is the fluctuating force designed to mimic the many collisions between the fluid which is assumed to be in thermodynamic equilibrium at some temperature and the particle. In the absence of force F, the particle exhibits a random walk in position. Without the damping term it would also perform a random walk in velocity, and the kinetic energy would increase without bound. However, any amount of dissipation is sufficient to stabilize it and the particle's average kinetic energy will become equal to $\frac{1}{2}\mathrm{k}Td$, where d is the number of spatial dimensions, T is the fluid temperature, and k is Boltzmann's constant. If one wants to extend this treatment to an oscillator, a problem arises in that a position dependent force cannot be incorporated into a first order equation. The Hamilton equations are, of course, first order, but there are two of them. By introducing a complex variable $b = (p + ix)/\sqrt{2}$, $b^* = (p - ix)/\sqrt{2}$, one can write the Hamilton equations for one degree of freedom as a single complex equation:

$$\dot{b} = i\frac{\partial H}{\partial b^*}. \tag{2}$$

To explore these ideas in more detail, consider the system of a single harmonic oscillator interacting with a bath of other harmonic oscillators[12]. The simple harmonic oscillator in coordinates $x = \sqrt{m\omega}x'$, $p = p'/\sqrt{m\omega}$, where x' and p' are the usual coordinate and momentum has the Hamiltonian

$$H = \omega b^* b \tag{3}$$

Here, k is the spring constant and $\omega \equiv \sqrt{k/m}$. The Hamilton equation (2) becomes

$$\dot{b} = i\omega b. \tag{4}$$

Interestingly, this formalism can be easily extended to the damped oscillator[12, 13] by allowing ω to become complex. Replacing ω with $\omega + i\gamma$ gives the equation of motion for the damped oscillator,

$$\dot{b} = i\omega b - \gamma b. \tag{5}$$

Here, the complex formalism goes beyond the real formalism, since the Hamiltonian does not technically exist for the damped oscillator unless auxiliary fields are added[13]. Note that this is not a fully complex Hamiltonian function which would result in doubling the number of equations of motion and producing an overdetermined system. Rather, the Hamiltonian takes values along a ray other than the real axis. If we add a fluctuating force, one obtains a complex Langevin equation,

$$\dot{b} = i\omega b - \gamma b + \eta(t). \tag{6}$$

This equation can be derived as the equation of motion of a tagged oscillator interacting with a collection of "bath" oscillators whose behavior is averaged over[12]. The bath provides both the random force and the damping. It can be used, for instance, to describe the behavior of a single-mode laser interacting with a thermal medium and thermal mirror fluctuations[12, 14]. Similarly, it can also be used to describe propagating modes in dynamical critical phenomena[6]. Thus the complex Langevin equation is a well-established equation for describing oscillating or propagating modes in a random medium.

If the Parisi-Wu quantization is applied to the Minkowski field theory directly, it has been shown[9, 10] that the correlation functions derived from the path integral

$$\int D\phi \exp(iS(\phi)/\hbar) \tag{7}$$

can be obtained from the long-time behavior of the Langevin equation

$$\dot{\phi} = i\delta S/\delta\phi^* - \epsilon\phi + \eta(x,t) \tag{8}$$

where t is a fictitious "Langevin time" unrelated to the real time in the path integral, x represents the four space-time variables, x_i, with $i = 1..4$, and the Gaussian noise term has the following correlation function:

$$< \eta^*(x,t)\eta(x',t') >= 2\hbar\delta^4(x-x')\delta(t-t'). \tag{9}$$

Field correlations are computed at equal Langevin times and the damping, ϵ, is taken to zero after correlation functions are calculated. Equation 8 appears to be a multivariate version of equation 6 (the first term being generalized to the RHS of equation 2) with the Minkowski action S playing the role of a Hamiltonian. For example, for the complex scalar field,

$$S = \int (|\partial_\mu\phi|^2 - m^2\phi^*\phi)d^4x \tag{10}$$

one obtains the complex Langevin equation

$$\dot{\phi} = i(-\partial^2\phi/\partial x_4^2 + \nabla^2\phi - m^2\phi) - \epsilon\phi + \eta(t) \tag{11}$$

One can understand the difference in sign between the spatial and local-temporal (x_4) derivative terms in relation to the different dynamic behavior of the interface in these directions. If one thinks of ϕ as a displacement field of elementary atoms from their quiescent-crystal locations, one expects oscillatory behavior in the spatial directions. The sign of the ∇^2 term is such as to provide the usual restoring force from neighboring atoms making this possible. A negative restoring force, as exists in the time direction, leads to an instability, as occurs in a ϕ^4 theory with a negative mass-squared term, for example. This will result in translational motion (a soft mode). If we think of the membrane as the physically relevant object, it is in translational motion in the temporal direction. Therefore the "Minkowski signature" of the D' Alembertian operator would appear to be directly related to the dynamics of the phase boundary, which is itself, of course, controlled by the Lagrangian of the "base-theory" of the elementary atoms. The fact that the instability that resulted in the motion of the phase boundary is a phase transition of the base-theory, which is likely driven by a spontaneous symmetry breaking, suggests that the Minkowski space we are familiar with is due to a spontaneous symmetry breaking from original space-time symmetry of the base-theory. Such a dynamical origin for Minkowski space, and the consequences of special relativity that result, is in rather distinct contrast to the kinematical origin postulated by Einstein. Indeed it is more like the view held by Lorentz and others who clung to the idea of a cosmic ether, even if invisible. The crystal and liquid in the picture presented here is a form of ether, which, however, is only invisible at low energies. When photon wavelengths get close to the elementary lattice spacings, then the deviation from linearity of their phonon-like dispersion relations in this theory will become apparent, and the existence of the crystal will have observable effects. These ideas are expanded in secs. III & IV.

Getting back to the Langevin equation under discussion, we now consider the consequences of our somewhat different interpretation of the Langevin time coordinate. The usual treatment calculates correlation functions at equal Langevin times, whereas we are essentially locking the Langevin time to the ordinary time through the presumed uniform motion of the phase boundary. It is important to see whether this will make any difference in the relationship to the quantum field theory. One notices a peculiarity in equation 11 when subjected to dimensional analysis. Taking the usual dimension of $[\ell^{-1}]$ for the ϕ field and $[\ell]$ for x_i and t variables leads to different dimensions for the $\dot{\phi}$ and $\Box\phi$ terms. One common solution is to let the fictitious time have dimensions $[\ell^2]$[9, 11]. Then dimensional consistency is obtained and $\hbar$ comes out dimensionless. However, since we want the fictitious time to become the real time, another solution must be taken. Introducing a parameter a with dimensions of length, which can be taken to be the lattice spacing, rewrite equation 11 as

$$\dot{\phi} = ia(-\partial^2\phi/\partial x_4^2 + \nabla^2\phi - m^2\phi) - a\epsilon\phi + \eta(x,t) \tag{12}$$

where

$$< \eta^*(x,t)\eta(x',t') >= 2a\hbar\delta^4(x-x')\delta(t-t'). \tag{13}$$

The two times now have the same dimensionality, the equation is dimensionally consistent and the two factors of a, one multiplying S and one multiplying $\hbar$ will cancel in the path integral ($\hbar$ will now be set to unity). For the free field theory we are considering here, the Langevin equation can be solved[8, 9, 11], with a long time stationary correlation function

$$D(x-x',t-t') \equiv \lim_{t,t'\to\infty} < \phi^*(x,t)\phi(x',t') > \tag{14}$$

(with $t-t'$ fixed) given by

$$D(x-x',t-t') = \frac{2a}{(2\pi)^5}\int d^4k \int d\omega \frac{e^{-ik(x-x')-i\omega(t-t')}}{\omega^2 + a^2(k^2-m^2+i\epsilon)^2}. \tag{15}$$

Setting $t-t' = x_4 - x_4'$, we get a free propagator of

$$D(x-x') = \frac{1}{(2\pi)^4}\int d^4k \frac{e^{-ik(x-x')}e^{-a|(k^2-m^2)(x_4-x_4')|}}{k^2-m^2+i\epsilon}. \tag{16}$$

This is slightly modified from the usual field-theory propagator which results from taking equal Langevin times, $t = t'$. However, the extra exponential factor affects only the off mass-shell propagator, and even for that is highly suppressed by the factor of the lattice spacing, a reasonable guess for which might be around 10^{-16} $(\text{eV})^{-1}$. It thus seems unlikely that this extra factor would affect calculations at today's accelerator energies. It breaks Lorentz invariance explicitly. As mentioned before, this theory is only approximately Lorentz invariant. Lorentz invariance is good at energies small compared to the inverse lattice spacing. From a more fundamental point of view, the rest frame of the crystal is a preferred frame and calculations should be performed in that frame. However, observable effects of this frame dependence are limited to very high energies. These are possibly accessible through studies of cosmic rays (see sec. III).

To sum, building on the known equivalence of the Minkowski path integral to a stochastic process involving a complex Langevin equation, it has been shown that ordinary quantum field theory may result from the dynamical critical behavior of an expanding phase boundary in a four-dimensional Euclidean space. In this picture, quantum fluctuations are actually thermal fluctuations in the higher dimensional space.

3 Special Relativity Realized Dynamically

The underlying theory pictured above is a classical dynamical system lying in the 4-D Euclidean space, governed by a universal Newtonian time. It is proposed that Minkowski space is the result of restricting attention to the hypersurface representing the phase boundary, and choosing local time to be the spatial coordinate perpendicular to the moving boundary. In calling it a Minkowski space, we are considering only a small portion of the surface which can be taken to be approximately flat. Globally, the spatial geometry is hyperspherical, and the space is a positively curved pseudo-Riemannian space similar to the positive-curvature case of the Robertson-Walker metric of standard General-Relativity-based big-bang cosmology.

To show the emergence of Minkowski space locally, a more detailed model is needed. If the phase boundary is considered the boundary between a liquid and crystalline solid, with the solid growing into the liquid, then a reasonable model for the photon is the surface acoustic wave, and for elementary fermions, screw dislocations in the crystal. The surface acoustic wave is a propagating solution within the surface that decays exponentially away from the surface. It obeys a phonon-like dispersion relation, with a speed somewhat below that of shear bulk waves. It is well known from the 1938 work of Frenkel and Kontorova [15] and of Frank and Eshelby in 1949 [16] that screw dislocations obey the Lorentz contraction formula with the speed of light replaced by the speed of transverse sound. In other words, the pattern of crystal distortion that surrounds the dislocation becomes elliptical for a moving dislocation, with the strain pattern in the direction of motion shrinking according to the Lorentz contraction formula. An "object" made from an array of such dislocations really does shrink in the direction of motion. In addition, the effective mass of the dislocation grows with velocity according to the relativistic formula (more precisely the energy and crystal-momentum transform according to the Lorentz transformation)[16, 17, 18]. Therefore, screw dislocations are prohibited from being accelerated beyond the velocity of transverse sound in a crystal, because the kinetic energy becomes infinite in that limit. In a real crystal, however, this limit can be exceeded by introducing a moving dislocation from an adjacent compatible medium where the sound speed is higher. The supersonic dislocation rapidly decelerates to subsonic velocities by emitting "vacuum Cerenkov radiation" [17, 18]. It is also conceivable to exceed the limit by violating the approximations of continuum linear elasticity theory on which these results are based. This relativistic behavior appears to be followed for any reasonable dislocation model for which perturbations are subject to continuum linear elasticity theory[18]. It is not immediately clear what the minimum requirements are[19], but coupling to a single type of phonon with a relativistic dispersion relation is necessary, and may be sufficient. Coupling to other types of phonons is possible only if these either have the same velocity or have an energy gap. The main point here is there can not be more than one limiting velocity for low-energy excitations. For instance, ordinary edge dislocations obey a more complicated set of contraction equations involving both the longitudinal and transverse sound velocities[16, 17].

Considering again the phase boundary universe model, if all matter is made up of screw dislocations then the above considerations strongly suggest that measuring rods constructed from "dislocation arrays" will obey the Lorentz contraction. For now, consider only observations made from the rest frame of the crystal. A measuring rod will *physically* shrink if moving with respect to this frame along the rod's direction. The Lorentz transformation also involves time, however. The Lorentz contraction and mass increase certainly will have physical effects on clocks that are constructed from moving dislocations. Günther [20] has investigated using the breather solution of the sine-Gordon equation as a clock (sine-Gordon soliton kinks are a lower-dimensional dislocation model). He finds such a clock slows with velocity in accordance with the usual time-dilation formula. If length and time standards are both based on solitons, full Lorentz invariance ensues.

For our case, assuming only length contraction and observing from the crystal rest frame, a simple light-clock where a flash of light is given off and bounces off a mirror held by a rigid frame to the light source, then back to a detector near the source, in either transverse or longitudinal orientations, exhibits time dilation following the usual treatment in special relativity. However, although the argument is the same, the assumptions are different. At this point we have not assumed anything about moving frames of reference. We are simply observing a moving rod and a moving clock from the rest frame of the crystal, where we know the speed of sound (light), and know it is isotropic (we are always assuming an isotropic crystal). This is all that is needed to demonstrate time dilation from Lorentz contraction of the light-clock. We notice that when observed from this frame, rods shrink, and clocks slow down due to physical, dynamical effects. Energy and crystal-momentum of dislocations also obey relativistic equations[16, 17, 18].

Now we ask what coordinate system is a reasonable one for a moving observer to use? Of course, the moving observer will use the shrunken rod to measure distance and the slow clock to measure time - what other reasonable choice does s/he have? It is also natural for moving observers to choose their local time coordinate to be along their own world line, and spatial hypersurfaces to consist of points all with the same time coordinate, with synchronization performed using light signals. The full forward Lorentz transformation, which consists not only of scale changes inherent in Lorentz contraction and time dilation, but also in the aforementioned axis rotations, ensues. This now allows us to transform coordinates between the crystal rest frame, and the natural frame of a moving observer. Inverting this transformation is simply a matter of mathematics. As is well known but seems to have been initially unappreciated by Lorentz, this inverse Lorentz transformation has the same form as the forward transformation, with the relative frame velocity negated. The point is that once the full forward Lorentz transformation is realized, fully reciprocal special relativity results simply due to the mathematics of the Lorentz transformation. In Einstein's special relativity, this is due to the symmetry of the underlying Minkowski space - a kinematical symmetry. All frames are exactly equivalent. Although our result is the same, conceptually it is very different, since the Minkowski space has resulted from a dynamical symmetry of the moving boundary solution. Unlike in the Einstein picture, the Lorentz contraction and time dilation have different causes in different frames in this picture. From the crystal rest frame, the shrinking of a moving rod and slowing of a clock are physical effects, caused by motion within the stationary crystal. From the moving frame, the observation that a rod and clock in the crystal rest frame *also* appear to be shortened and slowed are more of an illusion, created by using moving instruments, and a bent reference frame, with its different notion of simultaneity. Because these points of view are conceptually different (kinematic vs. dynamic symmetry), Lorentz, Larmor, Langevin and others held on to the latter view for some time after special relativity won acceptance[21]. In fact, the view of relativity given above is very similar to that of Lorentz, who introduced the concept of local time given above. The unobservability of the ether in this continuum theory eventually led to the demise of this viewpoint. However, if the underlying medium is not a continuum, but a lattice (which itself may lie in a continuum), then at high enough energies differences between the stationary and moving observer must eventually show up. This is because unlike the photon, the phonon dispersion relation is not a straight line. For a linear isotropic material in three dimensions it is given by

$$\omega^2(k) = (2c/a)^2(\sum_{i=1}^{3} \sin^2(k_i a/2)). \tag{17}$$

Surface phonons follow a similar dispersion relation. One has to get to within about 20% of the maximum frequency before the phonon curve differs from the photon curve by more than 1%. Above this point significant dispersion occurs. A very fast-moving light clock which blue-shifted the light into this frequency region would show measurable deviations. The crystal rest frame will be the only frame in which the speed of light at these high frequencies remains isotropic. It therefore becomes an observable preferred frame - the ether is detectable.

These considerations suggest a number of ways that this theory could be checked experimentally. Of course, the lattice spacing could always be made impossibly small, erasing all observable effects. Observations of very-high-energy cosmic rays can put a lower bound on the lattice spacing. Assuming the Plank relation $E = \hbar\omega$ (a possible origin of which is given in the next section) and setting $\hbar = 1$, the definite identification of cosmic ray photons at energies of a few times 10^{13} eV [22] means the high-energy cutoff of the photon dispersion relation must lie above this, so probably $a < 10^{-14}$ $(\text{eV})^{-1}$. An interesting enigma in Cosmic Ray physics is the presence of an "ankle" in the spectrum around 10^{18} eV, where the drop in intensity with energy becomes less steep, along with the apparent absence of the expected cutoff due to interactions with the cosmic microwave background radiation (CMB)[23]. This Greisen-Zatsepin-Kuzmin (GZK) cutoff [24] is due to pion photo-production from interactions between the cosmic ray particle (assumed a proton or light nucleus) and CMB photons. This effectively limits cosmic rays of energy above

5×10^{19} eV to a relatively short travel distance - within the local supercluster (photons and heavy nuclei are also limited by similar mechanisms involving starlight). However, the number of cosmic rays at this energy and higher, although small in absolute event counts, does not show any diminution from the earlier trend. In other words, there is no observational evidence for the GZK cutoff. Another puzzle is that if the very high energy cosmic rays do come from nearby sources, then, they would be expected to point to within a few degrees of their sources, despite the deflection of magnetic fields, due to the high momentum of the particles. However, there appears to be no correlation with possible nearby sources. A photon energy cutoff in the range 10^{16} to 10^{19} eV could invalidate the Lorentz transformation which is used to derive the GZK cutoff from the known behavior in the center-of-mass frame[25]. It would also affect the decays of other high-energy particles, such as neutral pions. A 10^{20} eV neutral pion could not decay into two photons, but at minimum into 10,000 photons for a photon energy cutoff of 10^{16} eV. This would be highly suppressed due to the large power of the fine structure constant required. If the weak bosons had similar cutoffs, then it seems the neutral pion could be made almost stable above a certain energy. Decay of the neutron could be similarly suppressed. If one or more of these neutral particles could travel cosmological distances above a certain energy threshold, it could possibly explain the ankle, due to the addition of a new species to the particle mix. High-energy neutral particles should point toward their sources even at great distances, since they are not much affected by magnetic fields (there is still some effect through magnetic moments). Interestingly, Farrar and Biermann have shown that the observed directions of some of the highest energy events can be correlated with distant quasars[26]. This would be consistent with the scenario sugested here.

4 Photons as Surface Acoustic Waves

Generically, surface acoustic waves (SAW's) traveling in the x-direction on a solid surface at $z = 0$, with the solid occupying the half-space $z < 0$ takes the form[27, 28]

$$u_j = u_{0j} e^{i(kx - \omega t) + \kappa z} \tag{18}$$

with k, ω, and κ real, and $\kappa \propto k$ at least for small k. Here u_j is the elastic displacement field for the solid, which is defined only for $z \leq 0$. Typically most of the energy in surface waves is confined to a region within a few wavelengths of the surface. The most common SAW is the Rayleigh wave, first described by Lord Rayleigh in 1885[27, 28]. It is polarized in the saggital plane (perpendicular to the surface), and consists of motion that is both transverse and longitudinal. It is dispersionless in the continuum version and has a typical phonon dispersion law on the lattice[29]. This wave does not seem to be a promising one to model the photon after, however, since it has only one polarization, regardless of the dimensionality of the surface. The Rayleigh wave is the only type of surface wave for the simplest case of a flat linear elastic half-space. However, if the surface is allowed to have properties different from the bulk, such as a different density, elastic constant, surface tension, curvature, roughness, piezoelectricity, magnetoelasticity, etc. then another surface wave will usually exist, a Love wave[30, 31, 32, 33]. This wave, originally derived for a finite slab of different material deposited on the half-space[34], has a shear-horizontal (SH) polarization, thus for a three dimensional surface would have two transverse polarizations. The Love wave also exists for a thin surface layer such as a thermodynamic phase boundary[31, 35, 36]. It can be seen as a perturbation of the SH surface skimming bulk wave (SSBW) that exists even for the simple half-space. The SSBW is a wave that does not decay below the surface. This solution is unstable with respect to virtually any surface property that retards the wave speed near the surface, which will turn it into an SH surface wave with exponential decay away from the surface, i.e. a Love wave[30, 31]. The Love wave is somewhat dispersive, due to the introduction of a quantity with dimensions of length that characterizes the surface-layer thickness. However if this is no more than a few lattice spacings, then the dispersion is similar to that of an ordinary phonon. Adding a liquid to the external space complicates but does not significantly change the situation. However, the case being envisioned here has a more complicated boundary condition than has been considered in the surface-wave literature, since the boundary is growing, perhaps rapidly. This can perhaps be treated by the method of virtual power[37], and is briefly considered by Maugin[31] and also by Kosevich and Tutov [36]. The transfer of momentum to the "new surface" of the growing crystal will modify the usual traction-free boundary condition of the Raleigh-wave solution. It is this latter boundary condition which prevents the SH polarization from existing in the simple half-space[28]. The violation of this boundary condition by the growing crystal is further evidence that SH waves probably do exist in this case.

Since the weak interactions also need to be accounted for, probably more structure needs to be incorporated into the model. If we imagine the elementary molecules to be non-spherical, then they have their own non-trivial symmetry group, compatible with but distinct from that of the crystal. This basis symmetry group could account for internal symmetries. For instance, if the molecule can be represented by a 4-d vector with nearest-neighbor Heisenberg-like interactions, then an SO(4) symmetry (isomorphic to SU(2) $\times$ SU(2)), which may be partially broken by other interactions, will exist. The spontaneous breaking of this symmetry will result in surface magnons[38]. These come in both acoustic and optical varieties. The surface magnons can also mix with surface elastic waves through the magneto-elastic effect, reminiscent of electroweak unification. These possibilities need to be examined in detail - they are mentioned here to indicate the rich possibilities for model building that occur in surface modes. It is also worth noting that a promising approach to incorporating chiral fermions on the lattice, necessary for a lattice approach to the weak interactions, incorporates a fourth spatial dimension, with the chiral fermions living on a domain wall[2, 39]. The picture of the universe presented here seems ideal for the realization of this mechanism.

A universal property of all surface modes is the exponential decay as the bulk is entered, characterized by a decay length proportional to the wavelength. This property can be used to derive the Plank relation $E = \hbar\omega$, perhaps the most fundamental equation of quantum mechanics, from the equipartition theorem. If we assume that all elementary degrees of freedom are thermally excited (actually not a completely good assumption due to conservation laws and partial non-ergodicity - see discussion below), then the equipartition theorem will give equal energy to each harmonic degree of freedom of amount kT_4, where k is Boltzmann's constant and T_4 is the 4-d temperature. For a surface wave with decay length $\kappa = bk$, where b is a constant, taking into account the energy of a wave being proportional to its square, one has an energy depth profile (1-d energy density)

$$E = E_0 e^{2\kappa x_4} \tag{19}$$

where x_4 is taken to be zero at the surface, and becomes negative inside the medium. The energy in the monatomic surface layer itself is given by $E_0 a$. The total energy can be computed by

$$E_{\text{tot}} = \int_{-\infty}^{0} E dx_4 = E_0/(2\kappa) \tag{20}$$

Setting $\kappa = bk$ (proportionality of decay length to wavelength), $\omega = ck$, and the total energy to kT_4, one can solve for the surface layer energy, $E_0 a$, now denoted E_3

$$E_3 = (2ba\text{k}T_4/c)\omega \tag{21}$$

which is the Plank relation if $\hbar = 2ba\text{k}T_4/c$. This is consistent with the thermal explanation of quantum mechanics given above, namely that $\hbar$ is essentially the 4-d temperature, with the necessary factors of a and c to fix the dimensions. The essential feature which gives higher frequency excitations higher energy on the surface is the higher concentration of SAW energy near the surface, compared to lower frequency excitations which are more spread out in the fourth dimension. Thus equal sharing of energy in four dimensions naturally leads to unequal energies on the 3-d surface, as embodied in the Plank relation.

Not all surface modes will necessarily become excited for two reasons. First is the effect of global conservation laws, and second is the probable lack of full thermodynamic equilibrium. Consider the liquid degrees of freedom directly above the growing surface. These are presumably in thermal equilibrium in their liquid environment. When the surface arrives, they are rather suddenly thrust into a new environment with modified interactions due to the translational symmetry breaking of the crystallization. They therefore do not have much time to adjust to these new conditions by the time they can be considered part of the new crystal surface. Eventually they reach a new equilibrium state well after the surface has passed and they become part of the bulk. Thus the surface degrees of freedom are in a transitional state. With the arrival of crystalline order comes a new conserved quantity, the crystal momentum. It is a consequence of the remaining discrete translational invariance but technically is a permutation invariance of the atomic position variables, resulting in conservation of wave number for phonons[40]. Since this is only a single global constraint (or three constraints in three dimensions) it would not appear to limit the allowed random excitations much. However, satisfying global conservation laws requires global correlations, and these take a long time to establish. Therefore one expects to have to satisfy conservation laws locally. This means that one should not expect widely separated thermal excited phonons whose wave vectors happen to add to zero. Rather one expects standing waves or standing wave packets where the zero net wave vector requirement is met pairwise and locally. Thus the random vibrational thermal energy of the liquid will re-organize on the surface primarily as such standing waves, with the amplitudes of constituent travelling waves locked, and phases randomly fluctuating. This may represent the zero-point energy of the photon field. Since each travelling wave mode is not independently excited, the net energy assigned to each is one-half that of the standing wave, i.e $\frac{1}{2}\hbar\omega$.

However, if a travelling propagating wave already exists on the surface (perhaps excited by dislocation interactions etc.) then it will be preserved by the crystal momentum law, and, since it is now an allowed excitation can exist independently of the standing wave and be given the full equipartition energy of $\hbar\omega$, on top of what it gets from the zero-point excitation. One wonders how multiple photon excitations can arise in such a picture, which will lead to a discussion of resonances in coupled oscillator systems. Before embarking on that, it is worth noting that all of the discussion here concerning zero-point energies and photon excitations is in one sense unnecessary, since once the formal equivalence of the stochastic evolution of the phase boundary and the quantum system is accepted, one can merely plug in the QED or standard model Lagrangian and obtain the equivalent Langevin equation for the phase boundary evolution, which will, due to the above equivalence, necessarily include all of the known particle excitations and quantum effects. The discussion here is therefore not to prove that each feature of quantum mechanics is included, but rather to illustrate how each quantum feature might be manifested in the phase boundary evolution. In a similar sense, energy quantization is not as apparent in the path integral formulation of quantum mechanics as it is in the canonical formulation, but it has to be there, and can be seen from multiple poles of the propagator.

A collection of coupled oscillators, even if somewhat non-linear, is generally not ergodic. This was demonstrated by the famous computer simulation of Fermi, Pasta and Ulam in which they coupled 64 harmonic oscillators with non-linear couplings, expecting to see the approach to equilibrium[41, 42, 43]. Instead they found that most modes remained unexcited with energy pouring back and forth between a few modes as in the Wilberforce pendulum - in other words a limit cycle as opposed to chaos. The only modes that participated were those that met or were very close to the resonance condition

$$\sum_i n_i \omega_i = 0 \tag{22}$$

where the n_i are integers[42, 43]. This was later understood in terms of the KAM theorem (Kolmogorov, Arnol'd, Moser), which essentially states that for small non-linearities only regions near resonant surfaces in phase space will get occupied. Full chaos only ensues when these resonant regions grow large enough to be overlapping[44]. Phenomena such as down-conversion or harmonic generation in the presence of small non-linearities can be understood in terms of the resonance condition. The n-photon state takes the form of an nth order resonance from this point of view. The integers in the resonance condition are the correspondents to energy quantization. The degree of excitation is consistent with that of a single n^{th} harmonic photon with which the state is resonantly linked. Therefore it seems plausible that the Plank relation and full photon spectrum, including zero point energy, does have a realization in the propagating modes of the phase boundary as it moves through the random medium.

5 Quantum Superposition and Measurement - Zitterbwegung and Spontaneous Symmetry Breaking

The picture described above treats quantum fluctuations as thermal fluctuations in the 4+1 dimensional space. In such a picture quantum tunneling is explained classically as thermal activation, i.e. due to a random kick of extra energy which results from thermal contact with the liquid and solid phases. Due to such thermal fluctuations, energy is not conserved over short time periods; it is conserved only in the average over time. Thermal fluctuations may create a kind of zitterbewegung – very rapid variation at small scales, that enforces the uncertainty principle and allows for superpositions. The ensemble average implied in a quantum expectation value is replaced by a time average. For rapid fluctuations which cover the ergodic subspace in times short compared to the time between measurements, these should yield identical results. Over very short time periods, additional correlations may appear in the time-averaged case, since the classical system is in a particular state at any one time, so subsequent states will retain some memory of previous states.

This more classical evolution affords the opportunity to explain the quantum measurement process as a spontaneous symmetry breaking event. Anderson has suggested that measuring devices incorporate spontaneous symmetry breaking in their operation[45]. Ne'eman has also espoused this viewpoint. In addition, he has shown that EPR type correlations can occur in classical systems with gauge symmetries, with the gauge connection enforcing long-distance correlations among fluctuating variables[46]. More detailed models have been considered in [47] and [48].

When a classical statistical mechanical system undergoes a spontaneous symmetry breaking, the ergodic phase space splits into non-communicating subspaces. From that point on, the system remains trapped in one of the subspaces. Which subspace is chosen is simply determined by the subspace the system happened to be in at the time of symmetry breaking. A measuring device is postulated to be any device that can couple its order parameter to a quantum system and that includes a control that can initiate spontaneous symmetry breaking of that order parameter. The measuring device, originally with an unbroken symmetry, couples to the system under study becoming strongly correlated with it. Then an adjustment is made to the potential of the measuring device which initiates spontaneous symmetry breaking. The measurement takes place at this time, when the ensemble of possible future states of the combined system splits into non-ergodic subensembles corresponding to the possible values of the order parameter, also corresponding to possible values of the measured quantity. Future evolution is confined to a single subensemble in the usual manner of a classical symmetry-breaking phase transition. In this picture measurements are well defined, the collapse is a physical event, and a clear distinction exists between what constitutes a measuring device and what does not. The symmetry breaking barrier does not even have to be infinitely high - all that is required is that the tunneling time of the post measurement state to be long compared to the time scale of the experiment. This is in contrast to what occurs if the same concept of spontaneous symmetry breaking is applied to explain measurement in standard quantum mechanics. Here it is difficult to see how even spontaneous symmetry breaking can break the superposition, especially if measuring devices are finite so tunneling probabilities are not quite zero (e.g. ref. [47] still uses the Everett interpretation to deal with the "collapse"). Nevertheless it is assumed that when the universe undergoes a cosmological phase transition it does not end up in a superposition of the possible outcomes but rather "measures itself" so as to fall into a single vacuum. In the new picture given here, an event like this is in the same category as a measurement and the result of both is a physical collapse of the available future phase space.

Because the "current past" is continuously undergoing 4-d thermal fluctuations, it is only frozen to the extent that the ensemble is limited due to spontaneous symmetry breaking. Thus questions such as "which slit did the electron go through" or "which direction was the spin pointing" are as meaningless here as they are in standard quantum mechanics. This is because the details of history are continuously being rewritten as both the current past and present fluctuate. Only to the extent that the ensemble is limited by spontaneous symmetry breaking can one make definite statements about past events. EPR (Einstein-Podolsky-Rosen) states, which consist of two separated spins in a net spin-0 state, can only undergo correlated fluctuations which obey the global angular-momentum conservation law. The spin direction of each particle will fluctuate in such a way that its partner fluctuates oppositely. Measurement of either spin is performed by spontaneously breaking the spin direction symmetry, after which a barrier will exist preventing further fluctuations of either particle. Such a process was envisioned in [46]. Such non-local correlations may seem odd, but they are formed by the causal process of separating the particles, and do not violate causality (causality is arbitrated from the crystal rest frame, where universal and local time are equivalent).

6 Dislocations as Candidates for Elementary Fermions

Dislocations, particularly screw dislocations and their variants, provide a rich building ground for models of elementary particles. In this section a detailed model will not be attempted, but rather the general problem of extending the screw dislocation into four dimensions will be discussed, which will result in a four-dimensional string.

The idea of representing elementary particles as dislocations in a medium is a rather old one. Burton talked of "strain-figures" that could move through a medium and interact[49]. Although most 19th century physicists considered matter to be separate from the ether, Larmor suggested the possibility of matter particles being singularities in the ether itself and sought a unified theory of matter and radiation through the properties of a single medium[50, 51]. In more recent times the modeling of elementary particles as topological solitons (a type of dislocation) has intrigued many, with the Skyrmion picture of the nucleon being perhaps the most successful.

The screw dislocation in three dimensions has a number of features that liken it to an elementary fermion. The left and right handed versions can be pair-produced or annihilated, and their elastic interactions have a number of electromagnetic analogies, the most often cited being to magnetostatics[18, 52, 53, 54]. Although double dislocations are not totally prohibited, they are very unfavorable energetically. Screw dislocations are, of course, line defects, so cannot be compared directly to point particles. One is tempted to interpret the line defect as a world-line. However, this implies an extension to four dimensions. An isolated screw dislocation is unfortunately not an option in four dimensions. This can be seen in a number of ways. If one circles a screw dislocation in three dimensions, then one finds after a single loop that one has advanced one lattice spacing to the next sheet of atoms in the direction of the dislocation. The degree of non-closure of the loop is represented by the Burgers vector of the dislocation, which for a screw dislocation on a cubic lattice is one lattice spacing long and in the same direction as the dislocation, or opposite for an oppositely-handed dislocation. Although the transition is gradual, the point on the loop at which one can be deemed to be on the next level can be arbitrarily defined - the set of these points for all possible loops is called the Volterra surface. The freedom of choice of the Volterra surface can be thought of as a form of gauge invariance. There are many atoms far from the dislocation which have moved some fraction of a lattice spacing from their original lattice positions, but there is not much stress associated with this since all of the neighboring atoms have moved a similar amount. Stresses are concentrated only around the dislocation line. If one tries to embed this structure into a non-dislocated 4-d lattice, then those atoms far from the dislocation which are shifted from their original lattice positions by near 1/2 of a lattice spacing will fit badly the undislocated lattices adjacent to them in the fourth dimension, where the atoms are all at their original undislocated positions. The energy of such a structure is proportional to the four-volume - it is no longer a one-dimensional dislocation. The other way one can see there is something wrong in simply promoting the screw dislocation to four dimensions, is that a loop apparently surrounding the dislocation can be moved into the fourth dimension, where the dislocation does not exist, and shrunk to a point. Thus there is no longer a consistent topological classification of this object.

The screw dislocation can be extended into the fourth dimension by copying it onto each successive 3-lattice as the fourth coordinate is changed. This produces a wall of identical dislocations. The solution is translational invariant in the fourth dimension and involves no new stresses, since each atom is exactly one lattice spacing away from its neighbor in the fourth direction. However, we now have a domain wall in 4-d or line in each 3-d slice. For an elementary particle we want something closer to a point in 3-d. An obvious solution would be to wrap the domain wall around onto itself into a small tube, the 3-d cross-section of which would be a string. The bending of the wall introduces stresses which favor a larger string, but this is opposed to the ordinary screw stress proportional to the string length, so there is the possibility of a stable equilibrium size. Going back to the 3-d cross-section, the Volterra surface is any surface bounded by the string. A loop that threads the string will pass through the Volterra surface and detect the dislocation.

Assuming a planar loop in the 3-d cross-section introduces a direction, the spatial normal to this plane (the temporal direction is also normal). This suggests the possible interpretation of a spin direction. Another strong possibility is that the constituent screw dislocations are not straight but form spiral helices. Ordinary screw dislocations often take helical form through a process that involves absorption of interstitials or vacancies[18, 53, 54]. The plane of the helix introduces another spatial direction which could be related to spin or spin precession. Interstitials are important in that they introduce curvature into the crystal[55]. Such curvature is absolutely necessary to produce the large-scale hyper-spherical spatial geometry inherent in the cosmological scenario outlined above. It also allows a connection between particle properties and gravitation.

One additional property of crystal dislocations that may provide an intriguing parallel to the strong interactions is the existence of partial dislocations[18, 53]. Under favorable conditions, a dislocation may split into two or more partial dislocations with fractional Burgers vectors. Such objects cannot exist in isolation since they would involve dislocating the entire lattice - resulting in infinite energy. These partial dislocations are linked by a sheet containing a stacking fault, which produces an attractive force proportional to the sheet area (the partial dislocations also repel each other through other elastic forces, resulting in an equilibrium separation). The possible analogy between quarks and partial dislocations, with gluons being related to the associated stacking faults is compelling. Confinement and fractional charge are inherent and linked properties of these configurations. Another common feature of dislocations in ordinary crystals is the formation of dislocation networks. These are ordered or disordered collections of either partial or full dislocations and anti-dislocations, with zero net Burgers vector. New kinds of dislocations can be defined from defects in an otherwise ordered dislocation network. For instance the chiral condensate could be modeled as a network of partial dislocations and associated stacking faults. Nucleons and mesons could then be modeled as dislocations and excitations of this underlying network, which is reminiscent of the Skyrmion approach. Ordered dislocation networks can have dislocations which can form an ordered network which can itself have higher-order dislocations, producing a possible hierarchy of dislocations several levels deep.

7 Gravity as Elasticity of Space

The similarities between the General Theory of Relativity and the theory of elasticity have been remarked upon by many authors. Sakharov spoke of relating General Relativity to a "metrical elasticity of space"[56]. Kokarev has likened space-time to a "strongly-bent plate" [57]. Several authors have developed three-dimensional continuum models of dislocations that resemble three-dimensional gravity[58].

Screw dislocations themselves do not result in curvature - rather they introduce torsion into the lattice, since an observer circling a screw dislocation finds themselves transported forward, along the dislocation direction. Two sources of curvature have been put forward - disclinations and extra matter (primarily interstitials). Disclinations are large angular defects. For example, the pentagons in a geodesic dome can be thought of as disclinations in an otherwise flat hexagonal tiling, and produces obvious curvature in the surface. The problem with disclinations is that they produce curvature only in large finite chunks, rather than building up from many small pieces. So, whereas a disclination is a good model for a cosmic string[59], it is not a good candidate for an elementary particle. We are therefore left with the extra matter concept, which has been championed by Kroner[55]. In Kroner's theory, the geometry of the resulting continuum model is characterized by both curvature, the source of which is extra matter, and torsion, which is caused by dislocations[54, 55, 60]. The obvious four-dimensional generalization would be the Einstein-Cartan-Sciama-Kibble theory of gravity, which supplements the usual Einstein equations with an equation relating spin density to the torsion tensor[61, 62]. Torsion effects are too small to be detected experimentally, so this theory is, so far, experimentally indistinguishable from General Relativity. In order to satisfy the equivalence principle, the absorption of interstitials by dislocations mentioned above would have to be a universal property, with the degree of absorption proportional to the energy, so that curvature could couple to the energy-momentum tensor. It is not clear that this would necessarily happen, however it could be forced by symmetries, since due to the Bianchi identity the Einstein tensor can only couple to a conserved quantity. Not all interstitials are necessarily absorbed. Unabsorbed interstitials are an intriguing dark-matter candidate. Unlike ordinary particles they do not persist, since they are true 4-d point defects. Their behavior is more like that of instantons. Their fleeting existence could make their detection difficult other than through their gravitational effects.

Regardless of the details of the gravitational theory that results, it will necessarily be a quantum theory of gravitation. This is because the evolution of the spatial hypersurface is influenced by the thermal fluctuations in the surrounding medium which are the source of quantum fluctuations in this picture. One certainly expects thermally induced curvature fluctuations. However, if the elementary lattice spacing is much larger than the Plank length, it is likely that such curvature fluctuations would be small, and gravitation would remain, in practical terms, a largely classical theory. As distances approached the lattice spacing, then the continuum theory (presumably a generalization of General Relativity) would have to be replaced with an appropriate lattice theory, just as continuum elasticity theory can be used for a crystal only for distances large compared to the lattice spacing. Of course, just as in ordinary crystallography, the lattice theory itself may be based on an underlying continuous space. The resolution of singularity problems in general relativity are more likely to come from the transition to an appropriate lattice theory than from the incorporation of quantum effects, unless the lattice spacing is of order the Plank length or smaller.

8 Cosmological Consequences

The model of an expanding phase boundary provides good explanations for some cosmological puzzles but introduces additional problems as well. Phase nucleation is a common way for structure to arise from chaos spontaneously. It naturally creates an expanding universe starting from a very small but not infinitesimal seed. Only if the initial fluctuation is above a certain minimum size, will the crystal grow - otherwise surface tension effects will remelt it back into the liquid. There would not seem to be a horizon problem because there is plenty of time before the big bang to establish causal contact, thermodynamic equilibrium etc. Also there is the "flatness problem" which, in a non-inflationary universe, requires a careful fine-tuning of parameters to create a universe as long lasting as ours which nevertheless has a reasonable matter density and is close to being spatially flat in the present era. Phase transitions only occur when there is a fine tuning between various terms in the Hamiltonian, so a system undergoing a phase transition is already *naturally fine tuned* between forces that favor the transition and those that don't. The other ingredient this model likely has which could reduce the need for fine-tuning would be dissipation, which could tame runaway solutions like inflation. In general, surface growth which is not diffusion-limited is controlled by the volume energy (which results in the liberation of latent heat), surface tension, and dissipation. The outward pressure from the volume energy takes the form of a repulsive cosmological constant and the 3-d surface tension may act like the ordinary spatial curvature term, but it is not immediately clear how to take dissipation into account within the standard Friedmann models. Comparison to ordinary phase transitions would suggest a period of slow growth at first, which accelerated as the surface term became less important, finally approaching a steady state constant growth rate. One can also consider the possibility that the background conditions responsible for the supercooling could vary over time. If this is allowed then a more complex growth-rate history could be accommodated.

An intriguing possibility for matter generation would be collisions between different crystal universes. Where crystals join, a lot of dislocations are formed. The join-boundary of two 3-d surfaces is a 2-d surface. Therefore, dislocations produced in such collisions would be distributed on 2-d surfaces within the combined 3-d surface of the joined crystals. Interestingly, matter in the universe is primarily distributed on a network of 2-d surfaces surrounding large voids. One can imagine this resulting from the twisting and folding of the join-boundary of a single cosmic collision or from a number of such events.

This scenario may have difficulty explaining both the uniformity of element abundances and of the cosmic background radiation. Helium could be produced in the cosmic collisions referred to above in much the same way as in the hot big-bang, but conditions would likely vary somewhat from place to place. A single large collision might be able to produce a fairly uniform result. Cosmic collisions, in addition to creating matter in the form of dislocations would also produce a lot of thermal radiation. Again, this could be fairly uniform for the case of a single large collision. This scenario shares some features with the colliding-branes string-model picture recently proposed by Khoury et. al.[63], although the geometry is rather different.

9 Discussion

At first glance, the idea that space could be crystalline would seem at odds with the notion of spatial isotropy. Wouldn't the axis directions create preferred directions in space? For distances large compared to the lattice spacing, this is not necessarily so. For instance, the long distance behavior of an isotropic crystal (one with isotropic elastic constants) is well approximated by isotropic linear elasticity theory which has full rotational invariance. Another example is lattice gauge theory. Here forces along axis directions differ from those along non-axis directions at short distances, but full rotational symmetry emerges at distances large compared to the lattice spacing. The longer lattice paths in diagonal directions are exactly compensated by the larger multiplicity of such paths. Also the surface of a growing crystal is more labile than the interior, resulting in features that are less "solid". For instance, even sessile dislocations can move through growth, via formation of kinks and jogs, though they are essentially locked in place once formed. Glissile dislocations (those that can move freely through the crystal) may themselves essentially stop in the bulk by transferring all of their momentum to the "growth tip" through a mechanism similar to a Newton's cradle onto which balls are added continuously, or a whip with a growing tip.

The similarities between condensed matter physics and particle physics are many. Phonons are surprisingly similar to photons. They can be thought of as Goldstone bosons resulting from the breaking of translation invariance, or as gauge fields relating to the remaining discrete translational invariance, which due to lattice periodicity, may be represented by an angular order parameter[45]. The counterpart to the Higgs mechanism is the plasma mechanism[64]. Even the chiral properties of the weak interaction may have an analog in the behavior of ^{3}He-A[65]. Several gauge theories of dislocations have been proposed[66, 67]. What is being proposed here can be thought of as going all the way with this program, namely hypothesizing that particle physics *is* condensed matter physics. The main experimental signature of such a proposal, regardless of the details, would be the effects of a finite lattice spacing. Besides the dramatic cutoff of gauge boson spectra above a certain energy, one can look for effects of dispersion near the cutoff. The lattice also makes all ultraviolet divergences finite, which will introduce small effects in higher order corrections. This also adds impetus to proposals that a serious effort be made to search experimentally for violations of Lorentz invariance[68]. The effects of living on a physical lattice are somewhat different from string-inspired Lorentz-invariance violations. The other experimental signature this scenario has in common with other extra-dimension scenarios is the possibility of energy conservation violation beyond the statistical violation already discussed and interpreted as quantum fluctuations[2]. One can imagine the possibility of a high-energy interaction radiating a longitudinal phonon into the bulk, for instance, which would look like a missing-energy event. This can be made rare by either a very weak coupling to these modes or by giving them a low frequency cutoff (a mass). Radiation forward into the liquid could be prevented by having the surface growing at a rate exceeding the sound speed in the liquid. Another possible experimental signature to look for would be effects of dissipation including lack of time reversal invariance. Although conservation laws may prevent dissipation on the surface itself, the bulk phases undoubtedly are dissipative. The moving phase boundary breaks time reversal invariance spontaneously. Both T and CPT invariance could be broken.

A final note concerning time in this theory is the special role played by the present. The edifice of the universe is constructed at the present surface from material provided by the undifferentiated current-future (liquid) state. The past, being a solid, is more fixed, though still can undergo some fluctuations. Present, past and future are different, distinguishable phases. This would seem to conform with our personal experience better than the picture presented in special relativity, where the present is not distinguished, and the future seems as well-formed as the past. Indeed, according to Einstein, "For us believing physicists, the distinction between past, present, and future is only an illusion, even if a stubborn one[69]." Davies states, "The four-dimensional space-time of physics makes no provision whatever for either a 'present-moment' or a 'movement' of time[70]." Quantum mechanics could play a possible role in blurring the future in the standard picture, but this depends on a definite resolution to the measurement problem. The phase boundary scenario, in contrast, matches well with the "process philosophy" concept of time as advanced by Whitehead, who talks of a "concresence" unfolding at the present where the indefinite future is molded into a definite past[71].

The preferred frame offered by the crystal rest frame also gives a different point of view for causality arguments. One can imagine the possibility of interactions that occur by faster-than-light mechanisms, just as a bullet can exceed the speed of sound in an ordinary crystal. Although in some frames of reference, cause may appear to precede effect, this will never occur in the crystal rest frame, regardless of interaction speed. Since Lorentz invariance is only approximate, all frames are not equivalent. The correct result is that observed in the preferred frame. Thus there is no longer a paradox created by faster-than-light interactions by which one could travel backwards in time and kill one's grandfather, for instance. Time always goes forward and effect follows cause in the crystal rest frame. Of course there is no evidence that any interaction or particle can exceed the speed of light, and ordinary dislocations probably can not, as previously discussed, but the removal of this causality paradox opens the door to such a possibility a crack wider.

10 Conclusion

At first glance this theory appears to be an anachronism - a neo-Lorentzian classical ether theory. Modifying special relativity and reintroducing an ether are probably the last thing that would enter the mind of a 20th or 21st century physicist, followed perhaps by a classical explanation of quantum mechanics. However, there are many ways in which this theory fits with modern ideas. The idea that we live on a membrane is becoming popular in string theory, and also for introducing chiral fermions into lattice gauge theory. Stochastic quantization, though never fully accepted in the realm of quantum mechanics, came very close to giving a classical statistical-mechanical explanation of quantum fluctuations. The very many analogies between particle physics and condensed-matter physics, especially in the realm of gauge field theories and spontaneous symmetry breaking, has led to tremendous sharing of ideas from one field to the other. The main difference between these is simply between relativistic and non-relativistic spacetime symmetries. One other difference is that, whereas in elementary particle physics gauge symmetries and Goldstone bosons are usually considered to be essentially separate mechanisms for producing massless particles (which conflict in the Higgs mechanism to give a mass), in condensed matter physics the gauge particles (phonons) can themselves be pictured as a type of Goldstone boson associated with the breaking of translational symmetry. A vector order parameter yields a vector Goldstone boson and a tensor order parameter (associated with breaking of rotational invariance) should produce a tensor Goldstone boson (the graviton). Gauge invariance is actually born from the ambiguities in defining the unperturbed lattice[45]. This economy of ideas (gauge bosons as Goldstone bosons) is appealing and may help to explain why some symmetries are gauged and others are not. To benefit from this analogy, however, an ether-type background would appear to be necessary to provide the required translational symmetry breaking.

The idea of a moving, expanding, phase boundary, where relativistic space-time emerges as a dynamical symmetry, integrates these ideas into a coherent picture; the big-bang and gravity, from the flexible geometry of the interface, are incorporated almost for free. The quantum measurement process is also vastly clarified in this picture as a consequence of spontaneous symmetry breaking. Ideas from chaos and ergodic theory, non-equilibrium statistical mechanics and dynamical critical phenomena play an important role. When these are added, a classical theory doesn't look so classical anymore.

Not much has been mentioned in this paper concerning the base-theory. What new sets of even more elementary particles (called elementary atoms above) and forces must be postulated for the underlying base-theory, upon which the dislocations and surface waves (our current set of elementary particles in this picture) can be built? Hopefully it is a simpler set than we currently have in the standard model. It seems possible that one or two types of elementary atoms, combined with a *short-range* force, repulsive at small distances and attractive at long, could be enough. Some model building seems to be in order, starting with extending simple crystal models to four dimensions.

In closing, one cannot help but speculate whether Einstein would have liked this idea. Certainly he may have disagreed with the reintroduction of the ether which he so strongly fought against, as well as the notion of a preferred frame. However from the point of view of the more fundamental base-theory there is no preferred frame – it is introduced through a spontaneous symmetry breaking, so the principle of relativity is safe there. Einstein's discomfort with the inherently probabilistic nature of quantum mechanics is well known, so perhaps he would be pleased with the application of his ideas on Brownian motion toward the explanation of quantum fluctuations, as well as the replacement of quantum mechanics with a deterministic (though chaotic) theory. Finally, there is the essentially geometric basis for electromagnetism and possibly all interactions through the picture of a dislocated crystal. This bears a rather strong resemblance to unified field theories that he worked on in his later years. On balance, this theory seems relatively in concert with the ideas of Einstein.

References

[1] These ideas were introduced in M. Grady, gr-qc/9805076; see also M. Chown, *New Scientist*, **161**, 42, 1999.

[2] The idea of the universe as a domain wall in a larger universe was discussed in V.A. Rubakov and M.E. Shaposhnikov, *Phys. Lett.* **125B**, 136, 1983.

[3] Another early domain-wall scenario is described in K. Akama, in *Lecture Notes in Physics, 176, Gauge Theory and Gravitation, Proceedings,* Nara, 1982; K. Kikawa, N. Nakanishi, and H. Nariai, Eds.; Springer-Verlag: Berlin, 1983, p 267, hep-th/0001113.

[4] String inspired domain wall universes are discussed in, e.g., A. Lukas et. al., *Phys. Rev. D* **59**, 086001, 1999; L. Randall and R. Sundrum, *Phys. Rev. Lett.* **83**, 4690, 1999.

[5] A. Reiss et. al., *Astron. J.*, **116**, 1009, 1998; S. Perlmutter et. al., *Nature*, **39**, 51, 1998; P.M. Garnavich et. al., *Ap. J.,* **493**, L53, 1998.

[6] P.C. Hohenberg and B.I. Halperin, *Rev. Mod. Phys.* **49**, 435, 1977; J. Zinn-Justin, *Quantum Field Theory and Critical Phenomena*, 3rd ed.; Clarendon Press: Oxford, 1996, pp 764-777.

[7] E. Nelson, *Quantum Fluctuations*; Princeton University Press: Princeton, 1985; E. Nelson, *Phys. Rev.* **150**, 1079, 1966; F. Guerra, *Phys. Rep.* **77**, 263, 1981; Ph. Blanchard, Ph. Comb and W. Zheng, *Mathematical and Physical Aspects of Stochastic Mechanics*; Springer-Verlag: Berlin, 1987.

[8] G. Parisi and Y. Wu, *Scientia Sinica* **24**, 483, 1981.

[9] H. Hüffel and H. Rumpf, *Phys. Lett.* **148B**, 104, 1984.

[10] E. Gozzi, *Phys. Lett.* **150B**, 119, 1985; H. Nakazato and Y. Yamanaka, *Phys. Rev. D* **34**, 492, 1986; H. Nakazato, *Prog. Theor. Phys.* **77**, 20, 1987.

[11] M. Namiki and S. Tanaka, in *Modern Problems of Theoretical Physics - Feschrift for Professor D. Ivanenko*; P.I. Pronin and Yu.N. Obukhov, Eds.; World Scientific: Singapore, 1991.

[12] H. Haken *Synergetics*; Springer-Verlag: Berlin, 1977, especially Ch. 6-8; *Rev. Mod. Phys.* **47**, 67, 1975.

[13] H. Dekker, Physica **95A**, 311, 1979; *Phys. Rep.* **80**, 1, 1981.

[14] H. Haken in *Encyclopedia of Physics, vol. XXV/2c: Laser Theory*; S. Flügge, Ed.; Springer-Verlag: Berlin, 1970.

[15] J. Frenkel and T. Kontorowa, *Phys. Z. Sowjet* **13**, 1 (1938).

[16] F.C. Frank, *Proc. Phys. Soc.* **A62**, 131, 1949; J. Eshelby, *Proc. Phys. Soc.* **A62**, 307, 1949.

[17] J. Weertman and J.R. Weertman, in *Dislocations in Solids - vol. 3: Moving Dislocations*; F.R.N. Nabarro, Ed.; North-Holland Publishing Co.: Amsterdam, 1980.

[18] F.R.N. Nabarro, *Theory of Crystal Dislocations*; Oxford University Press: London, 1967.

[19] This issue is addressed in A. Unzicker, gr-qc/0011064.

[20] H. Günther, *Phys. Stat. Sol.* **b149**, 101, 1988; H. Günther in *Proceedings of the Eighth International Symposium on Continuum Models and Discrete Systems,* Varna, Bulgaria, 1995; K.Z. Markov,Ed.; World Scientific: Singapore, 1996, p 507.

[21] A.I. Miller, in *Some Strangeness in the Proportion*; H. Woolf, Ed.; Addison-Wesley: Reading MA, 1980, p 66.

[22] M. Catanese and T.C. Weeks, *Publ. Astron. Soc. Pac.* **111**, 1193, 1999; B. Degrange and M. Punch, *C.R. Acad. Sci. Paris,* **1**, 189, 2000.

[23] A.V. Olinto, *Phys. Rep.* **333-334**, 329, 2000; A.A. Watson, *Phys. Rep.* **333-334**, 309, 2000; M. Boratav and A.A. Watson, *C.R. Acad. Sci. Paris*, **1**, 207, 2000.

[24] K. Greisen, *Phys. Rev. Lett.* **16**, 748, 1966; G.T. Zatsepin and V.A. Kuzmin, *Sov. Phys. JETP Lett.* **4**, 78, 1966.

[25] G. Amelino-Camelia and T. Piran, *Phys. Rev. D* **64**, 036005, 2001 and references therein.

[26] G.R. Farrar and P.L. Biermann, *Phys. Rev. Lett.* **81**, 3579, 1998.

[27] Lord Rayleigh (J.W. Strutt), *London Math. Soc. Proc.* **17**, 4, 1885, available in *Scientific Papers of Lord Rayleigh, vol II*; Dover: New York, 1964, p 441.

[28] G.W. Farnell, in *Physical Acoustics vol. VI*; Academic Press: New York, 1970, p 109; L.D. Landau and E.M. Lifshitz, *Theory of Elasticity*; Pergamon Press: London, 1959, p 105; A.A. Maradudin in *Nonequilibrium Phonon Dynamics*; W.E. Bron, Ed.; Plenum Press: New York, 1985, p 395.

[29] I.M. Lifshitz and A.M. Kosevich, in *Lattice Dynamics*; W.A. Benjamin: New York, 1969, p 53.

[30] I.A. Viktorov, *Sov. Phys. Acoustics*, **25**, 1, 1979.

[31] G.A. Maugin in *Recent Developments in Surface Acoustic Waves*; Springer-Veralg: Berlin, 1988, p. 158.

[32] A.I. Murdoch, *J. Mech. Phys. Solids*, **24**, 137, 1976.

[33] G.W. Farnell and E.L. Adler in *Physical Acoustics, Vol. IX*; W.P. Mason and R.N. Thurston, Eds.; Academic Press: New York, 1972.

[34] A.E.H. Love, *Some Problems of Geodynamics*; Cambridge University Press: London, 1911.

[35] H.F. Tiersten, *J. Appl. Phys.* **40**, 770, 1969; H.F. Tiersten, B.K. Sinha, and T.R. Mecker, *J. Appl. Phys.* **52**, 5614, 1981.

[36] A.M. Kosevich and A.V. Tutov in *Continuum Models and Discrete Systems - Proceedings of the 8th International Symposium,* June 11-16, 1995, Varna, Bulgaria; World Scientific: Singapore, 1996, p 444.

[37] N. Daher and G.A. Maugin, *Acta Mech.* **60**, 217, 1986. For additional information on excitations on moving interfaces see W. Kosiński, *Field Singularities and Wave Analysis in Continuum Mechanics*; Wiley: New York, 1986, and references therein.

[38] M.G. Cottam and D.R. Tilley, *Introduction to Surface and Superlattice Excitations*; Cambridge University Press: Cambridge, 1989.

[39] D.B. Kaplan, *Phys. Lett. B* **288**, 342, 1992; M. Creutz, *Rev. Mod. Phys.* **73**, 119, 2001, and references therein.

[40] N.W. Ashcroft and N.D. Mermin, *Solid State Physics*; Saunders College Publishing: Philadelphia, 1976, pp 784-789.

[41] E. Fermi, J. Pasta and S. Ulam, in *The Collected Papers of Enrico Fermi, vol. 2*; University of Chicago Press: Chicago, 1966, p 977 (original report dated 1955).

[42] M. Rasetti, *Modern Methods in Equilibrium Statistical Mechanics*; World Scientific: Singapore, 1986.

[43] H.S. Robertson, *Statistical Thermophysics*; Prentice Hall: Englewood Cliffs, NJ, 1993, pp 43-48.

[44] J. Ford and G.H. Lunsford, *Phys. Rev. A* **1**, 59, 1970.

[45] P.W. Anderson, *Basic Notions of Condensed Matter Physics*; Addison-Wesley: Reading, MA, 1984, pp 30-69.

[46] Y. Ne'eman, *Proc. Natl. Acad. Sci. USA* **80**, 7051, 1983; *Found. Phys.* **16**, 361, 1986.

[47] G.T. Zimányi and K. Vladár, *Phys. Rev. A* **34**, 3496, 1986; *Found. Phys. Lett.* **1**, 175, 1988.

[48] M. Grady, hep-th/9409049.

[49] C.V. Burton, London, Edinburgh, and Dublin *Philosophical Magazine and Journal of Science* **33**, 191, 1892.

[50] J. Larmor, *Aether and Matter*; Cambridge University Press: Cambridge, 1900; J. Larmor, *Phil. Trans. Roy. Soc. London* **A190**, 205, 1897.

[51] K.F. Schaffner, *Nineteenth Century Aether Theories*; Pergamon Press: Oxford, 1972.

[52] J.D. Eshelby, *Phys. Rev.* **90**, 248, 1953.

[53] J.P. Hirth and J. Lothe, *Theory of Dislocations*; McGraw Hill: New York, 1968.

[54] R. de Wit, in *Solid State Physics, vol. 10*; F. Sietz and D. Turnbull, Eds.; Academic Press: New York, 1960, p 249.

[55] E. Kröner, *Int. J. Theor. Phys.* **29**, 1219, 1990; E. Kröner in *Les Houches XXXV - Physics of Defects*; R. Balian, M. Kl'eman and J-P. Poirier, Eds.; North-Holland Publishing Company: Amsterdam, 1981, p 215.

[56] A.I. Sakharov, *Dokl. Akad. Nauk SSSR,* **177**, 70, 1967 (*Sov. Phys. Doklady* **12**, 1040, 1968).

[57] S.S. Kokarev, *Nuovo Cim.* **113B**, 1339, 1998; *Nuovo Cim.* **114B**, 903, 1999.

[58] See, e.g. C. Malyshev, *Ann. Phys.* **286**, 249, 2000, and references therein; also numerous articles in *RAAG Memoirs of the Unifying Study of Basic Problems in Engineering and Physical Sciences by Means of Geometry, vol. I-IV*; K. Kondo, Ed.; Gakujutsu Bunken Fukyu-Kai: Tokyo, 1968.

[59] A. Vilenkin and E.P.S. Shellard, *Cosmic Strings and Other Topological Defects*; Cambridge University Press: Cambridge, 1994.

[60] A. Trzesowski, *Int. J. Theor. Phys.* **33**, 931, 1994.

[61] F.W. Hehl, P. von der Hyde and G.D. Kerlick, *Rev. Mod. Phys.* **48**, 393, 1976; F.W. Hehl, J.D. McCrea, E.W. Mielke, and Y. Ne'eman, *Phys. Rep.* **258**, 1, 1995.

[62] M. Göckeler and T. Schücker, *Differential Geometry, Gauge Theories, and Gravity*; Cambridge University Press: Cambridge, 1987, ch. 5.

[63] J. Khoury, B.A. Ovrut, P.J. Steinhardt, and N. Turok, *Phys. Rev. D* **64**, 123522, 2001.

[64] P.W. Anderson, *Phys. Rev.* **130**, 439, 1963.

[65] G.E. Volovik, in *Topological Defects and Non-Equilibrium Dynamics of Symmetry Breaking Phase Transitions*; Y.M. Bunkov and H. Godfrin, Eds.; Kluwer Academic Publishers: Dordrecht, 2000, pp 353-387, cond-mat/9902171.

[66] A. Kadić and D.G.B. Edelen, *A Gauge Theory of Dislocations and Disclinations*; Springer-Verlag: Berlin, 1983.

[67] H. Kleinert, *Gauge Fields in Condensed Matter, Vol. II - Stresses and Defects*; World Scientific: Singapore, 1989.

[68] S. Coleman and S.L. Glashow, *Phys. Rev. D* **59**, 116008, 1999; D. Colladay and V.A. Kostelecký, *Phys. Rev. D* **55**, 6760, 1997; **58**, 16002, 1998.

[69] B. Hoffman (with H. Dukas), *Albert Einstein: Creator and Rebel*; Viking: New York, 1972, p 258.

[70] P.C.W. Davies, *The Physics of Time Asymmetry*; University of California Press: Berkely, 1976, p 21.

[71] A. N. Whitehead, *Process and Reality*; corrected edition, D.R. Griffin and D.W. Sherburne Eds.; Free Press: New York, 1978; *Physics and the Ultimate Significance of Time*; D. R. Griffin, Ed.; SUNY Press: Albany, 1986.

In: Trends in Quantum Gravity Research
Editor: David C. Moore, pp. 139-196
ISBN: 1-59454-670-3

Chapter 4

KINETIC QUANTUM THEORY OF GRAVITY

Fran De Aquino
Maranhao State University, Physics Department, S.Luis/MA, Brazil.

Abstract

Starting from the action function we have derived a theoretical background that leads to quantization of gravity and the deduction of a correlation between the gravitational and inertial masses, which depends on the *kinetic momentum* of the particle. We show that there is a reaffirmation of the strong equivalence principle and consequently the Einstein's equations are preserved. In fact such equations are deduced here directly from this kinetic approach to Gravity. Moreover, we have obtained a generalized equation for inertial forces, which incorporates the Mach's principle into Gravitation. Also, we have deduced the equation of Entropy; the Hamiltonian for a particle in an electromagnetic field and the reciprocal fine structure constant. It is possible to deduce the expression of the *Casimir force* and also to explain the *Inflation Period* and the *Missing Matter* without assuming the existence of *vacuum fluctuations*. This new approach for Gravity will allow us to understand some crucial matters in Cosmology. An experiment has been carried out to check the theoretical correlation between the gravitational and inertial masses. The experiment and results are presented on appendix A. The experimental data are in strongly accordance with the theory.

1 Introduction

Quantum Gravity was originally studied, by Dirac and others, as the problem of quantizing General Relativity. This approach has many difficulties, detailed by Isham [1]. In the 1970's physicists tried an even more conventional approach: simplify the Einstein's equations by pretending that they are *almost linear*, and then apply the standard methods of quantum field theory to the thus-oversimplified equations. But this method, too, failed. In the 1980's a very different approach, known as string theory, became popular. For a while there are many enthusiasts of string theory. But the mathematical difficulties in string theory are formidable, and it is far from clear that they will be resolved any time soon. At the end of 1997 Isham [2] pointed out several "Structural Problems Facing Quantum Gravity Theory". At the beginning of this new century, the problem of quantizing the gravitational field was still open.

In this work we propose a new approach to Quantum Gravity. Starting from the generalization of the *action function* we have derived a theoretical background that leads to quantization of gravity. The Einstein's equations of the General Relativity are deduced directly from this theory of Quantum Gravity. Also, it leads to a complete description of the Electromagnetic Field, providing a consistent unification of gravity with electromagnetism.

2 Theory

We start with the *action* for a free-particle that, as we know, is given by:

$$S = -\alpha \int_a^b ds$$

where α is a quantity which characterize the particle.

In Relativistic Mechanics, the action can be written in the following form [3]:

$$S = \int_{t_1}^{t_2} L dt = -\int_{t_1}^{t_2} \alpha c\sqrt{1 - V^2/c^2}\, dt$$

where

$$L = -\alpha c\sqrt{1 - V^2/c^2}$$

is the Lagrange's function.

In Classical Mechanics the Lagrange's function for a free-particle is, as we know, given by: $L = aV^2$ where V is the speed of the particle and a a quantity *hypothetically* [4] given by :

$$a = m/2$$

where m is the mass of the particle. However, there is no distinction about the kind of mass (if *gravitational mass*, m_g, or *inertial mass* m_i)neither about its sign $(\pm)$.

The correlation between a and α can be established based on the fact that on the limit $c \rightarrow \infty$ the relativistic expression for L must be reduced to the classic expression $L = aV^2$ .The result [5] is: $L = \alpha V^2/2c$. Therefore, if $\alpha = 2ac = mc$ we obtain $L = aV^2$. Now, we must decide if $m = m_g$ or $m = m_i$. We will see in this work that the definition of m_g includes m_i. Thus the right option is m_g, i.e.,

$$a = m_g/2.$$

Consequently, $\alpha = m_g c$ and the generalized expression for the action for a free-particle will have the following form:

$$S = -m_g c \int_a^b ds \tag{1}$$

or

$$S = -\int_1^2 m_g c^2 \sqrt{1 - V^2/c^2}\, dt \tag{2}$$

where the Lagrange's function is

$$L = -m_g c^2 \sqrt{1 - V^2/c^2}. \tag{3}$$

The integral $S = \int_1^2 m_g c^2 \sqrt{1 - V^2/c^2}\, dt$ preceded by the *plus* sign cannot have a *minimum*. Thus, the integrand of Eq.(2) must be *always positive*. Therefore, if $m_g > 0$ then necessarily $t > 0$; if $m_g < 0$ then $t < 0$. The possibility of $t < 0$ is based on the well-known equation $t = \pm t_0 \big/ \sqrt{1 - V^2/c^2}$ of Einstein's Theory.

Thus if the *gravitational mass* of a particle is *positive* then t is also *positive* and therefore given by $t = +t_0 \big/ \sqrt{1 - V^2/c^2}$. This leads to the well-known relativistic prediction that the particle goes to the *future* if $V \to c$. However, if the *gravitational mass* of the particle is *negative* then t is *negative* and given by $t = -t_0 \big/ \sqrt{1 - V^2/c^2}$. In this case the prediction is that the particle goes to the *past* if $V \to c$. Consequently, $m_g < 0$ is the necessary condition for the particle to go to the *past*. Further on it will be derived a correlation between *gravitational* and *inertial* masses, which contains the possibility of $m_g < 0$.

The Lorentz's transforms follow the same rule for $m_g > 0$ and $m_g < 0$, i.e., the sign before $\sqrt{1 - V^2/c^2}$ will be $(+)$ when $m_g > 0$ and $(-)$ if $m_g < 0$.

The *momentum*, as we know, is the vector $\vec{p} = \partial L / \partial \vec{V}$. Thus from Eq.(3) we obtain

$$\vec{p} = \frac{m_g \vec{V}}{\pm\sqrt{1 - V^2/c^2}} = M_g \vec{V}. \tag{4}$$

The sign $(+)$ in the equation above will be used when $m_g > 0$ and the sign $(-)$ if $m_g < 0$. Henceforth, by simplicity the signs $(\pm)$ before $\sqrt{1-V^2/c^2}$ will be omitted.

The derivate $d\vec{p}\,/dt$ is the *inertial* force F_i which acts on the particle. If the force is perpendicular to the speed we have

$$\vec{F}_i = \frac{m_g}{\sqrt{1-V^2/c^2}}\frac{d\vec{V}}{dt}. \tag{5}$$

However, if the force and the speed have the same direction, we find that

$$\vec{F}_i = \frac{m_g}{\left(1-V^2/c^2\right)^{3/2}}\frac{d\vec{V}}{dt}. \tag{6}$$

From Mechanics [6] we know that $\vec{p}\cdot\vec{V} - L$ denotes the *energy* of the particle, thus we can write

$$E_g = \vec{p}\cdot\vec{V} - L = \frac{m_g c^2}{\sqrt{1-V^2/c^2}} = M_g c^2. \tag{7}$$

This fundamental equation presents the concept of *Gravitational Energy*, E_g, in addition to the well-known concept of *Inertial Energy*, E_i, and shows that E_g is not null for $V=0$, but it has the finite value

$$E_{g0} = m_g c^2 \tag{8}$$

This is the particle's gravitational energy *at rest*.

The Eq.(7) can be rewritten in the following form:

$$E_g = m_g c^2 - \frac{m_g c^2}{\sqrt{1-V^2/c^2}} - m_g c^2 =$$

$$= \frac{m_g}{m_i}\left[m_i c^2 + \underbrace{\left(\frac{m_i c^2}{\sqrt{1-V^2/c^2}} - m_i c^2\right)}_{E_{Ki}}\right] = \tag{9}$$

$$= \frac{m_g}{m_i}\left(E_{i0} + E_{Ki}\right) = \frac{m_g}{m_i}E_i$$

By analogy to the Eq.(8), $E_{i0} = m_i c^2$ into the equation above, is the inertial energy *at rest*. Thus, $E_i = E_{i0} + E_{Ki}$ is the *total* inertial energy, where E_{Ki} is the *kinetic inertial energy*. From the Eqs.(7) and (9) we thus obtain

$$E_i = \frac{m_i c^2}{\sqrt{1 - V^2/c^2}} = M_i c^2. \tag{10}$$

For small velocities $(V << c)$ we obtain

$$E_i \approx m_i c^2 + \tfrac{1}{2} m_i V^2 \tag{11}$$

where we recognize the classical expression for the kinetic inertial energy of the particle.

The expression for the *kinetic gravitational energy*, E_{Kg}, is easily deduced by comparing of the Eqs.(7) and (9). The result is

$$E_{Kg} = \frac{m_g}{m_i} E_{Ki}. \tag{12}$$

In the presented picture, we can say that the *gravity* , $\vec{g}$, into a gravitational field produced by a particle of gravitational mass m_g depends on the particle's gravitational energy, E_g (given by Eq.(7)), because we can write

$$g = -G\frac{E_g}{r^2 c^2} = -G\frac{M_g c^2}{r^2 c^2} = -G\frac{M_g}{r^2} \tag{13}$$

where $M_g = m_g \left(1 - V^2/c^2\right)^{-\frac{1}{2}}$ is the *relativistic* gravitational mass defined in the Eqs.(4)and (7).

On the other hand, as we know, the gravitational force is *conservative*. Thus, gravitational energy, in agreement with the energy conservation law, can be expressed by the *decrease* of the inertial energy, i.e.,

$$\Delta E_g = -\Delta E_i \tag{14}$$

This equation expresses the fact that the decrease of gravitational energy corresponds to an increase of the inertial energy.

Therefore a variation ΔE_i in E_i yields a variation $\Delta E_g = -\Delta E_i$ in E_g. Thus $E_i = E_{i0} + \Delta E_i$; $E_g = E_{g0} + \Delta E_g = E_{g0} - \Delta E_i$ and

$$E_g + E_i = E_{g0} + E_{i0} \tag{15}$$

Comparison between (7) and (10) shows that $E_{g0} = E_{i0}$. Consequently we have

$$E_g + E_i = E_{g0} + E_{i0} = 2E_{i0} \tag{16}$$

However $E_i = E_{i0} + E_{Ki}$. Thus (16) becomes

$$E_g = E_{i0} - E_{Ki}. \tag{17}$$

Note the *symmetry* in the equations of E_i and E_g .Substitution of $E_{i0} = E_i - E_{Ki}$ into (17) yields

$$E_i - E_g = 2E_{Ki} \tag{18}$$

Squaring the Eqs.(4) and (7) and comparing the result, we find the following correlation between gravitational energy and *momentum* :

$$\frac{E_g^2}{c^2} = p^2 + m_g^{\ 2} c^2. \tag{19}$$

The energy expressed as a function of the *momentum* is, as we know, called *Hamiltonian* or Hamilton's function:

$$H_g = c\sqrt{p^2 + m_g^{\ 2} c^2}. \tag{20}$$

It is known that starting from the Schrödinger equation we may obtain the well-known expression for energy of a particle in periodic motion inside a cubical box of edge length L [7]. The result now is

$$E_n = \frac{n^2 h^2}{8m_g L^2} \qquad n = 1,2,3,... \tag{21}$$

Note that the term $h^2/8m_g L^2$ (energy) will be minimum for $L = L_{max}$ where L_{max} is the maximum edge length of a cubical box whose maximum diameter

$$d_{max} = L_{max}\sqrt{3} \tag{22}$$

is equal to *the maximum "diameter" of the Universe.*

The minimum energy of a particle is obviously its inertial energy at rest $m_g c^2 = m_i c^2$. Therefore we can write

$$\frac{n^2 h^2}{8 m_g L_{max}^2} = m_g c^2$$

Then from the equation above follows that

$$m_g = \pm \frac{nh}{c L_{max} \sqrt{8}} \qquad (23)$$

whence we see that there is a *minimum value* for m_g given by

$$m_{g(min)} = \pm \frac{h}{c L_{max} \sqrt{8}} \qquad (24)$$

The *relativistic* gravitational mass $M_g = m_g \left(1 - V^2/c^2\right)^{-\frac{1}{2}}$, defined in the Eqs.(4) and (7), shows that

$$M_{g(min)} = m_{g(min)} \qquad (25)$$

The *box normalization* leads to conclusion that the *propagation number* $k = |\vec{k}| = 2\pi/\lambda$ is restricted to the values $k = 2\pi n/L$. This is deduced assuming an *arbitrarily large but finite* cubical box of volume L^3 [8]. Thus we have

$$L = n\lambda$$

From this equation we conclude that

$$n_{max} = \frac{L_{max}}{\lambda_{min}}$$

and

$$L_{min} = n_{min} \lambda_{min} = \lambda_{min}$$

Since $n_{min} = 1$. Therefore we can write that

$$L_{max} = n_{max} L_{min} \tag{26}$$

From this equation we thus conclude that

$$L = nL_{min} \tag{27}$$

or

$$L = \frac{L_{max}}{n} \tag{28}$$

Multiplying (27) and (28) by $\sqrt{3}$ and reminding that $d = L\sqrt{3}$, we obtain

$$d = nd_{min} \qquad or \qquad d = \frac{d_{max}}{n} \tag{29}$$

Equations above show that the length (and therefore the *space*) is *quantized.*
By analogy to (23) we can also conclude that

$$M_{g(max)} = \pm \frac{n_{max} h}{cL_{min}\sqrt{8}} \tag{30}$$

since the relativistic gravitational mass, $M_g = m_g \left(1 - V^2/c^2\right)^{-1/2}$, is just a multiple of m_g.

Equation (26) tells us that $L_{min} = L_{max}/n_{max}$. Thus Eq.(30) can be written as follows

$$M_{g(max)} = \pm \frac{n_{max}^2 h}{cL_{max}\sqrt{8}} \tag{31}$$

Comparison of (31) with (24) shows that

$$M_{g(max)} = n_{max}^2 m_{g(min)} \tag{32}$$

which leads to following conclusion that

$$M_g = n^2 m_{g(min)} \tag{33}$$

This equation shows that *the gravitational mass is quantized.*
Substitution of (33) into (13) leads to quantization of *gravity*, i.e.,

$$g = -\frac{GM_g}{r^2} = n^2\left(-\frac{Gm_{g(min)}}{(r_{max}/n)^2}\right) = \qquad (34)$$
$$= n^4 g_{min}$$

From the Hubble's law follows that

$$V_{max} = \widetilde{H}l_{max} = \widetilde{H}(d_{max}/2)$$

$$V_{min} = \widetilde{H}l_{min} = \widetilde{H}(d_{min}/2)$$

whence

$$\frac{V_{max}}{V_{min}} = \frac{d_{max}}{d_{min}}$$

Equations (29) tell us that $d_{max}/d_{min} = n_{max}$. Thus the equation above gives

$$V_{min} = \frac{V_{max}}{n_{max}} \qquad (35)$$

which leads to following conclusion

$$V = \frac{V_{max}}{n} \qquad (36)$$

this equation shows that *velocity* is also quantized.

From this equation one concludes that we can have $V = V_{max}$ or $V = V_{max}/2$, but nothing in between. This shows clearly that V_{max} cannot be equal to c (speed of light in vacuum). Thus follows that

$$
\begin{array}{lll}
n=1 & V=V_{max} & \\
n=2 & V=V_{max}/2 & \\
n=3 & V=V_{max}/3 & Tachyons \\
\ldots\ldots & \ldots\ldots\ldots\ldots & \\
n=n_x-1 & V=V_{max}/(n_x-1) & \\
\hline
n=n_x & V=V_{max}/n_x=c \leftarrow & \\
n=n_x+1 & V=V_{max}/(n_x+1) & Tardions \\
n=n_x+2 & V=V_{max}/(n_x+2) & \\
\ldots\ldots & \ldots\ldots\ldots\ldots &
\end{array}
$$

where n_x is a big number.

Then c is the *upper limit* of speed of the *Tardions* and also the *lower limit* of speed of the *Tachyons*. Obviously that limit is *always the same in all inertial frames*. Therefore c can be used like a *reference speed*, which we may compare any speed V, as occurs in the relativistic factor $\sqrt{1-V^2/c^2}$. Thus in this factor c not refers to maximum propagation speed of the interactions such as suggest some authors; c is just a speed limit which is the same in any inertial frame.

The temporal coordinate x^0 of the space-time is now $x^0 = V_{max}t$ ($x^0 = ct$ is then obtained when $V_{max} \to c$). Substitution of $V_{max} = nV = n(\widetilde{H}l)$ into this equation yields $t = x^0/V_{max} = (1/n\widetilde{H})(x^0/l)$. On the other hand, since $V = \widetilde{H}l$ and $V = V_{max}/n$ then we can write that $l = V_{max}\widetilde{H}^{-1}/n$. Thus $(x^0/l) = \widetilde{H}(nt) = \widetilde{H}t_{max}$. Therefore we can finally write

$$t = (1/n\widetilde{H})(x^0/l) = t_{max}/n \tag{37}$$

which shows the quantization of *time*.

Now let us go back to Eq. (20) which will be called the *gravitational* Hamiltonian to distinguish it from the *inertial* Hamiltonian H_i:

$$H_i = c\sqrt{p^2 + m_i^2c^2}. \tag{38}$$

Consequently, the Eq. (18) can be rewritten in the following form:

$$H_i - H_g = 2\Delta H_i \tag{39}$$

where ΔH_i is the *variation on the inertial Hamiltonian* or *inertial kinetic energy*. A *momentum* variation Δp yields a variation ΔH_i given by:

$$\Delta H_i = \sqrt{(p+\Delta p)^2 c^2 + m_i^2 c^4} - \sqrt{p^2 c^2 + m_i^2 c^4} \tag{40}$$

Substituting Eqs.(20), (38) and (40) into (39) and making $p = 0$, we obtain

$$m_g c^2 - m_i c^2 = 2\left(\sqrt{\Delta p^2 c^2 + m_i^2 c^4} - m_i c^2\right).$$

From this equation we derive the general expression of correlation between the gravitational and inertial mass, i.e.,

$$m_g = m_i - 2\left[\sqrt{1 + \left(\frac{\Delta p}{m_i c}\right)^2} - 1\right] m_i . \tag{41}$$

Note that for $\Delta p > m_i c\left(\sqrt{5}/2\right)$ the value of m_g becomes *negative*.

Equation (41) can also be expressed in terms of velocity V of the particle. Starting from (4) we can write

$$(p + \Delta p) = \frac{(m_g - \Delta m_g)(V + \Delta V)}{\sqrt{1 - (V + \Delta V)^2 / c^2}}$$

For $V = 0$; $p = 0$. Thus the equation above reduces to:

$$\Delta p = (m_g - \Delta m_g)\Delta V \Big/ \sqrt{1 - (\Delta V / c)^2}$$

From the Eq.(16) we obtain:

$$E_g = 2E_{i0} - E_i = 2E_{i0} - (E_{i0} + \Delta E_i) = E_{i0} - \Delta E_i$$

However, Eq.(14) tells us that $-\Delta E_i = \Delta E_g$; it leads to $E_g = E_{i0} + \Delta E_g$ or $m_g = m_i + \Delta m_g$. Thus, in the expression of Δp we can replace $(m_g - \Delta m_g)$ by m_i, i.e.,

$$\Delta p = m_i \Delta V \Big/ \sqrt{1 - (\Delta V / c)^2}$$

We can therefore write

$$\frac{\Delta p}{m_i c}=\frac{V/c}{\sqrt{1-(V/c)^2}} \tag{42}$$

By substitution of the expression above into Eq.(41) we thus obtain:

$$m_g=m_i-2\left[\left(1-V^2/c^2\right)^{-\frac{1}{2}}-1\right]m_i \tag{43}$$

For $V=0$ the Eq.(43) gives

$$m_g=m_i$$

Therefore, in this case, the previously obtained quantized relation (33), $M_g=n^2 m_{g(min)}$, becomes

$$m_i=n^2 m_{i(min)} \tag{44}$$

which shows the quantization of *inertial mass.*

Finally, by dividing both members of Eq.(43)by $\sqrt{1-V^2/c^2}$ we readily obtain

$$M_g=M_i-2\left[\left(1-V^2/c^2\right)^{-\frac{1}{2}}-1\right]M_i \tag{45}$$

The *Lorentz's force* is usually written in the following form:

$$d\vec{P}/dt=q\vec{E}+q\vec{V}\times\vec{B}$$

where $\vec{P}=m_i\vec{V}/\sqrt{1-V^2/c^2}$. However, Eq.(4) tell us that $\vec{p}=m_gV/\sqrt{1-V^2/c^2}$.

Therefore, the expressions above must be corrected by multiplying its members by m_g/m_i ,i.e.,

$$\vec{P}\frac{m_g}{m_i}=\frac{m_g}{m_i}\frac{m_i\vec{V}}{\sqrt{1-V^2/c^2}}=\frac{m_g\vec{V}}{\sqrt{1-V^2/c^2}}=\vec{p}$$

and

$$\frac{d\vec{p}}{dt} = \frac{d}{dt}\left(\vec{P}\frac{m_g}{m_i}\right) = \left(q\vec{E} + q\vec{V} \times \vec{B}\right)\frac{m_g}{m_i} \tag{46}$$

That is now the general expression for Lorentz's force.

When the force is perpendicular to the speed, the Eq.(5) gives $d\vec{p}/dt = m_g\left(d\vec{V}/dt\right)/\sqrt{1-V^2/c^2}$ .By comparing with Eq.(46) we thus obtain

$$\left(m_i/\sqrt{1-V^2/c^2}\right)\left(d\vec{V}/dt\right) = q\vec{E} + q\vec{V} \times \vec{B}$$

Starting from this equation, well-known experiments have been carried out in order to verify the relativistic expression: $m_i/\sqrt{1-V^2/c^2}$.

In general, the *momentum* variation Δp is expressed by $\Delta p = F\Delta t$ where F is the applied force during a time interval Δt. Note that there is no restriction concerning the *nature* of the force F , i.e., it can be mechanical, electromagnetic, etc.

For example, we can look on the *momentum* variation Δp as due to absorption or emission of *electromagnetic energy* by the particle (by means of *radiation* and/or by means of *Lorentz's force* upon the *charge* of the particle).

In the case of radiation (any type), Δp can be obtained as follows. It is known that the *radiation pressure* , dP, upon an area $dA = dxdy$ of a volume $dV = dxdydz$ of a particle(the incident radiation normal to the surface dA)is equal to the energy dU absorbed per unit volume (dU/dV).i.e.,

$$dP = \frac{dU}{dV} = \frac{dU}{dxdydz} = \frac{dU}{dAdz} \tag{47}$$

Substitution of $dz = vdt$ (v is the speed of radiation) into the equation above gives

$$dP = \frac{dU}{dV} = \frac{(dU/dAdt)}{v} = \frac{dD}{v} \tag{48}$$

Since $dPdA = dF$ we can write:

$$dFdt = \frac{dU}{v} \tag{49}$$

However we know that $dF = dp/dt$, then

$$dp = \frac{dU}{v} \tag{50}$$

From Eq.(48) follows that

$$dU = dPdV = \frac{dVdD}{v} \tag{51}$$

Substitution into (50) yields

$$dp = \frac{dVdD}{v^2} \tag{52}$$

or

$$\int_0^{\Delta p} dp = \frac{1}{v^2}\int_0^D \int_0^V dVdD$$

whence

$$\Delta p = \frac{VD}{v^2} \tag{53}$$

This expression is general for all types of waves. Including *non-electromagnetic waves* like *sound waves*. In this case, v in Eq.(53), will be the speed of sound in the medium and D the *intensity* of the sound radiation.

In the case of *electromagnetic waves*, the Electrodynamics tells us that v will be given by

$$v = \frac{dz}{dt} = \frac{\omega}{\kappa_r} = \frac{c}{\sqrt{\frac{\varepsilon_r \mu_r}{2}\left(\sqrt{1+(\sigma/\omega\varepsilon)^2}+1\right)}}$$

Where k_r is the real part of the *propagation vector* $\vec{k}$; $k = \left|\vec{k}\right| = k_r + ik_i$; ε, μ and σ, are the electromagnetic characteristics of the medium in which the incident (or emitted) radiation is propagating ($\varepsilon = \varepsilon_r \varepsilon_0$ where ε_r is the *relative dielectric permittivity* and $\varepsilon_0 = 8.854 \times 10^{-12} F/m$; $\mu = \mu_r \mu_0$ where μ_r is the *relative magnetic permeability* and $\mu_0 = 4\pi \times 10^{-7} H/m$; σ is the *electrical conductivity*). For an *atom* inside a body , the incident(or emitted) radiation on this atom will be propagating inside the body , and consequently , $\sigma=\sigma_{body}$, $\varepsilon=\varepsilon_{body}$, $\mu=\mu_{body}$.

It is then evident that the *index of refraction* $n_r = c/v$ will be given by

$$n_r = \frac{c}{v} = \sqrt{\frac{\varepsilon_r \mu_r}{2}\left(\sqrt{1+(\sigma/\omega\varepsilon)^2}+1\right)} \tag{54}$$

On the other hand, from Eq.(50) follows that

$$\Delta p = \frac{U}{v}\left(\frac{c}{c}\right) = \frac{U}{c}n_r$$

Substitution into Eq.(41) yields

$$m_g = \left\{1 - 2\left[\sqrt{1+\left(\frac{U}{m_i c^2}n_r\right)^2} - 1\right]\right\}m_i \tag{55}$$

For $\sigma >> \omega\varepsilon$, the expression (54) gives

$$n_r = \frac{c}{v} = \sqrt{\frac{\mu\sigma c^2}{4\pi f}} \tag{56}$$

Substitution of (56) into (55) leads to

$$m_g = \left\{1 - 2\left[\sqrt{1+\frac{\mu\sigma}{4\pi f}\left(\frac{U}{m_i c}\right)^2} - 1\right]\right\}m_i$$

This equation shows that *atoms of ferromagnetic materials with very-high* μ can have its gravitational masses strongly reduced by means of *Extremely Low Frequency* (ELF) electromagnetic radiation. It also shows that atoms of *superconducting materials* (due to *very-high* σ) can also have its gravitational masses strongly reduced by means of ELF electromagnetic radiation.

Alternatively, we may put Eq.(55) as a function of the *power density* (or intensity), D, of the radiation. The integration of (51) gives $U = VD/v$. Thus we can write (55) in the following form:

$$m_g = \left\{1 - 2\left[\sqrt{1+\left(\frac{n_r^2 D}{\rho c^3}\right)^2} - 1\right]\right\}m_i \tag{57}$$

where $\rho = m_i/V$.

For $\sigma >> \omega\varepsilon$, n_r will be given by (56) and consequently (57) becomes

$$m_g = \left\{1 - 2\left[\sqrt{1 + \left(\frac{\mu\sigma D}{4\pi f\rho c}\right)^2} - 1\right]\right\}m_i \tag{58}$$

The vector $\vec{D} = (U/V)\vec{v}$, which we may define from (48), has the same direction of the propagation vector $\vec{k}$ and evidently corresponds to the *Poynting vector*. Then $\vec{D}$ can be replaced by $\vec{E} \times \vec{H}$ .Thus we can write $D = EH = E(B/\mu) = E[(E/v)/\mu] = (1/v\mu)E^2$.

For $\sigma >> \omega\varepsilon$ the Eq.(54) tells us that $v = \sqrt{4\pi f/\mu\sigma}$ consequently we obtain

$$D = E^2\sqrt{\frac{\sigma}{4\pi f\mu}}$$

This expression refers to the instantaneous values of D and E. The average value for E^2 is equal to $\frac{1}{2}E_m^2$ because E varies sinusoidaly (E_m is the maximum value for E). Consequently equation above tells us that the average density $\overline{D}$ is given by

$$\overline{D} = \tfrac{1}{2}E_m^2\sqrt{\frac{\sigma}{4\pi f\mu}}$$

Substitution of this expression into (58) yields the expression for $\overline{m}_g$. Substitution of the expression of D into (58) gives

$$m_g = \left\{1 - 2\left[\sqrt{1 + \frac{\mu}{c^2}\left(\frac{\sigma}{4\pi f}\right)^3\frac{E^4}{\rho^2}} - 1\right]\right\}m_i \tag{59a}$$

Note that for *extremely-low frequencies* the value of f^{-3} in this equation becomes highly expressive.

Since $E = vB$ the equation (59a) can also be putted as a function of B , i.e.,

$$m_g = \left\{1 - 2\left[\sqrt{1 + \left(\frac{\sigma}{4\pi f\mu c^2}\right)\frac{B^4}{\rho^2}} - 1\right]\right\}m_i \tag{59b}$$

For *conducting* materials with $\sigma \approx 10^7 S/m$; $\mu_r = 1$; $\rho \approx 10^3 kg/m^3$ the expression (59b) gives

$$m_g = \left\{1 - 2\left[\sqrt{1 + \left(\frac{\approx 10^{-12}}{f}\right)B^4} - 1\right]\right\}m_i$$

This equation shows that the decreasing in the *gravitational mass* of these conductors can become experimentally detectable for example, starting from 100Teslas at 10mHz.

One can then conclude that an interesting situation arises when a body penetrates a magnetic field in the direction of its center. The *gravitational mass* of the body decreases progressively. This is due to the intensity increase of the magnetic field upon the body while it penetrates the field. In order to understand this phenomenon we might, based on (45), think of the inertial mass as being formed by two parts: one *positive* and another *negative*. Thus, when the body penetrates the magnetic field its negative inertial mass increase, but its total inertial mass decreases, i.e., although there is an increase of inertial mass, the total inertial mass (which is equivalent to *gravitational mass*) will be reduced.

On the other hand, Eq.(4) shows that the *velocity of the body must to increase* as consequence of the gravitational mass decreasing since the *momentum* is conserved. Consider for example a spacecraft with velocity V_s and gravitational mass M_g. If M_g is reduced to m_g then the velocity becomes $V_s' = (M_g/m_g)V_s$ In addition, Eqs. 5 and 6 tell us that the *inertial forces* depend on m_g. Only in the particular case of $m_g = m_i$ the expressions (5) and (6) reduce to the well-known Newtonian expression $F = m_i a$. Consequently, one can conclude that the *inertial effects* on the spacecraft will also be reduced due to the *decreasing of its gravitational mass*. Obviously this leads to a new concept of aerospace flight.

Now consider an electric current $i = i_0 sin 2\pi ft$ through a conductor. Since the current density, $\vec{J}$, is expressed by $\vec{J} = di/d\vec{S} = \sigma\vec{E}$, then we can write that $E = i/\sigma S = (i_0/\sigma S)sin 2\pi ft$. Substitution of this equation into (59a) gives

$$m_g = \left\{1 - 2\left[\sqrt{1 + \frac{i_0^4 \mu}{64\pi^3 c^2 \rho^2 S^4 f^3 \sigma} sin^4 2\pi ft} - 1\right]\right\}m_i \qquad (59c)$$

If the conductor is a *supermalloy* rod $(1\times1\times400mm)$ then $\mu_r = 100{,}000$ (initial); $\rho = 8770 kg/m^3$; $\sigma = 1.6\times10^6 S/m$ and $S = 1\times10^{-6} m^2$. Substitution of these value into equation above yields the following expression for the *gravitational mass* of the supermalloy rod:

$$m_{g(sm)} = \left\{1 - 2\left[\sqrt{1 + \left(5.71\times10^{-12} i_0^4/f^3\right)\sin^4 2\pi ft} - 1\right]\right\}m_{i(sm)}$$

Some oscillators like the HP3325A (Op.002 High Voltage Output) can generate sinusoidal voltages with *extremely-low* frequencies down to $f = 1\times10^{-6}\,Hz$ and amplitude up to 20V (into 50Ω load). The maximum output current is $0.08A_{pp}$.

Thus, for $i_0 = 0.04A$ $\left(0.08A_{pp}\right)$ and $f < 2.25\times10^{-6}\,Hz$ the equation above shows that the *gravitational mass* of the rod becomes *negative* at $2\pi ft = \pi/2$; for $f \cong 1.7\times10^{-6}\,Hz$ at $t = 1/4f = 1.47\times10^5\,s \cong 40.8h$ it shows that $m_{g(sm)} \cong -m_{i(sm)}$.

It is important to realize that this is not the unique way of decreasing the gravitational mass of a body. It was noted earlier that the expression (53) is general for all types of waves including non-electromagnetic waves like sound waves for example. In this case, the velocity v in (53) will be the *speed of sound in the body* and D the *intensity* of the sound radiation. Thus from (53) we can write that

$$\frac{\Delta p}{m_i c} = \frac{VD}{m_i c} = \frac{D}{\rho c v^2}$$

It can easily shown that $D = 2\pi^2 \rho f^2 A^2 v$ where $A = \lambda P/2\pi\rho v^2$; A and P are respectively the amplitude and maximum pressure variation of the sound wave. Therefore we readily obtain

$$\frac{\Delta p}{m_i c} = \frac{P^2}{2\rho^2 c v^3}$$

Substitution of this expression into (41) gives

$$m_g = \left\{1 - 2\left[\sqrt{1 + \left(\frac{P^2}{2\rho^2 c v^3}\right)^2} - 1\right]\right\}m_i$$

This expression shows that in the case of sound waves the decreasing of gravitational mass is relevant for *very strong pressures* only.

It is known that in the nucleus of the Earth the pressure can reach values greater than $10^{13}\,N/m^2$. The equation above tells us that sound waves produced by pressure variations of this magnitude can cause strong decreasing of the *gravitational mass* at the surroundings of the point where the sound waves were generated. This obviously must cause an abrupt decreasing of the pressure at this place since pressure = weight /area = $m_g g$/area). Consequently a local instability will be produced due to the opposite internal pressure. The conclusion is that this effect may cause Earthquakes.

Consider a sphere of radius r around the point where the sound waves were generated (at $\approx 1{,}000km$ depth; the Earth's radius is $6{,}378km$). If the *maximum* pressure, at the explosion place (sphere of radius r_0), is $P_{max} \approx 10^{13}\,N/m^2$ and the pressure at the distance $r = 10km$ is $P_{min} = (r_0/r)^2 P_{max} \approx 10^9\,N/m^2$ then we can consider that in the sphere $P = \sqrt{P_{max}P_{min}} \approx 10^{11}\,N/m^2$.Thus assuming $v \approx 10^3\,m/s$ and $\rho \approx 10^3\,kg/m^3$ we can calculate the variation of gravitational mass in the sphere by means of the equation of m_g, i.e.,

$$\Delta m_g = m_{g(initial)} - m_g =$$

$$= m_i - \left\{1 - 2\left[\sqrt{1 + \left(\frac{P^2}{2\rho^2 cv^3}\right)^2} - 1\right]\right\} m_i =$$

$$= 2\left[\sqrt{1 + \left(\frac{P^2}{2\rho^2 cv^3}\right)^2} - 1\right]\rho V \approx 10^{11}\,kg$$

The *transitory* loss of this great amount of gravitational mass may produce evidently a strong pressure variation and consequently a strong Earthquake.

Finally, we can evaluate the energy necessary to generate that sound waves. From (48) we can write $D_{max} = P_{max}v \approx 10^{16}\,W/m^2$. Thus, the released power is $P_0 = D_{max}\left(4\pi r_0^2\right) \approx 10^{21}\,W$ and the energy ΔE released at the time interval Δt must be $\Delta E = P_0 \Delta t$. Assuming $\Delta t \approx 10^{-3}\,s$ we readily obtain

$$\Delta E = P_0 \Delta t \approx 10^{18}\,joules \approx 10^4\,Megatons$$

This is the amount of energy released by a magnitude 9 earthquake $(M_s = 9)$, i.e., $E = 1.74 \times 10^{(5+1.44M_s)} \cong 10^{18}\,joules$. The maximum magnitude in the *Richter* scale is 12. Note that the sole releasing of this energy at 1000km depth (without the effect of gravitational mass decreasing) cannot produce an Earthquake, since the sound waves reach 1km depth with pressures less than 10N/cm^2.

Let us now return to the principal development of the Gravity Theory.

The equivalence between frames of non-inertial reference and gravitational fields presupposed $m_g \equiv m_i$ because the inertial forces was given by $\vec{F}_i = m_i\vec{a}$, while the equivalent gravitational forces, by $\vec{F}_g = m_g\vec{g}$. Thus, to satisfy the equivalence ($\vec{a} \equiv \vec{g}$ and

$\vec{F}_i \equiv F_g$) it was *necessary that* $m_g \equiv m_i$. Now, the inertial force, $\vec{F}_i$, is given by Eq.(6), and from the Eq.(13) we can obtain the gravitational force, $\vec{F}_g$. Thus, $\vec{F}_i \equiv \vec{F}_g$ leads to

$$\frac{m_g}{\left(1-V^2/c^2\right)^{3/2}}\vec{a} \equiv G\frac{m'_g}{\left(r'\sqrt{1-V^2/c^2}\right)^2}\frac{m_g}{\sqrt{1-V^2/c^2}} \equiv$$

$$\equiv \left(G\frac{m'_g}{r'^2}\right)\frac{m_g}{\left(1-V^2/c^2\right)^{3/2}} \equiv \vec{g}\frac{m_g}{\left(1-V^2/c^2\right)^{3/2}}$$

whence results

$$\vec{a} \equiv \vec{g}$$

Consequently, the equivalence is evident, and therefore the Einstein's equations from the General Relativity continue obviously valid.

The new expression for F_i (Eqs.(5) and (6)) shows that the inertial forces are proportional to the *gravitational mass*, m_g. This means that these forces result from the gravitational interaction between the particle and the other gravitational masses of the Universe, just as *Mach's principle* predicts. Therefore the new expression for the inertial forces incorporates the Mach's principle into Gravitation Theory, and furthermore reveals that the inertial effects upon a particle can be reduced because, as we have seen, the gravitational mass may be reduced. When $m_g = m_i$ the *nonrelativistic* equation for inertial forces, $\vec{F}_i = m_g\vec{a}$, reduces to $\vec{F}_i = m_i\vec{a}$. This is the well-known *Newton's second law* for motion.

In Einstein's Special Relativity Theory the motion of a free-particle is described by means of $\delta S = 0$ [9]. Now based on Eq.(1), $\delta S = 0$ will be given by the following expression

$$\delta S = -m_g c\delta \int ds = 0.$$

which also describes the motion of the particle inside the gravitational field. Thus, the Einstein's equations from the General Relativity can be derived starting from $\delta\left(S_m + S_g\right) = 0$, where S_g and S_m refer to *action* of the gravitational field and the action of the matter, respectively [10].

The variations δS_g and δS_m can be written as follows[11]:

$$\delta S_g = \frac{c^3}{16\pi G}\int\left(R_{ik} - \tfrac{1}{2}g_{ik}R\right)\delta g^{ik}\sqrt{-g}\,d\Omega \qquad (60)$$

$$\delta S_m = -\frac{1}{2c}\int T_{ik}\delta g^{ik}\sqrt{-g}d\Omega \qquad (61)$$

where R_{ik} is the Ricci's tensor; g_{ik} the metric tensor and T_{ik} the matter's energy-momentum tensor:

$$T_{ik} = (P + \varepsilon_g)\mu_i\mu_k + Pg_{ik} \qquad (62)$$

where P is the pressure and $\varepsilon_g = \rho_g c^2$ is now, the density of gravitational energy, E_g, of the particle; ρ_g is then the density of gravitational mass of the particle, i.e., m_g at the volume unit.

Substitution of (60) and (61) into $\delta S_m + \delta S_g = 0$ yields

$$\frac{c^3}{16\pi G}\int\left(R_{ik} - \tfrac{1}{2}g_{ik}R - \tfrac{8\pi G}{c^4}T_{ik}\right)\delta g^{ik}\sqrt{-g}d\Omega = 0$$

whence,

$$\left(R_{ik} - \tfrac{1}{2}g_{ik}R - \tfrac{8\pi G}{c^4}T_{ik}\right) = 0 \qquad (63)$$

because the δg_{ik} are arbitrary.

The Eqs.(63) in the following forms;

$$R_{ik} - \tfrac{1}{2}g_{ik}R = \tfrac{8\pi G}{c^4}T_{ik} \qquad (64)$$

or

$$R_i^k - \tfrac{1}{2}g\delta_i^k R = \tfrac{8\pi G}{c^4}T_i^k. \qquad (65)$$

are the Einstein's equations from the General Relativity.

Making on the obtained equations for the gravitational field, the transition to the Classical Mechanics, we obtain:

$$\Delta\Phi = 4\pi G\left(\frac{\varepsilon_g}{c^2}\right) = 4\pi G\rho_g \qquad (66)$$

This is the nonrelativistic equation for the gravitational field, whose general solution is

$$\Phi = -G\int\frac{\varepsilon_g dV}{rc^2} \tag{67}$$

This equation express the nonrelativistic potential of the gravitational field for any distribution of mass. In particular, for only one particle with gravitational energy $E_g = m_g c^2$, the result is

$$\Phi = -GE_g \big/ rc^2 \tag{68}$$

Thus, the gravity $\vec{g}$ into the gravitational field created by the particle is

$$\vec{g} = -\frac{\partial\Phi}{\partial r} = -G\frac{E_g}{\vec{r}^2 c^2} = -G\frac{m_g}{\vec{r}^2}. \tag{69}$$

Therefore, the gravitational force $\vec{F}_g$ which acts on that field, upon another particle of gravitational mass m'_g is then given by:

$$\vec{F}_g = m'_g\vec{g} = -G\frac{m_g m'_g}{\vec{r}^2} \tag{70}$$

If $m_g > 0$ and $m'_g < 0$, or $m_g < 0$ and $m'_g > 0$ the force will be *repulsive*; *the force will never be null* due to the existence of a *minimum value* for m_g (see Eq. (24)). However, if $m_g < 0$ and $m'_g < 0$, or $m_g > 0$ and $m'_g > 0$ the force will be *attractive*. Just for $m_g = m_i$ and $m'_g = m'_i$ we obtain the *Newton's attraction law*.

Since the gravitational interaction can be repulsive, besides attractive, such as the electromagnetic interaction, then the *graviton* must have spin 1 and not 2. Consequently, the gravitational forces are also *gauge* forces because they are yield by the exchange of so-called "virtual" *quanta* of spin 1, such as the electromagnetic forces and the weak and strong nuclear forces.

Let us now deduce the *Entropy Differential Equation* starting from the Eq.(55). Comparison of the Eqs.(55) and (41) shows that $Un_r = \Delta pc$. For small velocities $(V << c)$, $\Delta p << m_i c$, so that $Un_r << m_i c^2$. Under these circumstances, the development of the Eq.(55) in power of $\left(Un_r / m_i c^2\right)$, gives

$$m_g = m_i - \left(\frac{Un_r}{m_i c^2}\right)^2 m_i \tag{71}$$

In the particular case of *thermal radiation,* it is usual to relate the energy of the photons to temperature, through the relationship $\langle h\nu\rangle \approx kT$ where $k = 1.38\times10^{-23}\,J/K$ is the Boltzmann's constant. Thus, in that case, the energy absorbed by the particle will be $U = \eta\langle h\nu\rangle \approx \eta kT$, where η is a particle-dependent absorption/emission coefficient. Therefore, Eq.(71) may be rewritten in the following form:

$$m_g = m_i - \left[\left(\frac{n_r \eta k}{c^2}\right)^2 \frac{T^2}{m_i^{\,2}}\right] m_i \tag{72}$$

For electrons at T=300K, we have

$$\left(\frac{n_r \eta k}{c^2}\right)^2 \frac{T^2}{m_e^{\,2}} \approx 10^{-17}$$

Comparing Eq.(72) with Eq.(18), we obtain

$$E_{Ki} = \frac{1}{2}\left(\frac{n_r \eta k}{c}\right)^2 \frac{T^2}{m_i}. \tag{73}$$

The derivative of E_{Ki} with respect to temperature T is

$$\frac{\partial E_{Ki}}{\partial T} = \left(n_r \eta k/c\right)^2 \left(T/m_i\right) \tag{74}$$

Thus,

$$T\frac{\partial E_{Ki}}{\partial T} = \frac{\left(n_r \eta k T\right)^2}{m_i c^2} \tag{75}$$

Substitution of $E_{Ki} = E_i - E_{i0}$ into Eq.(75) gives

$$T\left(\frac{\partial E_i}{\partial T} + \frac{\partial E_{i0}}{\partial T}\right) = \frac{\left(n_r \eta k T\right)^2}{m_i c^2} \tag{76}$$

By comparing the Eqs.(76) and (73) and considering that $\partial E_{i0}/\partial T = 0$ because E_{i0} does not depends on T, the Eq.(76) reduces to

$$T(\partial E_i/\partial T)=2E_{Ki} \tag{77}$$

However, Eq.(18) shows that $2E_{Ki}=E_i-E_g$ therefore Eq.(77) becomes

$$E_g=E_i-T(\partial E_i/\partial T) \tag{78}$$

Here, we can identify the energy E_i with the *free-energy* of the system-F and E_g with the *internal energy* of the system-U, thus we can write the Eq.(78) in the following form:

$$U=F-T(\partial F/\partial T) \tag{79}$$

This is the well-known equation of Thermodynamics. On the other hand, remembering $\partial Q=\partial\tau+\partial U$ (1st principle of Thermodynamics) and

$$F=U-TS \tag{80}$$

(Helmholtz's function), we can easily obtain from (79), the following equation:

$$\partial Q=\partial\tau+T\partial S. \tag{81}$$

For *isolated systems*, $\partial\tau=0$, we thus have

$$\partial Q=T\partial S \tag{82}$$

which is the well-know *Entropy Differential Equation.*

Let us now consider the Eq.(55) in the *ultra-relativistic case* where the inertial energy of the particle $E_i=M_ic^2$ is very greater than its inertial energy at rest m_ic^2. Comparison between (4) and (10) leads to $\Delta p=E_iV/c^2$ which, in the ultra-relativistic case, gives $\Delta p=E_iV/c^2\cong E_i/c\cong M_ic$. On the other hand, comparison between (55) and (41) shows that $Un_r=\Delta pc$. Thus $Un_r=\Delta pc\cong M_ic^2>>m_ic^2$. Consequently, Eq.(55) reduces to

$$m_g=m_i-2Un_r/c^2 \tag{83}$$

Therefore, the *action* for such particle, in agreement with the Eq.(2), is

$$
\begin{aligned}
S &= -\int_1^2 m_g c^2 \sqrt{1 - V^2/c^2}\, dt = \\
&= \int_1^2 \left(-m_i + 2Un_r/c^2\right)c^2 \sqrt{1 - V^2/c^2}\, dt = \\
&= \int_1^2 \left[-m_i c^2 \sqrt{1 - V^2/c^2} + 2Un_r \sqrt{1 - V^2/c^2}\right] dt.
\end{aligned} \tag{84}
$$

The integrant function is the *Lagrangean*, i.e.,

$$L = -m_i c^2 \sqrt{1 - V^2/c^2} + 2Un_r \sqrt{1 - V^2/c^2} \tag{85}$$

Starting from the Lagrangean we can find the Hamiltonian of the particle, by means of the well-known general formula:

$$H = V(\partial L/\partial V) - L.$$

The result is:

$$H = \frac{m_i c^2}{\sqrt{1 - V^2/c^2}} + Un_r \left[\frac{\left(4V^2/c^2 - 2\right)}{\sqrt{1 - V^2/c^2}}\right]. \tag{86}$$

The second term on the right hand side of the Eq.(86) results from the particle's interaction with the *electromagnetic field*. Note the similarity between the obtained Hamiltonian and the well-known Hamiltonian for the particle in a electromagnetic field[12]:

$$H = m_i c^2 / \sqrt{1 - V^2/c^2} + Q\varphi. \tag{87}$$

in which Q is the electric charge and φ , the field's *scalar* potential. The quantity $Q\,\varphi$ expresses, as we know, the particle's interaction with the electromagnetic field. Such as the second term on the right hand side of the Eq.(86).

It is therefore evident that it is the same quantity, expresses by means of different variables.

Thus, we can conclude that, in ultra-high energy conditions $\left(Un_r \cong M_i c^2 > m_i c^2\right)$, the gravitational and electromagnetic fields can be described by the *same* Hamiltonian, i.e., in these circumstances they are *unified* !

It is known that starting from that Hamiltonian we may obtain a complete description of the electromagnetic field. This means that from the present theory for gravity we can also derive *the equations of the electromagnetic field.*

Due to $Un_r = \Delta pc \cong M_i c^2$ the second term on the right hand side of the Eq.(86) can be written as follows

$$\Delta pc\left[\frac{\left(4V^2/c^2-2\right)}{\sqrt{1-V^2/c^2}}\right]=$$

$$=\left[\frac{\left(4V^2/c^2-2\right)}{\sqrt{1-V^2/c^2}}\right]M_ic^2=$$

$$=Q\varphi=\frac{QQ'}{4\pi\varepsilon_0 R}=\frac{QQ'}{4\pi\varepsilon_0 r\sqrt{1-V^2/c^2}}$$

whence

$$\left(4V^2/c^2-2\right)M_ic^2=\frac{QQ'}{4\pi\varepsilon_0 r}$$

The factor $\left(4V^2/c^2-2\right)$ becomes equal to 2 in the ultra-relativistic case, then follows that

$$2M_ic^2=\frac{QQ'}{4\pi\varepsilon_0 r} \tag{88}$$

From (44) we know that there is a minimum value for M_i given by $M_{i(min)}=m_{i(min)}$. The Eq.(43) shows that $m_{g(min)}=m_{i(min)}$ and Eq.(23) gives $m_{g(min)}=\pm h/cL_{max}\sqrt{8}=\pm h\sqrt{3/8}/cd_{max}$. Thus we can write

$$M_{i(min)}=m_{i(min)}=\pm h\sqrt{3/8}/cd_{max} \tag{89}$$

According to (88) the value $2M_{i(min)}c^2$ is correlated to $\left(QQ'/4\pi\varepsilon_0 r\right)_{min}=Q_{min}^2/4\pi\varepsilon_0 r_{max}$, i.e.,

$$\frac{Q_{min}^2}{4\pi\varepsilon_0 r_{max}}=2M_{i(min)}c^2 \tag{90}$$

where Q_{min} is the *minimum electric charge* in the Universe (therefore equal to minimum electric charge of the quarks, i.e., $\frac{1}{3}e$); r_{max} is the *maximum distance* between Q and Q', which should be equal to the so-called "diameter", d_c, of the *visible* Universe ($d_c=2l_c$ where l_c is obtained from the Hubble's law for $V=c$, i.e., $l_c=c\tilde{H}^{-1}$). Thus from (90) we readily obtain

$$Q_{\min} = \sqrt{\pi\varepsilon_0 hc\sqrt{24}\left(d_c/d_{\max}\right)} =$$
$$= \sqrt{\left(\pi\varepsilon_0 hc^2\sqrt{96}\tilde{H}^{-1}/d_{\max}\right)} = \tag{91}$$
$$= \tfrac{1}{3}e$$

whence we find

$$d_{max} = 3.4\times10^{30}\,m$$

This will be the maximum "diameter" that the Universe will reach. Consequently, Eq.(89) tells us that the *elementary quantum* of matter is

$$m_{i(min)} = \pm h\sqrt{3/8}/cd_{max} = \pm 3.9\times10^{-73}\,kg$$

Now by combination of gravity and the *uncertainty principle* we will derive the expression of the *Casimir force*.

An uncertainty Δm_i in m_i produces an uncertainty Δp in p and therefore an uncertainty Δm_g in m_g, which according to Eq.(41), is given by

$$\Delta m_g = \Delta m_i - 2\left[\sqrt{1+\left(\frac{\Delta p}{\Delta m_i c}\right)^2} - 1\right]\Delta m_i \tag{92}$$

From the uncertainty principle for position and momentum, we know that the product of the uncertainties of the simultaneously measurable values of corresponding position and momentum components is at least of the order of magnitude of $\hbar$, i.e.,

$$\Delta p\Delta r \sim \hbar$$

Substitution of $\Delta p \sim \hbar/\Delta r$ into(92)yields

$$\Delta m_g = \Delta m_i - 2\left[\sqrt{1+\left(\frac{\hbar/\Delta m_i c}{\Delta r}\right)^2} - 1\right]\Delta m_i \tag{93}$$

Therefore if

$$\Delta r << \frac{\hbar}{\Delta m_i c} \tag{94}$$

then the expression (93) reduces to:

$$\Delta m_g \cong -\frac{2\hbar}{\Delta r c} \tag{95}$$

Note that Δm_g does not depend on m_g.

Consequently, the uncertainty ΔF in the gravitational force $F = -Gm_g m'_g / r^2$, will be given by

$$\begin{aligned} \Delta F &= -G\frac{\Delta m_g \Delta m'_g}{(\Delta r)^2} = \\ &= -\left[\frac{2}{\pi(\Delta r)^2}\right]\frac{hc}{(\Delta r)^2}\left(\frac{G\hbar}{c^3}\right) \end{aligned} \tag{96}$$

The amount $\left(G\hbar/c^3\right)^{1/2} = 1.61\times 10^{-35}\,m$ is called the *Planck length*, l_{planck} ,(the length scale on which quantum fluctuations of the metric of the space time are expected to be of order unity). Thus, we can write the expression of ΔF as follows

$$\begin{aligned} \Delta F &= -\left(\frac{2}{\pi}\right)\frac{hc}{(\Delta r)^4} l^2_{planck} = \\ &= -\left(\frac{\pi}{480}\right)\frac{hc}{(\Delta r)^4}\left[\left(\frac{960}{\pi^2}\right) l^2_{planck}\right] = \\ &= -\left(\frac{\pi A_0}{480}\right)\frac{hc}{(\Delta r)^4} \end{aligned} \tag{97}$$

or

$$F_0 = -\left(\frac{\pi A_0}{480}\right)\frac{hc}{r^4} \tag{98}$$

which is the expression of the *Casimir force* for $A = A_0 = \left(960/\pi^2\right) l^2_{planck}$.

This suggests that A_0 is an *elementary area* related to existence of a *minimum length* $d_{min} = \tilde{k}\, l_{planck}$. What is in accordance with the *quantization of space* (29) which point out the existence of d_{min}.

One can be easily shown that the *minimum area* related to d_{min} is the area of an *equilateral triangle* of side length d_{min} ,i.e.,

$$A_{min} = \left(\tfrac{\sqrt{3}}{4}\right)d^2_{min} = \left(\tfrac{\sqrt{3}}{4}\right)\widetilde{k}^2 l^2_{planck}$$

On the other hand, the *maximum area* related to d_{min} is the area of an *sphere* of radius d_{min} ,i.e.,

$$A_{max} = \pi d^2_{min} = \pi\widetilde{k}^2 l^2_{planck}$$

Thus, the elementary area

$$A_0 = \delta_A d^2_{min} = \delta_A \widetilde{k}^2 l^2_{planck} \tag{99}$$

must have a value between A_{min} and A_{max}, i.e.,

$$\tfrac{\sqrt{3}}{4} < \delta_A < \pi$$

The previous assumption that $A_0 = \left(960/\pi^2\right)l^2_{planck}$ shows that $\delta_A \widetilde{k}^2 = 960/\pi^2$ what means that $5.6 < \widetilde{k} < 14.9$

Therefore we conclude that

$$l_{min} = \widetilde{k} l_{planck} \approx 10^{-34} m. \tag{100}$$

The $n-esimal$ area after A_0 is

$$A = \delta_A \left(n d_{min}\right)^2 = n^2 A_0 \tag{101}$$

One can also be easily shown that the *minimum volume* related to d_{min} is the volume of an *regular tetrahedron* of edge length d_{min} , i.e.,

$$\Omega_{min} = \left(\tfrac{\sqrt{2}}{12}\right)d^3_{min} = \left(\tfrac{\sqrt{2}}{12}\right)\widetilde{k}^3 l^3_{planck}$$

The *maximum volume* is the volume of a *sphere* of radius d_{min} , i.e.,

$$\Omega_{max} = \left(\tfrac{4\pi}{3}\right) d^3_{min} = \left(\tfrac{4\pi}{3}\right) \widetilde{k}^3 l^3_{planck}$$

Thus, the elementary volume $\Omega_0 = \delta_V d^3_{min} = \delta_V \widetilde{k}^3 l^3_{planck}$ must have a value between Ω_{min} and Ω_{max}, i.e.,

$$\left(\tfrac{\sqrt{2}}{12}\right) < \delta_V < \tfrac{4\pi}{3}$$

On the other hand, the $n-esimal$ volume after Ω_0 is

$$\Omega = \delta_V (n d_{min})^3 = n^3 \Omega_0 \qquad n = 1,2,3,\ldots,n_{max}.$$

The existence of n_{max} given by (26), i.e.,

$$n_{max} = L_{max} / L_{min} = d_{max} / d_{min} = \\ = \left(3.4 \times 10^{30}\right) / \widetilde{k} l_{planck} \approx 10^{64}$$

shows that the Universe must have a *finite volume* whose value at the present stage is

$$\Omega_{Up} = n^3_{Up} \Omega_0 = \left(d_p / d_{min}\right)^3 \delta_V d^3_{min} = \delta_V d^3_p$$

where d_p is the present length scale of the Universe. In addition as $\left(\tfrac{\sqrt{2}}{12}\right) < \delta_V < \tfrac{4\pi}{3}$ we conclude that the Universe must have a *polyhedral* space topology with volume between the volume of a *regular tetrahedron* of edge length d_p and the volume of the *sphere of* diameter d_p.

A recent analysis of astronomical data suggests not only that the Universe is *finite*, but also that it has a *dodecahedral* space topology [13,14], what is in strong accordance with the theoretical predictions above.

From (22) and (26) we have that $L_{max} = d_{max} / \sqrt{3} = n_{max} d_{min} / \sqrt{3}$. Since (100) gives $d_{min} \cong 10^{-34} m$ and $n_{max} \cong 10^{64}$ we conclude that $L_{max} \cong 10^{30} m$. From the *Hubble's law* and (22) we have that $V_{max} = \widetilde{H} l_{max} = \widetilde{H}\left(d_{max}/2\right) = \left(\sqrt{3}/2\right)\widetilde{H} L_{max}$ where $\widetilde{H} = 1.7 \times 10^{-18} s^{-1}$. Therefore we obtain $V_{max} \cong 10^{12} m/s$.

Now multiplying (98) by n^2 the expression of F_0 we obtain

$$F = n^2 F_0 = -\left(\frac{\pi n^2 A_0}{480}\right)\frac{hc}{r^4} = -\left(\frac{\pi A}{480}\right)\frac{hc}{r^4} \tag{102}$$

which is the general expression of the Casimir force.

Thus we conclude that *the Casimir effect* is just a gravitational effect related to the *uncertainty principle.*

Note that the Eq.(102) arises only when Δm_i and $\Delta m_i'$ satisfy Eq.(94). If only Δm_i satisfies Eq.(94), i.e., $\Delta m_i << \hbar/\Delta rc$ but $\Delta m_i' >> \hbar/\Delta rc$ then Δm_g and $\Delta m_g'$ will be respectively given by

$$\Delta m_g \cong -2\hbar/\Delta rc \quad and \quad \Delta m_g' \cong \Delta m_i$$

Consequently, the expression (96) becomes

$$\Delta F = \frac{hc}{(\Delta r)^3}\left(\frac{G\Delta m_i'}{\pi c^2}\right) = \frac{hc}{(\Delta r)^3}\left(\frac{G\Delta m_i' c^2}{\pi c^4}\right) = \\ = \frac{hc}{(\Delta r)^3}\left(\frac{G\Delta E'}{\pi c^4}\right) \tag{103}$$

However, from the uncertainty principle for *energy* and *time* we know that

$$\Delta E \sim \hbar/\Delta t \tag{104}$$

Therefore we can write the expression (103) in the following form:

$$\Delta F = \frac{hc}{(\Delta r)^3}\left(\frac{G\hbar}{c^3}\right)\left(\frac{1}{\pi \Delta t' c}\right) = \\ = \frac{hc}{(\Delta r)^3} l_{planck}^2 \left(\frac{1}{\pi \Delta t' c}\right) \tag{105}$$

From the General Relativity Theory we know that $dr = cdt/\sqrt{-g_{00}}$. If the field is *weak* then $g_{00} = -1 - 2\phi/c^2$ and $dr = cdt/(1+\phi/c^2) = cdt/(1 - Gm/r^2c^2)$. For $Gm/r^2 c^2 << 1$ we obtain $dr \cong cdt$. Thus, if $dr = dr'$ then $dt = dt'$. This means that we may change $(\Delta t'c)$ by (Δr) into (105). The result is

$$\Delta F = \frac{hc}{(\Delta r)^4}\left(\frac{1}{\pi} l^2_{planck}\right) =$$
$$= \left(\frac{\pi}{480}\right)\frac{hc}{(\Delta r)^4}\underbrace{\left(\frac{480}{\pi^2} l^2_{planck}\right)}_{\frac{1}{2}A_0} =$$
$$= \left(\frac{\pi A_0}{960}\right)\frac{hc}{(\Delta r)^4}$$

or

$$F_0 = \left(\frac{\pi A_0}{960}\right)\frac{hc}{r^4}$$

whence

$$F = \left(\frac{\pi A}{960}\right)\frac{hc}{r^4} \tag{106}$$

Now the Casimir force is *repulsive*, and its intensity is the half of the intensity previously obtained (102).

Consider the case when both Δm_i and $\Delta m_i'$ do not satisfy Eq.(94), and

$$\Delta m_i >> \hbar/\Delta rc$$
$$\Delta m_i' >> \hbar/\Delta rc$$

In this case, $\Delta m_g \cong \Delta m_i$ and $\Delta m_g' \cong \Delta m_i'$. Thus,

$$\Delta F = -G\frac{\Delta m_i \Delta m_i'}{(\Delta r)^2} = -G\frac{\left(\Delta E/c^2\right)\left(\Delta E'/c^2\right)}{(\Delta r)^2} =$$
$$= -\left(\frac{G}{c^4}\right)\frac{(\hbar/\Delta t)^2}{(\Delta r)^2} = -\left(\frac{G\hbar}{c^3}\right)\frac{hc}{(\Delta r)^2}\left(\frac{1}{c^2\Delta t^2}\right) =$$
$$= -\left(\frac{1}{2\pi}\right)\frac{hc}{(\Delta r)^4} l^2_{planck} =$$
$$= -\left(\frac{\pi}{1920}\right)\frac{hc}{(\Delta r)^4}\left(\frac{960}{\pi^2} l^2_{planck}\right) = -\left(\frac{\pi A_0}{1920}\right)\frac{hc}{(\Delta r)^4}$$

whence

$$F = -\left(\frac{\pi A}{1920}\right)\frac{hc}{r^4} \tag{107}$$

The force will be *attractive* and its intensity will be the *fourth part* of the intensity given by the first expression (102) for the Casimir force.

There is a crucial cosmological problem to be solved: the problem of the *hidden mass*. Most theories predict that the amount of known matter, detectable and available in the universe, is only about 1/10 to 1/100 of the amount needed to close the universe. That is, to achieve the density sufficient to close-up the universe by maintaining the gravitational curvature (escape velocity equal to the speed of light) at the outer boundary.

The Eq.(45) may solve this problem. We will start by substituting the well-known expression of *Hubble's* law for velocity, $V = \tilde{H}l$, into Eq.(45). ($\tilde{H} = 1.7\times10^{-18}s^{-1}$ is the Hubble constant). The expression obtained shows that particles which are at distances $l = l_0 = \left(\sqrt{5}/3\right)\left(c/\tilde{H}\right) = 1.3\times10^{26}m$ have *quasi null* gravitational mass $m_g = m_{g(min)}$; beyond this distance, the particles have *negative* gravitational mass. Therefore, there are two well-defined regions in the Universe; the region of the bodies with *positive* gravitational masses and the region of the bodies with *negative* gravitational mass. The total gravitational mass of the first region, in accordance with Eq.(45), will be given by

$$M_{g1} \cong M_{i1} = \frac{m_{i1}}{\sqrt{1-\overline{V}_1^2/c^2}} \cong m_{i1}$$

where m_{i1} is the total *inertial mass* of the bodies of the mentioned region; $\overline{V}_1 << c$ is the average velocity of the bodies at region 1. The total gravitational mass of the second region is

$$M_{g2} = M_{i2} - 2\left(\frac{1}{\sqrt{1-\overline{V}_2^2/c^2}} - 1\right)M_{i2}$$

where $\overline{V}_2$ is the average velocity of the bodies ; $M_{i2} = m_{i2}\Big/\sqrt{1-\overline{V}_2^2/c^2}$ and m_{i2} is the total *inertial mass* of the bodies of region 2.

Now consider that from Eq.(7), we can write

$$\xi = \frac{E_g}{V} = \frac{M_g c^2}{V} = \rho_g c^2$$

where ξ is the *energy density* of matter.

Note that the expression of ξ only reduces to the well-known expression ρc^2, where ρ is the sum of the inertial masses per volume unit, when $m_g = m_i$. Therefore, in the derivation of the well-known difference

$$\frac{8\pi G\rho_U}{3} - \tilde{H}^2 \tag{108}$$

which gives the *sign of the curvature* of the Universe [15], we must use $\xi = \rho_{gU} c^2$ instead of $\xi = \rho_U c^2$.The result obviously is

$$\frac{8\pi G\rho_{gU}}{3} - \tilde{H}^2 \tag{109}$$

where

$$\rho_{gU} = \frac{M_{gU}}{V_U} = \frac{M_{g1} + M_{g2}}{V_U} \tag{110}$$

M_{gU} and V_U are respectively the total gravitational mass and the volume of the Universe.

Substitution of M_{g1} and M_{g2} into expression (110) gives

$$\rho_{gU} = \frac{m_{iU} + \left[\left(\frac{3}{\sqrt{1-\overline{V}_2^2/c^2}} - \frac{2}{1-\overline{V}_2^2/c^2}\right)m_{i2} - m_{i2}\right]}{V_U}$$

where $m_{iU} = m_{i1} + m_{i2}$ is the total inertial mass of the Universe.

The volume V_1 of the region 1 and the volume V_2 of the region 2, are respectively given by $V_1 = 2\pi^2 l_0^3$ *and* $V_2 = 2\pi^2 l_c^3 - V_1$ where $l_c = c/\tilde{H} = 1.8\times 10^{26}\,m$ is the so-called "radius" of the visible Universe. Moreover, $\rho_{i1} = m_{i1}/V_1$ and $\rho_{i2} = m_{i2}/V_2$. Due to the *hypothesis of the uniform distribution of matter in the space*, follows that $\rho_{i1} = \rho_{i2}$.Thus, we can write

$$\frac{m_{i1}}{m_{i2}} = \frac{V_1}{V_2} = \left(\frac{l_0}{l_c}\right)^3 = 0.38$$

Similarly,

$$\frac{m_{iU}}{V_U} = \frac{m_{i2}}{V_2} = \frac{m_{i1}}{V_1}$$

Therefore, $m_{i2} = \frac{V_2}{V_U} m_{iU} = \left[1 - \left(\frac{l_0}{l_c}\right)^3\right] m_{iU} = 0.62 m_{iU}$ and $m_{i1} = 0.38 m_{iU}$.

Substitution of m_{i2} into the expression of ρ_{gU} yields

$$\rho_{gU} = \frac{m_{iU} + \left[\left(\frac{1.86}{\sqrt{1-\overline{V}_2^2/c^2}} - \frac{1.24}{1-\overline{V}_2^2/c^2} - 0.62\right) m_{iU}\right]}{V_U}$$

Due to $\overline{V}_2 \cong c$, we conclude that the term between bracket (hidden mass) *is very greater than* $10 m_{iU}$. The amount m_{iU} is the mass of known matter in the universe (1/10 to 1/100 of the amount needed to close the Universe).

Consequently, the total mass

$$m_{iU} + \left[\left(\frac{1.86}{\sqrt{1-\overline{V}_2^2/c^2}} - \frac{1.24}{1-\overline{V}_2^2/c^2} - 0.62\right) m_{iU}\right]$$

must be sufficient to close the Universe. This solves therefore the problem of the *hidden mass*.

There is another cosmological problem to be solved: the problem of the *anomalies* in the spectral red-shift of certain galaxies and stars.

Several observers have noticed red-shift values that cannot be explained by the Doppler-Fizeau effect or by the Einstein effect (the gravitational spectrum shift, supplied by Einstein's theory).

This is the case of the so-called *Stefan's quintet* (a set of five galaxies which were discovered in 1877), whose galaxies are located at approximately the same distance from the Earth, according to very reliable and precise measuring methods. But, when the velocities of the galaxies are measured by its red-shifts , the velocity of one of them is much greater than the velocity of the others.

Similar observations have been made on the *Virgo constellation* and spiral galaxies. Also the Sun presents a red-shift greater than the predicted value by the Einstein effect.

It seems that some of these anomalies can be explained if we consider the Eq.(45) in the calculation of the *gravitational mass* of the point of emission.

The expression of the gravitational spectrum shift, supplied by Einstein's theory [16] is given by

$$\Delta\omega = \omega_1 - \omega_2 = \frac{\phi_2 - \phi_1}{c^2}\omega_1 = = \frac{-Gm_{g2}/r_2 + Gm_{g1}/r_1}{c^2}\omega_1 \tag{111}$$

where ω_1 is the frequency of the light at the point of emission ; ω_2 is the frequency at the point of observation; ϕ_1 and ϕ_2 are respectively, the Newtonian gravitational potentials at the point of emission and at the point of observation.

This expression has been deduced from $t = t_0\sqrt{-g_{00}}$ [17] which correlates *own time* (real time), t, with the temporal coordinate x^0 of the space-time ($t_0 = x^0/c$).

When the gravitational field is *weak*, the temporal component g_{00} of the metric tensor is given by $g_{oo} = -1 - 2\phi/c^2$[18].Thus we readily obtain

$$t = t_0\sqrt{1 - 2Gm_g/rc^2} \tag{112}$$

Curiously, this equation tell us that we can have $t < t_0$ when $m_g > 0$; and $t > t_0$ for $m_g < 0$. In addition, if $m_g = c^2r/2G$, i.e., if $r = 2Gm_g/c^2$ (*Schwarzschild radius*) we obtain $t = 0$.

Let us now consider the well-known process of stars' *gravitational contraction*. It is known that the destination of the star is directly correlated to its mass. If the star's mass is less than $1.4M_\odot$ (Schemberg-Chandrasekhar's limit), it becomes a *white dwarf.* If its mass exceeds that limit, the pressure produced by the degenerate state of the matter no longer counterbalances the gravitational pressure, and the star's contraction continues. Afterwards occur the reactions between protons and electrons (capture of electrons), where *neutrons* and anti-neutrinos are produced.

The contraction continues until the system regains stability (when the pressure produced by the neutrons is sufficient to stop the gravitational collapse). Such systems are called *neutron stars*.

There is also a critical mass for the stable configuration of neutron stars. This limit has not been fully defined as yet, but it is known that it is located between $1.8M_\odot$ and $2.4M_\odot$. Thus, if the mass of the star exceeds $2.4M_\odot$, the contraction will continue.

According to Hawking[19] collapsed objects cannot have mass less than $\sqrt{\hbar c/4G} = 1.1\times10^{-8}kg$. This means that, with the progressing of the compression, the neutrons cluster must become a cluster of *superparticles* where the *minimal inertial mass* of the superparticle is

$$m_{i(sp)} = 1.1 \times 10^{-8}\,kg. \tag{113}$$

Symmetry is a fundamental attribute of the Universe that enables an investigator to study particular aspects of physical systems by themselves. For example, the assumption that space is homogeneous and isotropic is based on *Symmetry Principle*. Also here, by symmetry, we can assume that there are only *superparticles* with mass $m_{i(sp)} = 1.1 \times 10^{-8}\,kg$ in the cluster of *superparticles*.

Based on the mass-energy of the superparticles ($\sim 10^{18}$ GeV) we can say that they belong to a putative class of particles with mass-energy beyond the *supermassive* Higgs bosons (the so-called X bosons). It is known that the GUT's theories predict an entirely new force mediated by a new type of boson, called simply X (or X boson). The X bosons carry both electromagnetic and color charge, in order to ensure proper conservation of those charges in any interactions. The X bosons must be extremely massive, with mass-energy in the unification range of about 10^{16} GeV.

If we assume the superparticles *are not hyper*massive Higgs bosons then the possibility of the *neutrons cluster* to become a *Higgs bosons cluster* before becoming a *superparticles cluster* must be considered. On the other hand, the fact that superparticles must be so massive also means that it is not possible to create them in any conceivable particle accelerator that could be built. They can exist as free particles only at a very early stage of the Big Bang from which the universe emerged.

Let us now imagine the Universe coming back for the past. There will be an instant in which it will be similar to a *neutrons cluster*, such as the stars at the final state of gravitational contraction. Thus, with the progressing of the compression, the *neutrons* cluster becomes a superparticles cluster. Obviously, this only can occur before 10^{-23}s(after the Big-Bang).

The temperature T of the Universe at the 10^{-43}s< t < 10^{-23}s period can be calculated by means of the well-known expression[20]:

$$T \approx 10^{22}\left(t/10^{-23}\right)^{-1/2} \tag{114}$$

Thus at $t \cong 10^{-43}\,s$ (at the *first* spontaneous breaking of symmetry) the temperature was $T \approx 10^{32}\,K$ (~10^{19}GeV).Therefore, we can assume that the absorbed electromagnetic energy by each *superparticle*, before $t \cong 10^{-43}\,s$, was $U = \eta kT > 1 \times 10^{9}\,J$ (see Eqs.(71) and (72)). By comparing with $m_{i(sp)}c^2 \cong 9 \times 10^{8}\,J$, we conclude that $U > m_{i(sp)}c$. Therefore, *the unification condition* $\left(U\eta_r \cong M_i c^2 > m_i c^2\right)$ is satisfied. This means that, *before* $t \cong 10^{-43}\,s$, *the gravitational and electromagnetic interactions were unified.*

From the *unification condition* $\left(U\eta_r \cong M_i c^2\right)$, we may conclude that the superparticles' *relativistic inertial mass* $M_{i(sp)}$ is

$$M_{i(sp)} \cong \frac{U\eta_r}{c^2} = \frac{\eta n_r kT}{c^2} \approx 10^{-8}\,kg \tag{115}$$

Comparing with the superparticles' *inertial mass at rest* (113), we conclude that

$$M_{i(sp)} \approx m_{i(sp)} = 1.1\times 10^{-8}\,kg \tag{116}$$

From Eqs.(83) and (115), we obtain the superparticle's *gravitational mass at rest*, i.e.,

$$\begin{aligned} m_{g(sp)} &= m_{i(sp)} - 2M_{i(sp)} \cong \\ &\cong -M_{i(sp)} \cong -\frac{\eta n_r kT}{c^2} \end{aligned} \tag{117}$$

and consequently, the superparticle's *relativistic gravitational mass*, is

$$M_{g(sp)} = -\frac{\eta n_r kT}{c^2\sqrt{1-V^2/c^2}} \tag{118}$$

Thus, the gravitational forces between two *superparticles* , according to (13), is given by:

$$\begin{aligned} \vec{F}_{12} &= -\vec{F}_{21} = -G\frac{M_{g(sp)}M'_{g(sp)}}{r^2}\hat{\mu}_{21} = \\ &= \left[\left(\frac{M_{i(sp)}}{m_{i(sp)}}\right)^2\left(\frac{G}{c^5\hbar}\right)(\eta n_r \kappa T)^2\right]\frac{\hbar c}{r^2}\hat{\mu}_{21} \end{aligned} \tag{119}$$

Due to the *unification* of the gravitational and electromagnetic interactions at that period, we have

$$\begin{aligned} \vec{F}_{12} &= -\vec{F}_{21} = G\frac{M_{g(sp)}M'_{g(sp)}}{r^2}\hat{\mu}_{21} = \\ &= \left[\left(\frac{M_{i(sp)}}{m_{i(sp)}}\right)^2\left(\frac{G}{c^5\hbar}\right)(\eta \kappa T)^2\right]\frac{\hbar c}{r^2}\hat{\mu}_{21} = \\ &= \frac{e^2}{4\pi\varepsilon_0 r^2} \end{aligned} \tag{120}$$

From the equation above we can write

$$\left[\left(\frac{M_{i(sp)}}{m_{i(sp)}}\right)^2\left(\frac{G}{c^5\hbar}\right)(\eta\kappa T)^2\hbar c\right]=\frac{e^2}{4\pi\varepsilon_0} \tag{121}$$

Now assuming that

$$\left(\frac{M_{i(sp)}}{m_{i(sp)}}\right)^2\left(\frac{G}{c^5\hbar}\right)(\eta\kappa T)^2=\psi \tag{122}$$

the Eq.(121) can be rewritten in the following form:

$$\psi=\frac{e^2}{4\pi\varepsilon_0\hbar c}=\frac{1}{137} \tag{123}$$

which is the well-known *reciprocal fine structure constant* .

For $T=10^{32}K$ the Eq.(122) gives

$$\psi=\left(\frac{M_{i(sp)}}{m_{i(sp)}}\right)^2\left(\frac{G}{c^5\hbar}\right)(\eta n_r\kappa T)^2\approx\frac{1}{100} \tag{124}$$

This value has the same order of magnitude that the exact value(1/137) of the *reciprocal fine structure constant.*

From equation (120) we can write:

$$\left(G\frac{M_{g(sp)}M'_{g(sp)}}{\psi c\vec{r}}\right)\vec{r}=\hbar \tag{125}$$

The term between parenthesis has the same dimensions that the *linear momentum* $\vec{p}$. Thus (125) tells us that

$$\vec{p}\cdot\vec{r}=\hbar. \tag{126}$$

A component of the momentum of a particle cannot be precisely specified without loss of all knowledge of the corresponding component of its position at that time ,i.e., a particle cannot precisely localized in a particular direction without loss of all knowledge of its momentum component in that direction . This means that in intermediate cases the product of the uncertainties of the simultaneously measurable values of corresponding position and momentum components is *at least of the order of magnitude* of $\hbar$,i.e.,

$$\Delta p.\Delta r \geq \hbar \tag{127}$$

This relation, *directly obtained here from the Unified Theory*, is the well-known relation of the *Uncertainty Principle* for position and momentum.

According to Eq.(83), the gravitational mass of the superparticles at the *center* of the cluster becomes *negative* when $2\eta n_r kT/c^2 > m_{i(sp)}$, i.e., when

$$T > T_{critical} = \frac{m_{i(sp)}}{2\eta n_r k} \approx 10^{32}\,K.$$

According to Eq.(114) this temperature corresponds to $t_c \approx 10^{-43}\,s$.

With the progressing of the compression, more superparticles into the center will have *negative* gravitational mass. Consequently, there will have a critical point in which the *repulsive* gravitational forces between the superparticles with negative gravitational masses and the superparticles with positive gravitational masses will be so strong that an explosion will occur. This is the experiment that we call the Big Bang.

Now, starting from the Big Bang to the present time. Immediately after the Big Bang, the superparticles' *decompressing* begins. The gravitational mass of the most central superparticle will only be positive when the temperature becomes smaller than the critical temperature, $T_{critical} \approx 10^{32}\,K$. At the maximum state of compression (exactly at the Big Bang) the volumes of the superparticles was equal to the elementary volume $\Omega_0 = \delta_V d_{min}$ and the volume of the Universe was $\Omega = \delta_V (nd_{min})^3 = \delta_V d^3_{initial}$ where $d_{initial}$ was the *initial* length scale of the Universe. At this very moment the *average* density of the Universe was equal to the *average* density of the superparticles, thus we can write

$$\left(\frac{d_{initial}}{d_{min}}\right)^3 = \frac{M_{i(U)}}{m_{i(sp)}} \tag{128}$$

where $M_{i(U)} \approx 10^{53}\,kg$ is the inertial mass of the Universe. It has already been shown that $d_{min} = \tilde{k} l_{planck} \approx 10^{-34}\,m$. Then, from the Eq.(128), we obtain:

$$d_{initial} \approx 10^{-14}\,m \tag{129}$$

After the Big Bang the Universe expands itself from $d_{initial}$ up to d_{cr} (when the temperature decreasing reaches the critical temperature $T_{critical} \approx 10^{32}\,K$, and the gravity becomes *attractive*). Thus, it expands by $d_{cr} - d_{initial}$, under effect of the *repulsive* gravity

$$\bar{g} = \sqrt{g_{max} g_{min}} =$$

$$= \sqrt{\left[G\left(\tfrac{1}{2} M_{g(U)}\right)/\left(\tfrac{1}{2} d_{initial}\right)^2\right]\left[G\tfrac{1}{2} M_{i(U)}/\left(\tfrac{1}{2} d_{cr}\right)^2\right]} =$$

$$= \frac{2G\sqrt{M_{g(U)}M_{i(U)}}}{d_{cr}d_{initial}} = \frac{2G\sqrt{\sum m_{g(sp)}M_{i(U)}}}{d_{cr}d_{initial}} =$$

$$= \frac{2G\sqrt{\chi\sum m_{i(sp)}M_{i(U)}}}{d_{cr}d_{initial}} = \frac{2GM_{i(U)}\sqrt{\chi}}{d_{cr}d_{initial}}$$

during to a time $t_c \approx 10^{-43} s$ .Thus,

$$d_{cr} - d_{initial} = \tfrac{1}{2}\bar{g}(t_c)^2 = \left(\sqrt{\chi}\right)\frac{GM_{i(U)}}{d_{cr}d_{initial}}(t_c)^2 \quad (130)$$

The Eq.(83), gives

$$\chi = \frac{m_{g(sp)}}{m_{i(sp)}} = 1 - \frac{2Un_r}{m_{i(sp)}c^2} = 1 - \frac{2\eta n_r kT}{m_{i(sp)}c^2} \approx 10^{-32}T$$

The temperature at the beginning of the Big Bang (t=0) should have been very greater than $T_{critical} \approx 10^{32} K$. Thus, χ must be a very big number. Then it is easily seen that during this period, the Universe expanded at an astonishing rate. Thus, there is an evident *inflation* period, which ends at $t_c \approx 10^{-43} s$.

With the progressing of the *decompression* the *superparticles* cluster becomes a neutrons cluster. This means that the neutrons are created *without its antiparticle*, the antineutron. Thus it solves the matter/antimatter dilemma that is unresolved in many cosmologies.

Now a question: How did the *primordial superparticles* appear at the beginning of the Universe? It is a proven *quantum* fact that a *wave function* Ψ may *collapse* and that at this moment all the possibilities that it describes are suddenly expressed in *reality*. This means that, through this process, particles can be suddenly *materialized.*

The materialization of the *primordial superparticles* into a critical volume denotes *knowledge* of what would happen starting from that *initial condition*, fact that points towards the *existence* of a Creator.

Conclusion

We have described a coherent way for the quantization of gravity, which provides a consistent unification of gravity with electromagnetism. As we have seen, this new approach will allow us to understand some crucial matters in Quantum Cosmology.

The equation of correlation between gravitational and inertial masses, which has been derived directly from the theory of gravity, has relevant technological consequences. We have seen that gravitational mass can be negative at specific conditions. This means that it will be possible to build gravitational binaries (*gravitational motors*), and to extract energy from any site of a gravitational field. Obviously, the Gravity Control will be also very important to Transportation Systems. On the other hand, negative gravitational mass suggests the possibility of dipole gravitational radiation. This fact is highly relevant because now we may build transceivers to operate with gravitational waves. Furthermore, the receiver would allow us to directly observe for the first time the Cosmic Microwave Background in Gravitational Radiation, which would picture the Universe at the beginning of the Big-Bang.

Appendix A

In the beginning of this work it was shown (Eq.59c) that when an alternating electric current passes through a conductor its *gravitational mass* is reduced in accordance with the following expression

$$m_g = \left\{1 - 2\left[\sqrt{1 + \left(i^4 \mu / 64\pi^3 c^2 \rho^2 S^4 f^3 \sigma\right)} - 1\right]\right\} m_i \qquad \text{(A1)}$$

In this equation i refers to the instantaneous electric current; $\mu = \mu_r \mu_0$ is the magnetic permeability of the conductor; c is the speed of light; ρ is the density (kg/m^3) of the conductor; S is the area of the cross section(m^2) of the conductor; σ is the electric conductivity of the conductor (S/m); m_i is the *inertial mass* of the conductor and f the frequency of the electric current (Hz).

If the conductor is a *mumetal* wire with the following characteristics: relative *magnetic permeability* = μ_r = 100,000; electrical *conductivity* = $\sigma = 1.9 \times 10^6\, S/m$; *density* = $\rho = 8740\, kg.m^{-3}$. Then (A1) gives

$$m_g = \left\{1 - 2\left[\sqrt{1 + 4.8 \times 10^{-36}\left(i^4 / S^4 f^3\right)} - 1\right]\right\} m_i \qquad \text{(A2)}$$

If $i \cong i_0 = 0.6A$, $f = 10mHz$ and the wire has diameter = 0.127mm then substitution of these values into (A2) yields

$$m_g = -6.98 m_i \qquad \text{(A3)}$$

Thus, if the length of the wire is $l = 10m$ then $m_i = \rho Sl = 1.1\times10^{-3}kg = 1.1g$ and consequently results $m_g = -7.7g$ The electrical resistance R of the wire is $R = l/\sigma S = 414.4\Omega$ and the *average* power required (*true power*) is $P = V_{rms}I_{rms}\cos\varphi = RI^2_{rms} = R\left(i_0/\sqrt{2}\right)^2 \cong 74.59W$ As we know in an AC circuit the *apparent power* is $P_{ap} = ZI^2_{rms}$ (volt-amp), where Z is the impedance; the *reactive* power is $P_q = XI^2_{rms}$, $X = Z\sin\varphi$ is the reactive impedance. These powers are related by means of the following equation: $P^2_{ap} = P^2 + P^2_q$. It is easy to see that the powers P_{ap} and P_q depend on the frequency f, but P is *independent* of f. Therefore if we decrease f the power P won't be altered.

Note that decreasing the frequency down to $1\mu Hz$ we can increase 1000 times the area S (mumetal wire with 4mm diameter). Then by keeping the same current ($i \cong 0.6A$) the result will be the same of (A3), i.e., $m_g = -6.98m_i$ However, due to the increasing of the area S the value of m_i increases 1000 times, ie., $m_i = \rho Sl = 1.1kg$ and consequently the gravitational mass becomes $m_g = -7.7kg$ By increasing the length of the wire from 10m up to 10,000m we obtain

$$m_g = -7{,}700kg$$

In this case the power required (*true power*) will be the same, i.e., $P \cong 74.59W$ since the resistance R of the wire will be the same: $R = l/\sigma S = 414.4\Omega$.

This means that we can produce a *gravitational lift force*

$$F = m_g g = -7{,}700\times 9.8 = -75{,}460N$$

with power requirements of 74.59W only.

Next, we will describe an experiment, in which there has been observed *strong* decreases in weight of a *thin mumetal* wire (0.127mm diameter; 10m length; 1.1g mass) when it has been subjected to sinusoidal currents of $1.20A_{pp}$ with extremely-low frequencies (ELF) of 5mHz, 10mHz , and 15mHz.

The mumetal wire has the characteristics previously mentioned.

The ELF sinusoidal currents were generated by a HP3325A (Op.002 High Voltage Output) which can generate sinusoidal *voltages* with *extremely-low* frequencies down to $f = 1\times10^{-6}Hz$ and amplitude up to 20V (40Vpp into 50Ω load). The maximum output current is $0.08A_{pp}$.

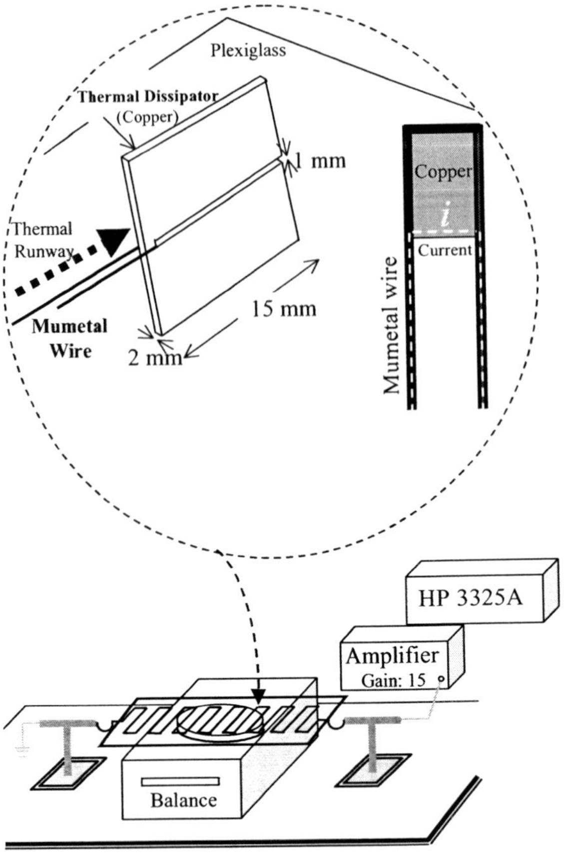

Note: The curve a hanging flexible wire assumes when supported at its ends and acted upon by a uniform gravitational field is called *catenary*. Note that the extreme parts of the Mumetal wire make catenaries which will be inverted after $t \sim 12$s if *f* = *5mHz*, after $t \sim 11$s if *10mHz*, and after $t \sim 10$s if *15mHz*.

Fig. 1. Experimental Set-up.

As shown in figure 1, an amplifier has been connected to the HP3325A in order to increase the maximum output current up to $1.20A_{pp}$. The mumetal wire sandwiched by 2 transparent Plexiglass plates (150×250×2mm) has been placed over a balance and connected to the system amplifier/ HP3325A.

The mass of mumetal wire has been measured while passing through it an ELF electric current. For the mass measurements we used a XL-500 Pan balance with maximum weight capability of 500g and resolution of 0.01g.

We have started applying through the wire a sinusoidal current of $1.20A_{pp}$ (amplitude $i_0 = 0.60A$) and frequency $f = 5mHz$. Then at $t = 5s$ the balance showed a mass *decrease* of ~ 27.3% with respect to *initial mass of the wire*. The mass decrease becomes equal to 100% at $t \cong 9s$.

For $10mHz$ at $t = 5s$ the balance showed a mass *decrease* of ~18.2%. The mass decrease becomes equal to 100% at $t \cong 8s$. For $15mHz$ at $t = 5s$ the balance showed a mass *decrease* of ~36.4%. The mass decrease becomes equal to 100% at $t \cong 7s$.

Since the weight is given by $\vec{P} = m_g \vec{g}$ and $\vec{g}$ is not altered, we conclude that the experimental results refers to m_g ,i.e., the *gravitational mass* of the mumetal wire is decreased (independently of its *inertial mass*) when an electric current with extremely-low frequency passes through it.

Table 1 - Experimental results of Δm_g and correlation $m_g / m_{g(initial)}$ for the Mumetal wire ($m_{g(initial)} = m_{i(initial)}$ = 1.10g; through the wire a sinusoidal current of $1.20A_{pp}$ (amplitude $i_0 = 0.60A$)). Experimental data are the average of 10 measurements. The standard deviation of the single data is between 3 and 5%.

t (s)	5mHz				10mHz				15mHz			
	Δm_g (g)		$m_g/m_{g(initial)}$		Δm_g (g)		$m_g/m_{g(initial)}$		Δm_g (g)		$m_g/m_{g(initial)}$	
	Exp	*The*	*Exp*	*The*	*Exp*	*The*	*Exp*	*The*	*Exp*	*The*	*Exp*	*The*
5	0.3	0.21	0.7	0.81	0.2	0.31	0.8	0.72	0.4	0.38	0.6	0.65
10	1.3	1.42	-0.2	-0.29	2.1	2.04	-0.9	-0.85	2.1	2.13	-0.9	-0.94
15	4.2	4.16	-2.8	-2.78	4.7	4.84	-3.3	-3.40	3.7	3.68	-2.4	-2.35
20	7.9	7.90	-6.2	-6.18	7.4	7.21	-5.7	-5.55	3.5	3.33	-2.2	-2.03
25	12.0	12.13	-9.9	-10.03	8.3	8.13	-6.5	-6.39	1.3	1.44	-0.2	-0.31
30	16.3	16.39	-13.8	-13.90	7.4	7.21	-5.7	-5.55	0.2	0.16	0.8	0.85
35	20.5	19.14	-17.6	-17.40	4.7	4.84	-3.3	-3.40	0.1	0.10	0.9	0.91
40	23.2	23.30	-20.1	-20.18	2.1	2.04	-0.9	-0.85	0.9	0.83	0.2	0.25
45	25.1	25.27	-21.8	-21.97	0.2	0.31	0.8	0.72	2.9	2.79	-1.6	-1.54
50	26.1	25.94	-22.7	-22.58	0.0	0.00	1.0	1.00	3.8	3.81	-2.5	-2.46

- The mass of the plexiglass plates is greater than $26.1g$. See Fig.1.

Table1 presents the mass decreases, Δm_g, as function of t. The *theoretical* values of Δm_g, calculated as follows

$$\begin{aligned}\Delta m_g &= m_{g(initial)} - m_g = \\ &= 2\left[\sqrt{1+\left(i_0^4 \mu / 64\pi^3 c^2 \rho^2 S^4 f^3 \sigma\right)\sin^4 2\pi f t} - 1\right] m_{g(initial)} \\ &= 2\left[\sqrt{1+1.86\times10^{-4}\left(i_0^4 / f^3\right)\sin^4 2\pi f t} - 1\right] m_{g(initial)}\end{aligned}$$

are also on that Table to be compared with those supplied by the experiment.

The values of the correlation $m_g/m_{(initial)}$(experimental and theoretical results) as function of the *instantaneous* currents i have been plotted on the Fig.2 to be compared. There is a strongly accordance with the experimental data and theory.

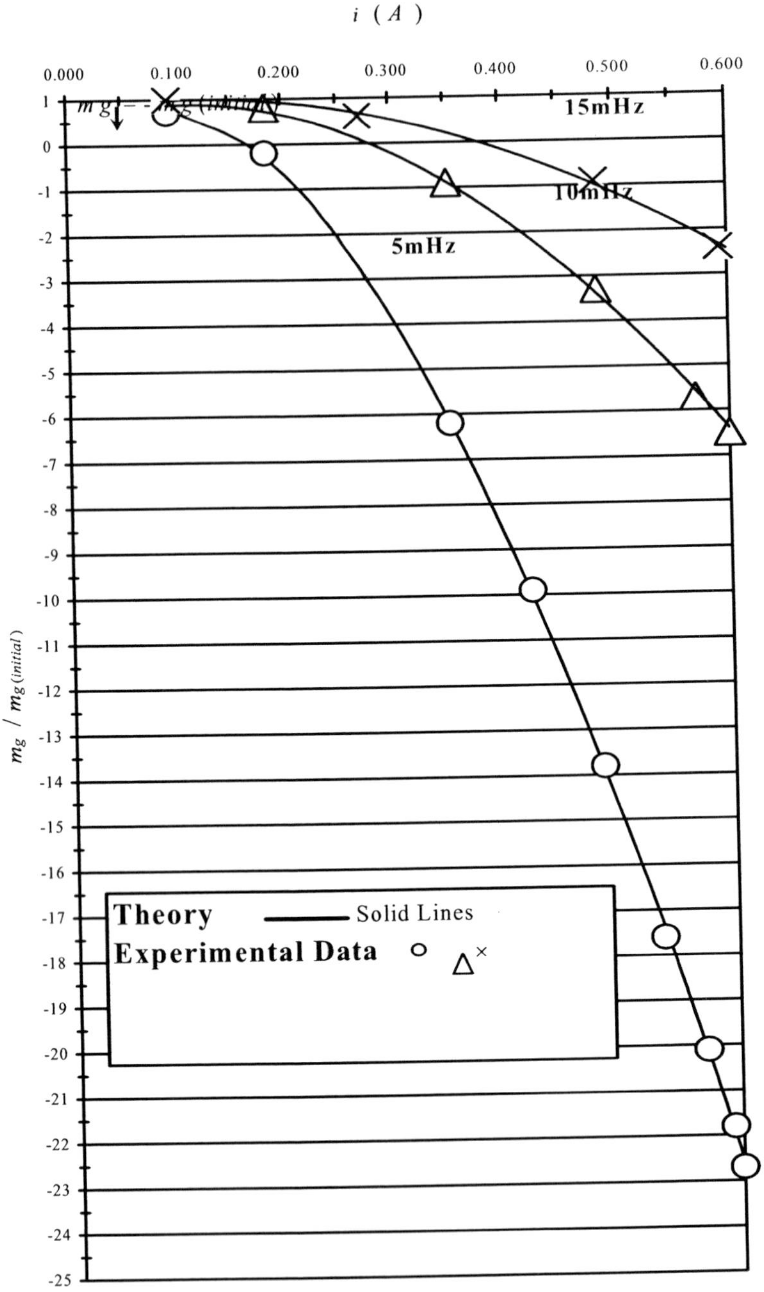

Fig. 2. Distribution of the correlation *mg/mg* (initial) as a function of *i*.

The strong variations in the gravitational mass of the mumetal wire correlated to the small values of electric current and applied voltages show that the observed phenomenon cannot be attributed to ion winds or other electrokinetic effects (Charge asymmetries, etc.). It tells us precisely about the theoretical correlation between gravitational and inertial mass deduced in this work.

Appendix B

It was noted earlier that when an alternating electric current passes through a conductor its gravitational mass is reduced in accordance with the following expression

$$m_g = \left\{1 - 2\left[\sqrt{1 + \left(i_0^4 \mu / 64\pi^3 c^2 \rho^2 S^4 f^3 \sigma\right)\sin^4 2\pi f t} - 1\right]\right\} m_i \qquad \text{(B1)}$$

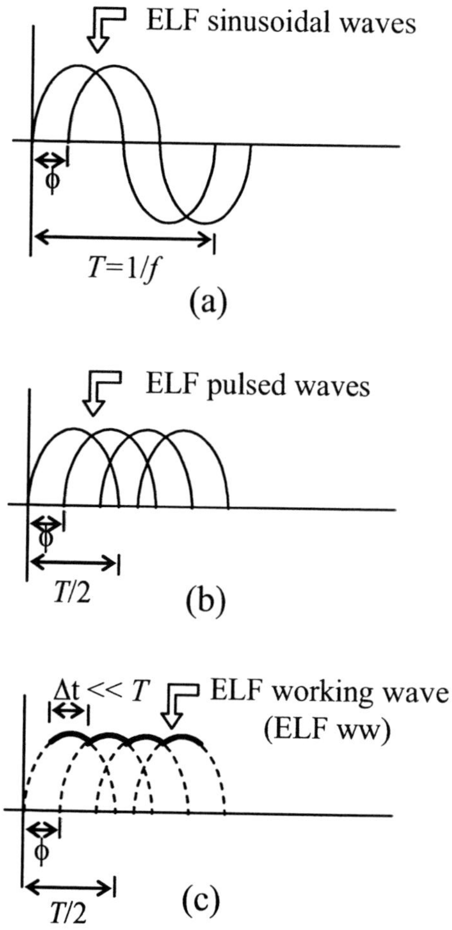

Fig. 3 - ELF working waves

In this equation i_0 refers to amplitude of the electric current; $\mu = \mu_r \mu_0$ is the magnetic permeability of the conductor; c is the speed of light; ρ is the density (kg/m^3) of the conductor; S is the area of the cross section(m^2) of the conductor; σ is the electric conductivity of the conductor (S/m); m_i is the *inertial mass* of the conductor and f the frequency of the electric current (Hz). We have shown that the *gravitational mass* of a *supermalloy* wire is strongly decreased when the electric current through the wire has extremely-low frequency. As we have seen, for $i_0 = 0.04A$ and $f \cong 1.7 \times 10^{-6} Hz$ at $t = 1/4f = 1.47 \times 10^5$ s $\cong 40.8h$ Eq.(59) gives $m_{g(sm)} \cong - m_{i(sm)}$.

The period of this wave is too long. In order to reduce the period of the wave we can reduce the diameter of the wire. For example, in the case of *supermalloy* or *mumetal* wire 0.005" diameter, the period will be strongly reduced down to ~100s. In addition, by digitizing the top of this ELF wave, as shown in Fig.3, we may produce a ELF digitized wave which obviously becomes much more adequate for practical use.

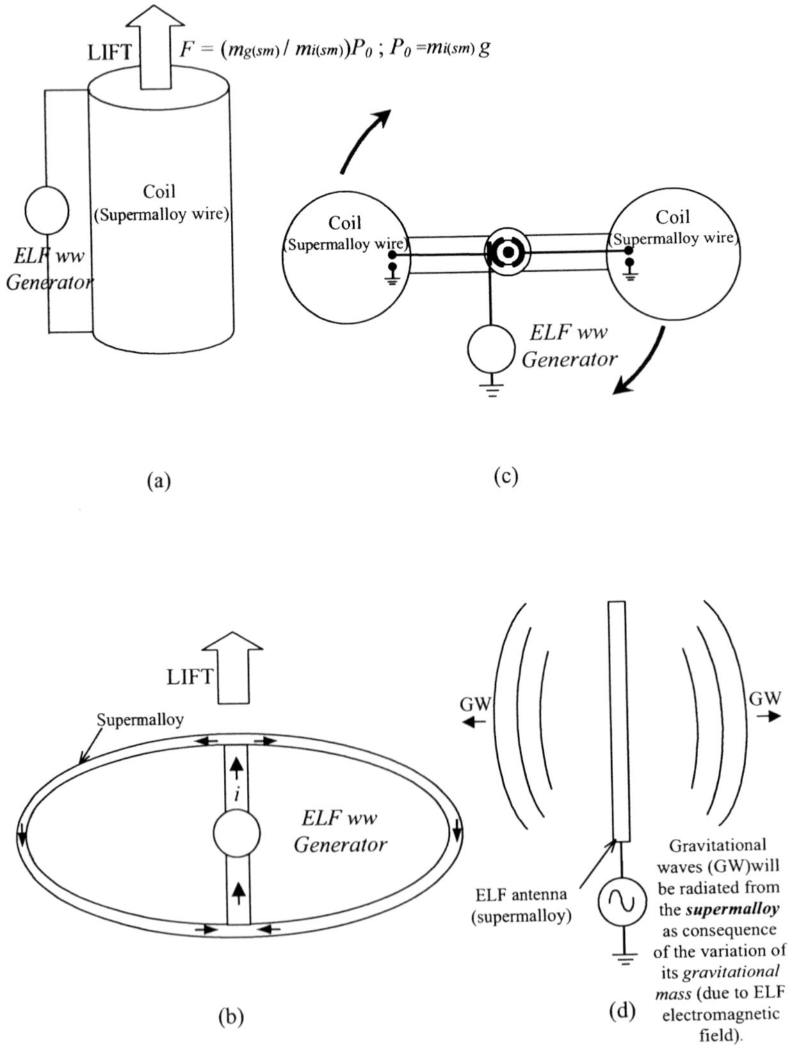

Fig. 4. Schematic diagram of system using gravity control: (a) and (b) The generation of Gravitational Lift Force. (c) The Gravitational Motor (d) The Gravitational Radiation Transmitter.

This possibility points to some interesting systems as shown in Fig. 4. Figures 4(a) and 4(b) show the *generation of Lift Force*. Figure 4(c) shows a new concept of motor : the *Gravitational Motor*, also based on the gravity control.

When the *gravitational field* of an object changes, the changes ripple outwards through spacetime. These ripples are called *gravitational radiation* or *gravitational waves*.

The existence of gravitational waves follows from the General Theory of Relativity. In Einstein's theory of gravity the gravitational waves propagate at the speed of light.

Just as electromagnetic waves (EM), gravitational waves (GW) too carry energy and momentum from their sources. Unlike EM waves, however, there is no dipole radiation in Einstein's theory of gravity. The dominant channel of emission is quadrupolar. But the existence of *negative* gravitational mass suggest the possibility of dipole gravitational radiation.

This fact is highly relevant because now we can build a gravitational wave transmitter to generate detectable levels of gravitational radiation. Gravitational waves are very suitable as a means of transmitting information because of their low interaction an therefore low scattering. In figure 4(d) we present the *Gravitational Radiation Transmitter*, a new concept of transmitter that arises from this new technology.

The phenomenon of gravitational mass decreasing in a conductor when an alternating electric current passes through it, such as described by (B1), is a macroscopic reflex of the gravitational mass decreasing of its atoms. Note that in (B1), the variables ρ, μ, σ are macroscopic characteristics of the *conductor* and consequently the *decreasing factor* (m_g/m_i) will be the same for all types of atoms inside the conductor. This phenomenon is general for any kinds of conductors including *plasmas*. Thus, if a plasma contains different types of ions then the Eq. (B1) can be used to express the gravitational mass of each one of these ions. As the decreasing factor is the same for all ions inside the plasma, the different gravitational masses of the ions arise only due to the difference among their inertial masses.

Consider a *Carbon Plasma* confined within a *toroidal chamber*. The plasma has density $\rho = 28Kg/m^3$ and electric conductivity $\sigma \approx 10^4 S/m$ at temperature $T = 16{,}000K$. An ELF electric current (ELF working wave) is induced through the Plasma. If this ELF working wave is a sequence of *tops* similar to region very close to the amplitude of the original wave, of ELF electric current, then we can assume $sin2\pi ft \cong 1$ in (B1), and consequently when this ELF current passes through the Carbon Plasma the *gravitational mass of the carbon ions*, m_{gC}, will be reduced and given by

$$m_{gC} = \left\{1 - 2\left[\sqrt{1 + 10^{-33}\left(i_0^4/S^4 f^3\right)} - 1\right]\right\} m_{iC}$$

Note that, the equation above shows that in practice it is possible to reduce strongly the gravitational masses of the Carbon ions, for example, if the diameter of the cross section of the plasma toroid is 5mm $\left(S = 1.96 \times 10^{-5} m^2\right)$, $f = 10mHZ$ and $i_0 \cong 116A$.

Usually a Carbon plasma is produced by evaporating a *thin graphite wire* within some hundred nanoseconds (exploding the wire by means of a strong electric current). To ensure a good energy input and a homogeneous plasma a preheating system is used.

Equation (4) shows that the velocities of the Carbon ions must increase as a consequence of the decreasing of its gravitational masses since the *momentum* is conserved. In practice the gravitational mass decreasing can easily reach 100 times. This means that the velocities of the ions can be increased a 100 times, which is equivalent to an increasing of 10,000 times in temperature (since $\Delta T \propto \Delta V^2$), i.e., it is equivalent to increase the initial temperature of the plasma from 16,000K up to 1.6×10^8 K.

In other words, in this case, the velocities of the Carbon ions inside the toroid should be the same as if the temperature of the plasma was to reach 1.6×10^8K.

As we know *nuclear reactions* can occur at temperatures greater than 10^7K. Consequently, if Hydrogen atoms are sprayed into this *Artificial Plasma* the well-known *CNO Cycle* may occur, i.e.,

$$
\begin{aligned}
{}_1H^1 + {}_6C^{12} &\rightarrow {}_7N^{13} + \gamma \\
{}_7N^{13} &\rightarrow {}_6C^{13} + e^+ + \nu \\
{}_1H^1 + {}_6C^{13} &\rightarrow {}_7N^{14} + \gamma \\
{}_1H^1 + {}_7N^{14} &\rightarrow {}_8O^{15} + \gamma \\
{}_8O^{15} &\rightarrow {}_7N^{15} + e^+ + \gamma \\
{}_1H^1 + {}_7N^{15} &\rightarrow {}_6C^{12} + {}_2He^4
\end{aligned}
$$

The total energy released in these reactions is, as we know, 24.7MeV. Note that the initial ${}_6C^{12}$ acts like a kind of catalyst in the process, since it appears again at the end of the process. On the other hand, it is easy to see that by controlling the amount of Hydrogen sprayed into the toroid it is possible to control the total energy produced in nuclear reactions.

The mass-13 isotope of Nitrogen $\left({}_7N^{13}\right)$ is unstable with half-life of 9.97 minutes. This means that there is a delay of 9.97 minutes to occur the beta decays reaction: ${}_7N^{13} \rightarrow {}_6C^{13} + e^+ + \nu$. However we can put a smaller amount of ${}_7N^{14}$ (stable) inside the toroidal chamber *before* the Carbon plasma is produced. When the plasma is produced and the ELF current applied, the decreasing factor $\left(m_g/m_i\right)$ for the ${}_7N^{14}$ ions will be the same of the Carbon ions and consequently the velocities of the ${}_7N^{14}$ ions will also be increased at the same proportion of the Carbon ions. Therefore the following reactions

$$
\begin{aligned}
{}_1H^1 + {}_7N^{14} &\rightarrow {}_8O^{15} + \gamma \\
{}_8O^{15} &\rightarrow {}_7N^{15} + e^+ + \gamma \\
{}_1H^1 + {}_7N^{15} &\rightarrow {}_6C^{12} + {}_2He^4
\end{aligned}
$$

of the CNO cycle can occur without delaying. Although we have illustrated the CNO cycle starting with Carbon, it is known that similar sequences are possible starting with ${}_7N^{14}$ or ${}_8O^{15}$.

If instead of Nitrogen we put Hydrogen inside the toroidal chamber before the plasma is produced, then the velocities of the ${}_1H^1$ ions will also be increased at the same proportion of the Carbon ions since the decreasing factor (m_g/m_i) for the ${}_1H^1$ ions will also be the same as the Carbon ions. Consequently, in this case reactions of the *Proton-Proton Cycle* can occur, i.e.,

$$
\begin{aligned}
&{}_1H^1 + {}_1H^1 \rightarrow {}_1H^2 + e^+ + \nu \\
&{}_1H^1 + {}_1H^2 \rightarrow {}_2He^3 + \nu \\
&{}_2He^3 + {}_2He^3 \rightarrow {}_2He^4 + {}_1H^1 + {}_1H^1
\end{aligned}
$$

The total energy released in these reactions is also 24.7MeV.

It is important to note that in this case, the reactions of the *Proton-Proton Cycle* and *CNO cycle* occur simultaneously inside the toroidal chamber.

On the other hand if, instead of ${}_1H^1$ we put ${}_1H^2$ (Deuterium) inside the toroidal chamber then the following reactions can occur

$$
\begin{aligned}
&{}_1H^2 + {}_1H^2 \rightarrow {}_2He^3 + {}_0n^1 + 3.3MeV \\
&{}_1H^2 + {}_1H^2 \rightarrow {}_1H^3 + {}_1H^1 + 4.0MeV \\
&{}_1H^3 + {}_1H^2 \rightarrow {}_2He^4 + {}_0n^1 + 17.6MeV
\end{aligned}
$$

Appendix C

In this appendix we will show that strong fluxes of ELF radiation upon electric/electronic circuits can suddenly increase the electric currents and consequently to damage these circuits.

Let us consider an electric current I through a conductor subjected to electromagnetic radiation with power density D and frequency f.

Under these circumstances the *gravitational mass* m_{ge} of the *electrons* of the conductor, according to Eq. (58), is given by

$$m_{ge} = \left\{1 - 2\left[\sqrt{1 + \left(\frac{\mu\sigma D}{4\pi f\rho c}\right)^2} - 1\right]\right\} m_e \qquad (C1)$$

where $m_e = 9.11 \times 10^{-31} kg$.

Note that m_{ge}, becomes less than the inertial mass, m_e. If the radiation upon the conductor has extremely-low frequency (ELF radiation) then m_{ge} can be strongly reduced. For example, if $f \approx 10^{-6} Hz$, $D \approx 10^5 W/m^2$ and the conductor is made of *copper* ($\mu \cong \mu_0$; $\sigma = 5.8\times10^7 S/m$ and $\rho = 8900 kg/m^3$) then

$$\left(\frac{\mu\sigma D}{4\pi f \rho c}\right) \approx 1$$

and consequently $m_{ge} \approx 0.1 m_e$.

According to Eq.(6) the force upon each *free electron* is given by

$$\vec{F}_e = \frac{m_{ge}}{\left(1 - V^2/c^2\right)^{3/2}}\frac{d\vec{V}}{dt} = e\vec{E} \qquad \text{(C2)}$$

where E is the applied electric field. Therefore the decreasing of m_{ge} produces an increase in the velocity V of the free electrons and consequently the *drift velocity* V_d is also increased. It is known that the density of electric current J through a conductor [21] is given by

$$\vec{J} = \Delta_e \vec{V}_d \qquad \text{(C3)}$$

where Δ_e is the density of the free electric charges (For cooper conductors $\Delta_e = 1.3\times10^{10} C/m^3$). Therefore increasing V_d produces an increase in the electric current I. Thus if m_{ge} is reduced 10 times $(m_{ge} \approx 0.1 m_e)$ the drift velocity V_d is increased 10 times as well as the electric current. This sudden increase in the electric currents of electric/ electronic circuits can cause damage.

In order for the ELF radiation to arrive at each electron, the flux density D must be greater than D_{min} given by

$$D_{min} = \frac{hf^2}{A_{electron}} \qquad \text{(C4)}$$

where $A_{electron}$ is the "area of cross section" of the electron. We know that the *leptons* should have length scale less than $10^{-19} m$ [22]. This means that an electron has a maximum, "radius" of $r_e \sim 10^{-19} m$. The plausible relation given by Brodsky and Drell [23] for the simplest composite theoretical model of the electrons, $|g-2| = r_e/\lambda\!\!\!^{-}_c$ or $|g - g_{DIRAC}| = r_e/\lambda\!\!\!^{-}_c$,where $\lambda\!\!\!^{-}_c = 3.9\times10^{-13} m$ and $|g-2| = 1.1\times10^{-10} m$ [24] gives an electron radius of

$$r_e \approx 10^{-22} m$$

Therefore assuming $A_{electron} \approx 10^{-45} m$ (C4) gives

$$D_{min} \approx 10^{12} f^2 \tag{C5}$$

Thus, for $f \approx 10^6 Hz$ we have $D_{min} \approx 1W/m^2$.

Since the *orbital electrons* moment of inertia is given by $I_i = \Sigma(m_i)_j r_j^2$, where m_i refers to *inertial mass* and not to gravitational mass, then the *momentum* $L = I_i\omega$ of the conductor *orbital electrons* are not affected by the ELF radiation. Consequently this radiation just affects the conductor *free electron* velocities.

Appendix D

Here we will show that the possible existence of ELF radiation into solar radiation can explain the anomalous acceleration which has been observed on the *Pioneer 10 and 11 spacecrafts* in the solar system [25] and also the anomalous behavior of mechanical systems during *solar eclipses* observed by Allais [26] with paraconical pendula and Saxl and Allen [27] with a torsion pendulum and measurements with gravimeters.

Equation (58) shows that the presence of ELF radiation (frequency ranging between $0.1\mu Hz$ down to $0.1nHz$) into solar radiation can slightly *reduce* the gravitational masses of any body in the solar system. The gravitational mass of these bodies become *less than* their inertial masses, m_i, as expressed by

$$m_g = \left\{1 - 2\left[\sqrt{1+\left(\frac{\mu\sigma D}{4\pi f\rho c}\right)^2} - 1\right]\right\}m \tag{D1}$$

The total energy of the spacecraft (Hamiltonian) according to (20), is $H = \sqrt{p^2c^2 + m_g^2c^2}$. Therefore the decreasing of m_g reduces the total energy of the spacecraft, and consequently its acceleration. This explains the fact that the *Pioneer 10* and *11* spacecrafts, launched by NASA, in the early 1970s, are receding from the sun slightly more slowly than they should be.

Similarly, the ELF solar radiation slightly reduces the gravitational mass of the Earth, $M_{g\oplus}$, and consequently it becomes smaller than its inertial mass $M_{i\oplus}$.

From Electrodynamics we know that radiation with frequency f propagating within a material with electromagnetic characteristics ε, μ and σ has the amplitudes of its waves attenuated by $e^{-1}=0.37$ (37%) when it penetrates a distance z, given by

$$z = \frac{1}{\omega\sqrt{\frac{1}{2}\varepsilon\mu\left(\sqrt{1+(\sigma/\omega\varepsilon)^2}-1\right)}} \tag{D2}$$

The radiation is mostly absorbed if it penetrates a distance $\delta \cong 5z$.

Based on this equation we can easily conclude that the ELF solar radiation is mostly absorbed by the moon, therefore during the eclipses, when the moon passes in front of the sun, the ELF solar radiation ceases to fall upon the Earth and, according to (D1), the gravitational mass of the Earth *increases* ($M_{g\oplus}$ becomes equal to $M_{i\oplus}$) slightly increasing the gravity $g = GM_{g\oplus}/r^2$. Similarly, the gravitational mass of the pendulum, m_g, also *increases* slightly during the eclipse. Since the period, T, of the pendulum is given by

$$T = 2\pi\sqrt{\left(\frac{m_i}{m_g}\right)\frac{l}{g}} \tag{D3}$$

one can conclude that during the eclipses the pendulum's periods are slightly *decreased.* This means that their motion becomes *faster during the eclipses*, such as has been observed in the experiments of Allais, Saxl and Allen.

Appendix E

Equation (70) shows that the gravitational interaction can be repulsive, besides attractive. Therefore, as with electromagnetic interaction, the gravitational interaction must be produced by the exchange of "virtual" *quanta of* spin 1 and mass null, i.e., the *gravitational* "virtual" *quanta* (gravitons) must have spin 1 and not 2.

It is known that the gravitational interaction is *instantaneously* communicated to all the particles of the Universe. This means that the velocity of the *gravitational* "virtual" *quanta* must be *infinite.*

Consider a Mumetal ELF antenna as showed in Fig.5. The ELF electric current through it is $i_e = i_0 sin\omega t = i_0 sin2\pi ft$. According to (59c) the gravitational mass of the antenna is given by

$$m_g = \left\{1 - 2\left[\sqrt{1 + \left(i_e^4 \mu / 64\pi^3 c^2 \rho^2 S^4 f^3 \sigma\right)} - 1\right]\right\} m_i \qquad \text{(E1)}$$

where ρ, μ, σ and S are respectively the density, the magnetic permeability, the electric conductivity and the area of the antenna cross section.

It is easy to see that the ELF electric current yields a variation in the gravitational mass of the antenna, which is detected instantaneously by all particles of the Universe, i.e., the gravitational "virtual" *quanta* emitted from the antenna will instantaneously reach all particles.

When a particle absorbs photons, the *momentum* of each photon is transferred to particle and, in accordance with (41), the *gravitational mass* of the particle is altered. Similarly to the photons the gravitational "virtual" *quanta* have mass null and *momentum*. Therefore the gravitational masses of the particles are also altered by the absorption of gravitational "virtual" *quanta*.

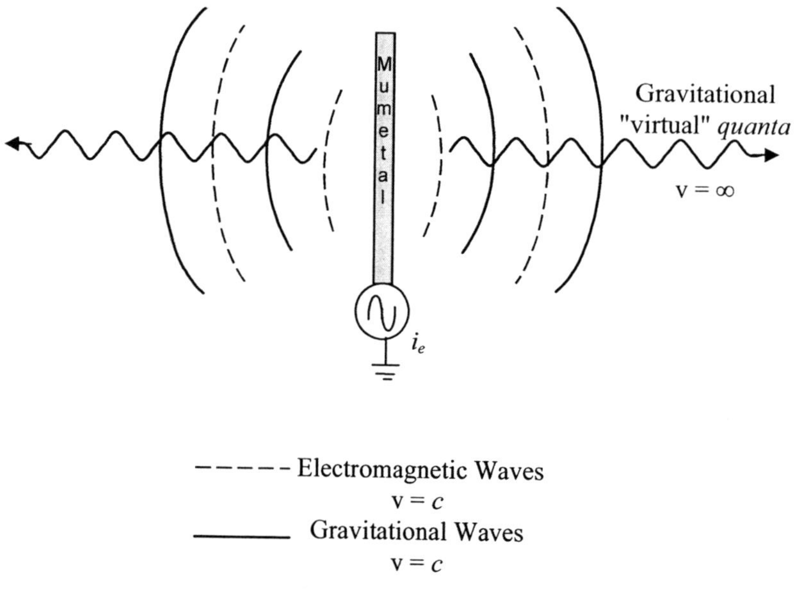

Fig. 5. Mumetal ELF antenna

If the gravitational "virtual" *quanta* are emitted by an antenna (like a Mumetal ELF antenna) and absorbed by a similar antenna, tunned to the same frequency f, the changes on the gravitational mass of the receiving antenna, in accordance with the *principle of resonance*, will be similar to changes occurred on the transmitting antenna, and consequently the induced current through the receiving antenna has the same frequency f and, in agreement with (E1), must be similar to electric current through the transmitting antenna. The Fig. 6 shows the emission and detection of gravitational "virtual" *quanta* by two Mumetal ELF antennas.

Note that the changes of gravitational mass of the antenna also produce the so-called *gravitational waves* which are *ripples in the geometry of the spacetime*. This is produced by the changes on the *gravitational field* of the antenna. When the *gravitational field* changes, the changes ripple outwards through space and take a *finite time* to reach other objects. In Einstein's theory of gravity these ripples (gravitational waves) propagate at the *speed of light* (c).

Therefore the velocity of the gravitational waves is much less than the velocity (∞) of the gravitational "virtual" *quanta* (gravitons). There is another fundamental difference between the gravitational waves and gravitons: the gravitational waves are *real* unlike the gravitons which are *virtual* .

Note that a Mumetal ELF antenna emits gravitons and gravitational waves simultaneously. Thus it is not only a gravitational antenna: it is a *macroscopic quantum gravitational antenna* because can also emit and detect gravitational "virtual" *quanta*, which can to transmit information *instantaneously* from any distance in the Universe *without* scattering.

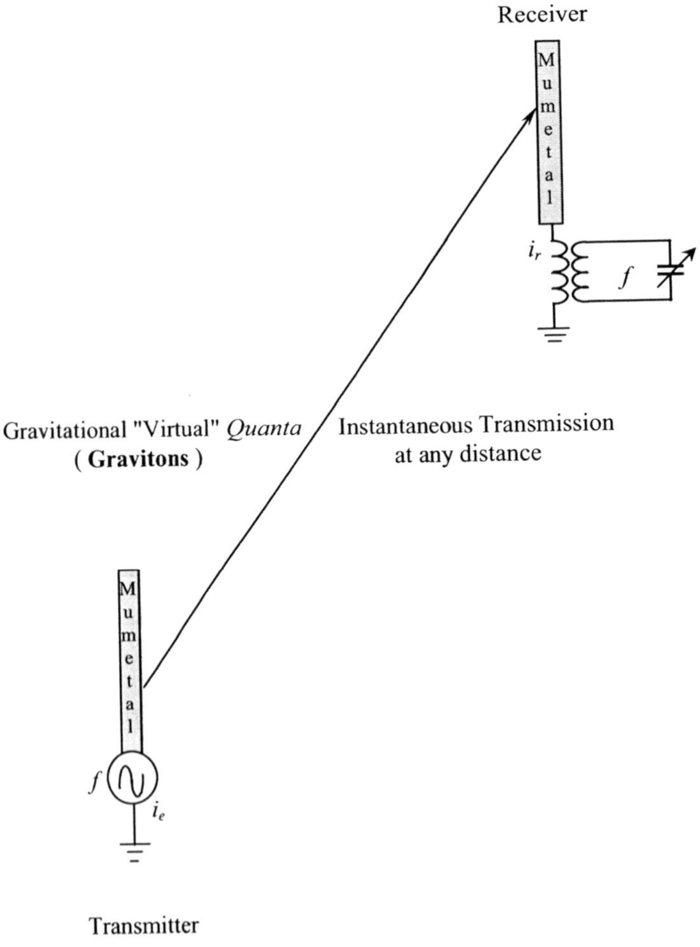

Fig. 6. Transmitter and Receiver of gravitational "Virtual" *Quanta*

Unlike the electromagnetic waves the gravitational waves have low interaction and consequently low scattering. Therefore gravitational waves are suitable as a means of transmitting information. However, when the distance between transmitter and receiver is too large, for example of the order of magnitude of several light-years, the transmission of information by means of gravitational waves becomes impracticable due to the long time necessary to receive the information. The velocity of the gravitational waves is equal to the speed of light (c) therefore the delay would be in the order of several years.

The velocity of the gravitational "virtual " *quanta* is infinite thus there is no delay during the transmissions. The scattering of this radiation is null. Therefore this gravitational "virtual" radiation or *gravitational "virtual" waves* are very suitable as a means of transmitting information at any distances including astronomical distances.

In order to check these theoretical predictions we propose the following experiment: A transmitter and a receiver both with Mumetal antennas will be placed in two very distant places, like Mars and Earth (the distance is $\sim 7.9 \mathrm{X} 10^{10}$m). Electromagnetic waves or gravitational *real* waves emitted from Mars will need ~ 4.4 minutes to arrive at Earth. There is no delay in the case of gravitational "virtual" waves due to their infinite velocity. Therefore simply checking that there is no delay during the transmission by using Mumetal antennas we can check the existence of the gravitons.

Since the gravitational masses of the antennas vary during the transmissions then another way to check the existence of the gravitons is to measure the weight of the receiving and transmitting antennas during the transmissions. In this case is not necessary to put the antennas in very distant places.

It is easy to see that the information transportation with infinite velocity by means of gravitons promises to be quite useful for the Internet (*Quantum* Internet) and also for the development of *Quantum* Teleportation Systems.

By operating with infinite velocity and not with the speed of light these systems will solve in the future the problem of the cosmic transportation of long range, since it is impracticable for spacecrafts – even with velocities very close to light speed – to reach places whose distances are greater than 100 light-years.

References

[1] Isham, C. J. (1975) *Quantum gravity, an Oxford Symposium,* OUP.

[2] Isham, C.J., (1997) "Structural Problems Facing Quantum Gravity Theory", in M, Francaviglia, G, Longhi, L, Lusanna, and E, Sorace, eds., *Proceedings of the 14th International Conference on General Relativity and Gravitation*, 167-209, (World Scientific, Singapore, 1997).

[3] Landau, L. and Lifchitz, E. (1969) *Theorie du Champ*, Ed.MIR, Moscow, Portuguese version (1974) Ed. Hemus, S.Paulo, p.35.

[4] Landau, L. and Lifchitz, E. (1969) *Mecanique*, Ed.MIR, Moscow, 15.

[5] Landau, L. and Lifchitz, E.[3], p.35.

[6] Landau, L. and Lifchitz, E.[3], p.36.

[7] Beiser, A. (1967) *Concepts of Modern Physics*, McGraw-Hill, Portuguese version (1969) Ed. Poligno, S.Paulo, p.151.

[8] Schiff, L.I. (1981) *Quantum Mechanics*, McGraw-Hill, p.54.

[9] Landau, L. and Lifchitz, E.[3], p.324.
[10] Landau, L. and Lifchitz, E.[3], p.357.
[11] Landau, L.and Lifchitz,E.[3],p.p.358-359.
[12] Landau, L. and Lifchitz, E.[3], p.64.
[13] Luminet, J. at al., (2003) *Nature* **425**, 593–595.
[14] Ellis, G.F.R (2003) *Nature* **425**, 566–567.
[15] Landau, L. and Lifchitz, E.[3], p.440-441.
[16] Landau, L. and Lifchitz, E.[3], p.333.
[17] Landau, L. and Lifchitz, E.[3], p.331.
[18] Landau, L. and Lifchitz, E.[3], p.327.
[19] Hawking, S.W (1971) *MNRAS.*, **152**, 75.
[20] Carr, B.J.(1976) *Astrophys.J.,* **206**,10.
[21] Quevedo, C.P.(1978) *Eletromagnetismo*, McGraw-Hill, p. 102.
[22] Fritzsch, H. (1984) *Quarks-Urstoff unserer Welt,* R. Piper GmbH&Co. KG,München, Portuguese version (1990), Ed. Presença, Lisboa, p.215.
[23] Brodsky, S . J., and Drell, S . D., (1980) Anomalous Magnetic Moment and Limits on Fermion Substructure, *Phys. Rev. D*, **22**, 2236.
[24] Dehmelt, H.G.,(1989) *Experiments with an isolated subatomic particle at rest*, Nobel Lecture, p.590.
[25] Anderson, J.D, Lang, P.A, Lau, E. L, Liu, A.S, Nieto, N.M and Turyshev, S.G. (2002) Study of the anomalous acceleration of Pioneer 10 and 11, *Phys. Rev. D* **65**, 082004. gr-qc/0104064.
[26] Allais, M.F.C (1957) Mouvement du pendule paraconique et éclipse totale de Soleil 30 jui 1954, *C.R. Acad. Sci*, **245**,2001.
[27] Saxl, E.J. and Allen, M.(1971) 1970Solar Eclipse as "Seen" by a Torsion Pendulum, *Phys.rev. D3*, **823**.

In: Trends in Quantum Gravity Research
Editor: David C. Moore, pp. 197-218

ISBN 1-59454-670-3

Chapter 5

SEMICLASSICAL REDUCTIVE QUANTUM GRAVITY

Vladimir S. Mashkevich*
Physics Department, Queens College, The City University of New York
65-30 Kissena Boulevard Flushing, New York 11367-1519

Abstract

It is shown that the inclusion of quantum jumps, i.e., state vector reduction, in the semiclassical gravity construction opens a new avenue for the solution, on the one hand, of the serious difficulties of the construction per se and, on the other hand, of the challenging puzzles of dark energy and dark matter. In the problem of quantum gravity, the simplest and most natural construction is that of semiclassical gravity. In the latter, the energy-momentum tensor entering into the Einstein equation is represented by the expectation value of the corresponding operator. In a conventional treatment, there exists no satisfactory generalization of normal ordering to curved spacetime. The renormalization of the energy-momentum tensor is based on a set of axioms; one of the latter is that the tensor must be four-divergence free. The results of the renormalization suffer from serious difficulties: an ambiguity and a nonlocal dependence on metric. In addition, the conventional treatment denounces the concept of particles and the Hamiltonian. It is commonly accepted that things look even worse when state reduction is involved in dynamics. In fact, the opposite situation occurs. The reduction, being nonlocal and instantaneous, implies a universal time and, as a consequence, the structure of spacetime as the direct product of cosmological time and space. This allows for introducing normal ordering, particles, and the Hamiltonian. The renormalized energy-momentum tensor is unique and involves at most second derivatives of metric. On that basis, semiclassical reductive quantum gravity is constructed—a theory in which metric is treated classically whereas a quantum treatment of matter includes state vector reduction. The theory is assumed to be fundamental. In the theory, the semiclassical Einstein equation is violated due to the following. First, the energy-momentum tensor is not divergence free. Second, the six space components of the Einstein tensor involve the second time derivative of metric, but the other four components involve only the first time derivative. Therefore the latter components must be continuous. The energy-momentum tensor should be complemented by a pseudo energy-momentum tensor with four degrees of freedom which would compensate for the breakdown both of the divergence freedom condition and of the continuity of the

*E-mail address: Vladimir_Mashkevich@qc.edu

four components of the energy-momentum tensor. The compensatory tensor is, by definition, the energy-momentum tensor of pseudomatter. The latter is represented by a pressural dust, i.e., a perfect fluid with a constant pressure, which has four degrees of freedom. The pressural dust comprises both dark energy (cosmological constant) and dark matter. So the presence of dark energy and dark matter in the real world provides an observable evidence of characteristically quantum gravitational effects. That is a challenge to a conventional opinion that there exists no such recognized evidence. The reductive semiclassical Einstein equation is composed of ten equations for six space components of metric and four pseudomatter variables (density and four-velocity). The elimination of the latter variables results in the metric equation. Dark matter is represented by a pseudodust, which implies the fruitlessness of efforts to represent dark matter by any kind of ordinary matter.

Introduction

Quantum Gravity in a broad sense is a theory that would combine, or unify General Relativity and Quantum Theory. To construct such a theory is today at the core of fundamental physics. The construction is faced with two problems: (i) The description and dynamics of spacetime structure, especially metric, under the quantum treatment of matter; (ii) The description and dynamics of quantum matter in the presence of gravity. In a narrow sense, the term quantum gravity refers to the problem (i).

The present study is dedicated mainly to the problem (i). The problem (ii) is considered only so far as it concerns the problem (i).

In general relativity, metric is a dynamical variable, so it seems quite natural to quantize it along with matter variables. A realization of such a program would result in totally quantum gravity [K], [BT]. But totally quantum gravity is confronted with a problem of principle—that of the interpretation of state vector. In particular, it is absolutely unclear what meaning may be assigned to state vector reduction, i.e., quantum jumps. A standard interpretation of a state vector pertaining to a quantum system is formulated in terms of the results of experiments which might be performed on the system. In totally quantum gravity where all dynamical variables are quantized, the universe represents a perpetually closed quantum system, on which no experiments might be performed.

In the problem (i), the simplest and most natural construction is semiclassical quantum gravity, in which metric is treated classically [Mø], [Ro], [St], [CH2], [V]. From the above it might be assumed that semiclassical gravity may be a fundamental theory. Classical metric removes the problem of the interpretation of the state vector of the universe.

The basic equation of general relativity is the Einstein equation: $G_{\mu\nu} = 8\pi\varkappa T_{\mu\nu}$, $\mu,\nu = 0,1,2,3$, where G is the Einstein tensor, $\varkappa$ is the gravitational constant ($c = 1$), and T is the energy-momentum tensor. In semiclassical gravity, the latter is represented by an effective quantity—the expectation value of the corresponding operator: $T_{\mu\nu} := (\Psi, \widehat{T}_{\mu\nu}\Psi)$ where Ψ is a state vector of quantum matter.

In a conventional treatment of semiclassical gravity [Wa], there exists no satisfactory generalization of the vacuum state and normal ordering to curved spacetime, which complicates significantly the problem of the renormalization of the effective tensor $T_{\mu\nu}$. The renormalization is based on a set of axioms [Wa]; one of those is that the tensor $T_{\mu\nu}$ must be divergence free: $T^{\mu\nu}{}_{;\nu} = 0$. The results of the renormalization suffer from serious difficulties [Wa], [V]: an ambiguity; a nonlocal dependence on metric; a failure on the chronology horizon. In addition, the conventional treatment denounces the concept of particles and the Hamiltonian [BD], [Fu].

It is commonly accepted that things look even worse when state vector reduction is involved in dynamics [CH1]: $T_{\mu\nu}$ and, as a consequence, $G_{\mu\nu}$ become discontinuous. In fact, the opposite situation occurs. The reduction, being nonlocal and instantaneous, implies a universal time and, as a consequence, the structure of spacetime as the direct product of cosmological time and space with metric $ds^2 = dt^2 + g_{ij}dx^i dx^j$ [M2]. That allows for introduction of the vacuum state, normal ordering, particles, and the Hamiltonian. The renormalized tensor $T_{\mu\nu}$ is unique and involves at most second derivatives of metric.

On that basis, semiclassical reductive quantum gravity is constructed—a theory in which metric is treated classically whereas a quantum treatment of matter includes state vector reduction, i.e., quantum jumps. The theory is assumed to be fundamental.

In semiclassical reductive quantum gravity, the semiclassical Einstein equation is violated due to the following. First, the tensor $T_{\mu\nu}$ is not divergence free, $T^{\mu\nu}{}_{;\nu} \neq 0$. Second, the six space components G_{ij} $(i, j = 1,2,3)$ of the Einstein tensor involve the second time derivative of metric, $\ddot{g}_{ij}$, but the other four components $G_{0\mu}$ involve only the first time derivative, $\dot{g}_{ij}$. Under state vector reduction, the discontinuity of $\ddot{g}_{ij}$ is feasible, but g_{ij} and $\dot{g}_{ij}$ must be continuous. Therefore $G_{0\mu}$ must be continuous. The tensor $T_{\mu\nu}$ should be complemented by a pseudo energy-momentum tensor $\Theta_{\mu\nu}$ with four degrees of freedom which would compensate for both the inequality $T^{\mu\nu}{}_{;\nu} \neq 0$ and the discontinuity of $T_{0\mu}$. The compensatory tensor $\Theta_{\mu\nu}$ is, by definition, the energy-momentum tensor of pseudomatter. The latter is represented by a perfect fluid with a constant pressure, which has just four degrees of freedom. We call such a fluid a pressural dust. Thus $\Theta_{\mu\nu} = (\sigma + \pi)\upsilon_\mu\upsilon_\nu - \pi g_{\mu\nu}$ where σ is the density, $\pi = \text{const}$ is the pressure, and υ_μ is the four-velocity. Now we put $\pi = -\Lambda/8\pi\varkappa$, $\varepsilon := \sigma + \pi$, so that the pressural dust comprises both dark energy (the cosmological constant Λ) and dark matter represented by an incoherent dust, $\varepsilon\upsilon_\mu\upsilon_\nu$. It follows that the presence of dark energy and dark matter in the real world provides an observable evidence of characteristically quantum gravitational effects. That is a challenge to a conventional opinion that "...to date, there is no recognized experimental evidence of characteristically *quantum* gravitational effects" [CH1].

The reductive semiclassical Einstein equation is of the form $G_{\mu\nu} - \Lambda g_{\mu\nu} = 8\pi\varkappa(T_{\mu\nu} + \varepsilon\upsilon_\mu\upsilon_\nu)$. It is composed of ten equations for the six components of metric, g_{ij}, and the four dark matter variables, ε and υ_μ. The elimination of those variables results in the metric equation $B_{00}B_{ij} - B_{0i}B_{0j} = 0$ where $B_{\mu\nu} := G_{\mu\nu} - \Lambda g_{\mu\nu} - 8\pi\varkappa T_{\mu\nu}$. The metric equation represents six equations for the six g_{ij}.

In the absence of dark matter, the metric equation reduces to the semiclassical Einstein equation with the cosmological constant. The latter equation is generally violated, so that dark matter is indispensable.

Generally, the conditions $\upsilon_0^2 \geq 1$ and $\upsilon_i^2 \geq 0$ may be violated, so that dark matter is represented by a pseudodust rather than by an ordinary dust. This implies the fruitlessness of efforts to represent dark matter by any kind of ordinary matter whatever.

Let us turn to the general problem of (the interpretation of) quantum reduction. The latter gives rise to a jump of the classical quantities $\ddot{g}_{ij}$. It is these classical jumps that provide an operationalistic meaning to quantum reduction.

Furthermore, a jump of $\ddot{g}_{ij}$ determines the exact time at which a state vector reduction happens. (Quantum mechanics does not determine the time.)

1 Quantum Gravity as a Unification of General Relativity and Quantum Theory

In this section, we cite some information on General Relativity and Quantum Theory which is used in what follows.

1.1 The Problem of Unification

Quantum Gravity in a broad sense is a theory that should combine, or unify General Relativity and Quantum Field Theory—the two theories which form the basis of contemporary physics. The construction of such a theory sought after by both relativists and elementary particle physicists is today at the core of fundamental physics. The construction is faced with two problems:

(i) The establishment of spacetime structure, especially metric, and its dynamics;

(ii) The description and dynamics of quantum fields in the presence of gravity.

In a narrow sense, the term quantum gravity refers to the problem (i). It is this problem that the present study is mainly devoted to. The problem (ii) is considered only so far as it has to do with the problem (i).

1.2 General Relativity

The classical theory of gravitation is General Relativity. In the latter, both spacetime structure, including metric, i.e., (M^4, g), and matter are treated classically. The basic equation of general relativity is the Einstein equation [S]

$$G_{\mu\nu} = 8\pi\varkappa T_{\mu\nu}, \quad \mu,\nu = 0,1,2,3 \tag{1}$$

where $G_{\mu\nu}$ is the Einstein tensor, $T_{\mu\nu}$ is the energy-momentum tensor of matter, and $\varkappa$ is the gravitational constant $(c = 1)$.

In view of the identity

$$G^{\mu\nu}{}_{;\nu} \equiv 0 \tag{2}$$

the energy-momentum tensor should be divergence free:

$$T^{\mu\nu}{}_{;\nu} = 0 \tag{3}$$

which provides the equation of motion for matter.

The simplest description of matter is that by means of a perfect fluid, whose energy-momentum tensor is

$$T_{\mu\nu} = (\rho + p)u_\mu u_\nu - p g_{\mu\nu} \tag{4}$$

where ρ is the density, p is the pressure, and u_μ is the four-velocity. The special case $p = 0$ is called (incoherent) dust, or incoherent matter. The equation of motion (1.2.3) for the perfect fluid reads

$$w(u^\nu{}_{;\nu}u^\mu + u^\mu{}_{;\nu}u^\nu) + w_{,\nu}u^\nu u^\mu - p^{;\mu} = 0 \tag{5}$$

where

$$w := \rho + p \tag{6}$$

In the case of

$$p = \text{const} \tag{7}$$

the equation of motion (1.2.5) contains p and ρ only in the combination (1.2.6). We will call a perfect fluid with $p = \text{const} \neq 0$ a pressural dust.

1.3 Quantum Theory

The basic concepts of Quantum Theory are these: Operator $\hat{O}$, in particular, a field operator $\hat{\phi}$ defined on spacetime; state vector Ψ; expectation value, or average value $(\Psi, \hat{O}\Psi)$; observable represented by a selfadjoint operator.

In the Schrödinger picture, dynamics, i.e., time evolution is determined by the Schrödinger equation

$$\frac{d\Psi}{dt} = -\mathrm{i}\hat{H}\Psi \tag{8}$$

$(\hbar = 1)$ where $\hat{H}$ is the Hamiltonian.

The interpretation of the state vector depends on measurement theory.

2 Difficulties of General Relativity and Quantum Theory

In this section, we touch upon and discuss some difficulties of General Relativity and Quantum Theory—ingredients of Quantum Gravity—which bear on the subject of our study.

2.1 The Existence of Physically Nonequivalent Solutions to the Cauchy Problem

It is commonly known that there are four degrees of freedom in the solution to the Cauchy problem in general relativity [W1], [LL]. On the other hand, a diffeomorphism $F : M^4 \to M^4$ involves exactly four degrees of freedom too. For a Ricci flat spacetime, the different solutions are diffeomorphic, i.e., physically equivalent [SW], [FM]. To fix a solution, four complementary conditions should be introduced [W1].

The situation changes dramatically when spacetime is not Ricci flat. What follows in this subsection is a concise refined account of the relevant results of [M1], [M2].

Let the complementary conditions be

$$g_{0i}(x) = 0, \quad i = 1,2,3 \tag{9}$$

$$R(x) = f(x) \tag{10}$$

where $x = (x^\mu)$ are (local) coordinates, $g_{\mu\nu}(x)$ is metric tensor, $R(x)$ is the scalar curvature, and $f(x)$ is a function on M^4. The equations

$$\begin{aligned} &R^i_j - \tfrac{1}{2}Rg^i_j = 8\pi\varkappa T^i_j, \quad i,j = 1,2,3 \\ &R = f \end{aligned} \tag{11}$$

(R^{μ}_{ν} is the Ricci tensor) form a system of seven equations for the seven metric components g_{ij}, g_{00}. The four equations

$$R^{\mu}_{0} - \frac{1}{2}Rg^{\mu}_{0} = 8\pi\varkappa T^{\mu}_{0}, \quad \mu = 0,1,2,3 \tag{12}$$

are constraints on initial conditions.

The system (2.1.3) may be rewritten in the form of

$$\begin{array}{l} R^{i}_{j} = 8\pi\varkappa T^{i}_{j}, \quad 1 \leq i < j \leq 3 \quad \text{(3 equations)} \\ R^{1}_{1} - R^{2}_{2} = 8\pi\varkappa(T^{1}_{1} - T^{2}_{2}) \\ R^{2}_{2} - R^{3}_{3} = 8\pi\varkappa(T^{2}_{2} - T^{3}_{3}) \\ R = f \\ 8\pi\varkappa T^{\mu}_{\mu} + f = 0 \end{array} \tag{13}$$

The highest time derivatives involved in R^{i}_{j}, $(R^{1}_{1} - R^{2}_{2})$, $(R^{2}_{2} - R^{3}_{3})$, and R are $\ddot{g}_{ij}$ and $\dot{g}_{00}$.

Let matter be represented by a spin 1 field A_{μ} with the energy-momentum tensor [BD]

$$T_{\mu\nu} = \frac{1}{\zeta}\left\{A_{\mu}\left(A^{\rho}{}_{;\rho}\right)_{,\nu} + A_{\nu}\left(A^{\rho}{}_{;\rho}\right)_{,\mu} - g_{\mu\nu}\left[A^{\sigma}\left(A^{\rho}{}_{;\rho}\right)_{,\sigma} + \frac{1}{2}\left(A^{\rho}{}_{;\rho}\right)^{2}\right]\right\} \tag{14}$$

where ζ is a parameter determining the choice of gauge. We find

$$T^{i}_{j} = -\frac{1}{2\zeta}g^{i}_{j}(A^{0})^{2}g^{00}\ddot{g}_{00} + \widetilde{T}^{i}_{j} \tag{15}$$

$$T^{\mu}_{0} = \frac{1}{2\zeta}A^{\mu}A^{0}g^{00}\ddot{g}_{00} + \widetilde{T}^{\mu}_{0} \tag{16}$$

where the terms involving $\ddot{g}_{00}$ are singled out. Turning back to the system (2.1.5), we see that the quantities

$$T^{i}_{j} = \widetilde{T}^{i}_{j} \quad \text{for } j \neq i \tag{17}$$

and

$$T^{1}_{1} - T^{2}_{2} = \widetilde{T}^{1}_{1} - \widetilde{T}^{2}_{2}, \quad T^{2}_{2} - T^{3}_{3} = \widetilde{T}^{2}_{2} - \widetilde{T}^{3}_{3} \tag{18}$$

do not involve $\ddot{g}_{00}$, but the quantity

$$T^{\mu}_{\mu} = -\frac{1}{\zeta}(A^{0})^{2}g_{00}\ddot{g}_{00} + \widetilde{T}^{\mu}_{\mu} \tag{19}$$

does involve $\ddot{g}_{00}$.

Thus there are seven quasilinear equations (2.1.5) for (g_{ij}, g_{00}), the highest time derivatives being $(\ddot{g}_{ij}, \ddot{g}_{00})$.

Now let g and $\bar{g}$ be solutions to the system (2.1.5) with the functions f and $\bar{f}$, respectively. If the ranges of those functions are different,

$$\text{ran}\,\bar{f} \neq \text{ran}\,f \tag{20}$$

then g and $\bar{g}$ are nondiffeomorphic. Indeed,

$$\bar{g} = F^{*}g \tag{21}$$

implies

$$\bar{R} = F^* R \tag{22}$$

so that

$$\operatorname{ran} \bar{R} = \operatorname{ran} R \tag{23}$$

which does not hold.

Thus there are solutions to the Cauchy problem which are physically non-equivalent. It is the degree of freedom described by the function f that brings about the nonequivalence.

2.2 The Problem of Dark Energy and Dark Matter

General relativity and especially cosmology are confronted with the problems of dark energy (or cosmological constant) and dark matter [D], [Ri], [St], [R], [Fr], [C], [MR]. To wit, the Einstein equation should be extended:

$$G_{\mu\nu} - \Lambda g_{\mu\nu} = 8\pi\varkappa(T_{\mu\nu} + T_{\mu\nu}^{\text{dark matter}}) \tag{24}$$

where Λ is the cosmological constant. In terms of a perfect fluid, the treatment of the problem is this:

$$\text{for dark energy} \quad p = -\rho\,,\; p = -\frac{\Lambda}{8\pi\varkappa} \tag{25}$$

$$\text{for dark matter} \quad p \approx 0,\; |u_i u^i| \ll 1 \text{ (cold dark matter)} \tag{26}$$

The existence of dark energy and the nature of dark matter are challenging puzzles.

2.3 The Measurement Problem

The measurement problem in quantum theory per se is beyond the scope of our study; but some issues of the problem will be touched upon in what follows. So we restrict ourselves here to quoting d'Espagnat: "The problem of measurement in quantum mechanics is considered as nonexistent or trivial by an impressive body of theoretical physicists and as presenting almost insurmountable difficulties by a somewhat lesser but steadily growing number of their colleagues" [E1]; "...measurement constitutes a riddle, and a great many theories were put forward as attempts to solve this riddle. The number is so considerable indeed that to try and review them all would be an almost impossible undertaking" [E2].

2.4 Indeterminacy of the Time of State Vector Reduction

In most formulations of quantum mechanics, there appears the following problem: At which precise time does a state vector reduction happen? [Rv]. The answer to this question is that quantum mechanics without a dynamical theory of reduction does not determine the time of the latter. We quote von Neumann [N]: "...we must always divide the world into two parts, the one being the observed system, the other the observer. In the former, we can follow up all physical processes (in principle at least) arbitrary precisely. In the latter, this is meaningless. The boundary between the two is arbitrary to a very large extent." The point is that the unitary time evolution (1.3.1) of the resulting component of the state vector after the reduction is the same as before.

According to this view, the question "when does the measurement, i.e., the reduction happen" is meaningless.

Rovelli [Rv] has claimed that "contrary to that view, quantum mechanics *does* give a precise answer to this question, although a peculiar answer." The idea is based on measuring an observable represented by a projection operator related to the coupled system formed by an observed system and an observer system. The range of the operator is the state ripe for registering possible results, i.e., the reduction.

But we argue that the measurement of the projection operator poses the same question: What is the time when the measurement result occurs? So it is impossible to tear apart the vicious circle.

3 Semiclassical Quantum Gravity as a Fundamental Theory

In this section, we argue that there is reason to consider semiclassical quantum gravity as a fundamental theory.

3.1 Totally Quantum Gravity: The Universe as a Perpetually Closed Quantum System and the Problem of Interpretation of State Vector

In general relativity, metric is a dynamical variable, so that it seems quite natural to quantize it along with matter field variables. A realization of such a program would result in totally quantum gravity [K], [BT], [GSR], [HI], [DN], [CH1], [IPS], [P1].

We argue that totally quantum gravity is confronted with a problem of principle—the problem of the interpretation of state vector. A standard interpretation of a state vector pertaining to a quantum system may be either epistemological or ontological; but in either case it is formulated in terms of results of experiments which might be performed on the system. We quote Penrose [P2]: "Suppose the state-vector is $|\Psi\rangle$. Then we can consider making an observation corresponding to the observable $Q = |\Psi\rangle\langle\Psi|$. The state $|\Psi\rangle$ is the only one, up to proportionality, for which the observable Q yields the result 1 with certainty. The state must 'know' that it has to produce this result in the event that the observation Q is actually performed. This is a completely objective property, so the fact that the state is (proportional to) $|\Psi\rangle$ is an objective property of the world.

I am expressing a point of view here which, for some reason, is not often maintained. One frequently hears the opposing view that 'the state-vector merely expresses our state of knowledge about a system', or 'the state-vector is expressing a property of ensembles of systems rather than of a single system.' However, the point of view that I am expressing is that the state-vector is clearly defining an objective property of a *single* system. It is not a 'testable' property of that system in the sense that we can perform an experiment on it to determine what its state $|\Psi\rangle$ actually is, but it is an 'objective' property of that individual system in the sense that the state of that one system is characterized by the results of experiments that one *might* perform on it."

Thus even if the interpretation is ontological, i.e., objective, it is operationalistic. In totally quantum gravity, the universe represents a perpetually closed quantum system, on which no experiments might be performed. We quote Goldstein and Teufel [GT]: "One of the most fascinating applications of quantum gravity is to quantum cosmology. Orthodox quantum theory attains physical meaning only via its predictions about the statistics of outcomes of measurements of observables, performed by observers that are not part of the system under consideration, and seems to make no clear physical statements about the behaviour of a closed system, not under observation. The quantum formalism concerns the interplay between—and requires for its very meaning—two kinds of objects: a quantum system and a more or less classical apparatus. It is hardly imaginable how one could make any sense out of this formalism for quantum cosmology, for which the system of interest is the whole universe, a closed system if there ever was one."

In particular, it is absolutely unclear what meaning may be assigned to state vector reduction, i.e., quantum jumps.

Another point that should be taken into account is the principle of equivalence [Wh]: "...on the one hand, the principle of equivalence provides the possibility to geometrisize gravitational fields and, on the other, it gives limits,..., under which gravitation cannot be interpreted as a usual field."

At the conclusion of this subsection, we quote Weinstein [We]: "Gravity, however, has resisted quantization. There exist several current research programmes in this area, including superstring theory and canonical quantum gravity. One often comes across the claim that the gravitational field must be quantized, and that quantization will give rise to a ...local uncertainty in the gravitational field. Here we will examine this claim, and see how the very things that make general relativity such an unusual 'field' theory not only make the quantization of the theory so technically difficult, but make the very idea of a 'fluctuating gravitational field' so problematic."

3.2 Interpretation of Quantum Theory and Classical Concepts: Advocating Semiclassical Gravity

Orthodox quantum theory requires, for its physical formulation, classical concepts relating to a classical world, in which experimental results occur. We quote Butterfield and Isham [BI]: "We dub our first approach to interpreting quantum theory, 'instrumentalism'...It includes views that apply to quantum theory some general instrumentalism about all scientific theories; and views that advocate instrumentalism only about quantum theory, based on special considerations about that subject. We will not comment on the first group, since we see no special connections with quantum gravity...

On the other hand, some views in the second group do have connections with quantum gravity, albeit 'negative' ones. Thus consider the Copenhagen interpretation of quantum theory: understood, not just as the minimal statistical interpretation of the quantum formalism in terms of frequencies of measurement result, but as insisting on a classical realm external to the quantum system, with a firm 'cut' between them, and with no quantum description of former. In so far as this classical realm is talking about 'quantum gravity', we are making a category error by trying to apply quantum theory to something that forms part of the classical background of that theory: 'what God has put assunder, let no man bring together'...

If it is indeed wrong to quantize the gravitational field, it becomes an urgent question how matter—which presumably is subject to the laws of quantum theory—should be incorporated in the overall scheme. To discuss this, we shall focus on the so-called 'semiclassical quantum gravity' approach. Here, one replaces the right-hand side of Einstein's field equations by a quantum expectation value, so as to couple a classical spacetime metric... to quantized matter..."

3.3 Semiclassical Gravity

Thus there is good reason to believe that semiclassical gravity [Mø], [Ro], [S], [CH1], [V], [BI] may be a fundamental theory.

The semiclassical Einstein equation retains the form of (1.2.1):

$$G_{\mu\nu} = 8\pi\varkappa T_{\mu\nu} \tag{27}$$

but now

$$T_{\mu\nu} := (\Psi, \widehat{T}_{\mu\nu}\Psi) \tag{28}$$

where $\widehat{T}_{\mu\nu}$ is the energy-momentum tensor operator and Ψ is a state vector of quantized matter, i.e., quantum fields $\hat{\phi}$. Thus

$$G_{\mu\nu} = G_{\mu\nu}[g] \tag{29}$$

$$\widehat{T}_{\mu\nu} = \widehat{T}_{\mu\nu}[g, \hat{\phi}] \tag{30}$$

where $g = \{g_{\mu\nu}\}$.

It should be realized that in spite of the fact that the quantity (3.3.2) is of the form of a quantum expectation value (the average value), the interpretation of it is by no means statistical: the quantity (3.3.2) does not relate to the averaging over a set of any measurement results. As long as Ψ is a state vector of the matter of the whole universe, (3.3.2) is, in fact, not an "average value" but rather an "effective quantity". It is (3.3.2) and (3.3.1) that give an ontological, objective meaning to the state vector Ψ.

The above argument lifts the following objection by Belinfante against (3.3.1), (3.3.2) [B]: "This would be most unusual, to equate a c-number to an expectation value of a q-number, and I think this violates the principles of quantum theory. Suppose, we had a situation which were a superposition of states, say 40% probability for one state of the matter and 60% probability for a different state. Suppose one did a million experiments; then one would expect a correlation between the measured actual state of the matter and the surrounding gravitational field. That is, one would in 400000 cases find one gravitational field and in 600000 cases a different gravitational field. If, however, the gravitational field were given by the expectational value, one should in all 1000000 cases find the same gravitational field, obtained by a 40-60 average." See also [I].

4 Conventional Semiclassical Construction and its Difficulties

In this section, we describe a conventional construction of semiclassical gravity and serious difficulties that it runs into [S], [CH1], [BI], [Wa], [V].

4.1 Spacetime Structure

In the conventional semiclassical construction, the spacetime structure is the same as in general relativity. Spacetime is defined as a pair

$$(M,g) \tag{31}$$

where M is a connected four-dimensional Hausdorff C^∞ manifold and g is a Lorentzian metric on M [HE].

It is the poorness of the structure (4.1.1) that results in serious difficulties of the conventional construction.

4.2 Renormalization of the Energy-Momentum Tensor

In this and the next subsection we closely follow [Wa]. Since quantum fields $\hat{\phi}$ are well defined only as distributions on spacetime, the operator $\widehat{T_{\mu\nu}}$ (3.3.4) involves taking the product of distributions at the same spacetime point. Consequently, some regularization is needed to define the effective energy-momentum tensor $T_{\mu\nu}$ (3.3.2).

Normal ordering gives an entirely satisfactory prescription for defining the latter. However, there is no satisfactory straightforward generalization of normal ordering to curved spacetime (4.1.1). Namely, in a generic spacetime (4.1.1), there is no preferred vacuum state.

It is useful to take an axiomatic approach and seek to determine $T_{\mu\nu}$ by the properties one would expect it to satisfy. There are four such properties [Wa] including (1.2.3):

$$T^{\mu\nu}{}_{;\nu} = 0 \tag{32}$$

4.3 Difficulties

There are two principal difficulties which confront any attempt to calculate matter effects in (3.3.1). First, there is a two-parameter ambiguity in the definition of $T_{\mu\nu}$. The second difficulty is this: The classical Einstein equation (1.2.1) is of second order in derivatives of the metric, but $T_{\mu\nu}$ (3.3.2) contains terms of fourth order in derivatives of the metric. In addition, except in some simple, special cases, $T_{\mu\nu}$ is a highly nonlocal functional of the metric.

4.4 Failure on the Chronology Horizon

The best way to present the issue is to quote Vissar [V]:

"*There are points on the chronology horizon where semiclassical Einstein equations fail to hold.*"

Note that the semiclassical Einstein equations [(3.3.1)] fail for a subtle reason; they fail simply because at some points the RHS [(3.3.2)] fails to exist, not necessary because the RHS is infinite...

The physical interpretation is that semiclassical quantum gravity fails to hold (at some points) on the chronology horizon; a fact which can be read in two possible ways:

1. If you assume that semiclassical quantum gravity is the fundamental theory (at best a minority opinion, and there are very good reasons for believing that this is not the case), then by *reductio ad absurdum* the chronology horizon must fail to form. Chronology is protected, essentially by *fiat*.

2. If you are willing to entertain the possibility that semiclassical quantum gravity is not the whole story (the majority opinion), then it follows from the above that issues of chronology protection cannot be settled at the semiclassical level. Chronology protection must then be settled (one way or another) at the level of a full theory of quantum gravity."

It must be emphasized that the case in point is the *conventional* semiclassical construction based on the poor spacetime structure (4.1.1).

4.5 Nonexistence of a Universal Vacuum State and Consequences

The poorness of the spacetime structure in the case of a generic curved spacetime results in rather specific features of the conventional semiclassical quantum gravity [BD], [Fu], [F]. There exists no well defined universal vacuum state, and, as a consequence, there are no universal creation/annihilation operators. Those facts imply the denouncement of the concept of particles. Finally, the Hamiltonian and the Schrödinger picture are abandoned.

Here we turn our attention to the problem of the greation/annihilation operators. We quote Weinberg [W2]: "...there is a deeper reason for constructing the Hamiltonian out of creation and annihilation operators, which goes beyond the need to quantize any preexisting field theory like electrodynamics, and has nothing to do with whether particles can actually be produced or destroyed. The great advantage of this formalism is that if we express the Hamiltonian as a sum of products of creation and annihilation operators, with suitable non-singular coefficients, then the *S*-matrix will automatically satisfy a crucial physical requirement, the cluster decomposition principle, which says in effect that distant experiments yield uncorrelated results. Indeed, it is for this reason that formalism of creation and annihilation operators is widely used in non-relativistic quantum statistical mechanics, where the number of particles is tipically fixed. In relativistic quantum theories, the cluster decomposition principle plays a crucial part in making field theory inevitable."

4.6 The Problem of State Vector Reduction

The status of the conventional semiclassical gravity looks even worse when state vector reduction is involved in dynamics [CH1]. The effective energy-momentum tensor $T_{\mu\nu}$ (3.3.2) and, by (3.3.1), the Einstein tensor $G_{\mu\nu}$ become discontinuous. The four components $G_{0\mu}$ involve only the first time derivatives of metric components, $\dot{g}_{ij}$ [W1], [LL]. Therefore a quantum jump of the $T_{0\mu}$ would result in that of $\dot{g}_{ij}$, which is inadmissible.

5 Reductive Semiclassical Construction: Semiclassical Reductive Quantum Gravity

In this section, we advance a novel semiclassical construction—including state vector reduction.

5.1 Quantum Jumps and Spacetime Structure

Be that as it may, quantum jumps, i.e., state vector reduction should be incorporated into quantum gravity. In the case of semiclassical gravity, which is the theme of our study, the incorporation of quantum jumps results first of all in the enrichment of spacetime structure. To wit, the jumps, being nonlocal and instantaneous, imply the existence of a universal time. The latter, in its turn, implies the structure of spacetime as the direct product of cosmological time and cosmological space:

$$\begin{aligned} &M = M^4 = T \times S, \quad M \ni p = (t,s) \\ &t \in T \subset \mathbb{R}, \ s \in S \end{aligned} \tag{33}$$

The one-dimensional manifold T is the universal cosmological time, the three-dimensional manifold S is a cosmological space. The tangent space M_p at a point $p \in M$ is

$$M_p = T_t \oplus S_s, \ p = (t,s) \tag{34}$$

Now metric may be made into the form of

$$g = dt \otimes dt - h_t \tag{35}$$

where h_t is a Riemannian metric on S depending on time, i.e.,

$$g(x) = dt^2 + g_{ij}(x)dx^i dx^j, \ x = (t,\vec{x}), \ g_{ij} = -h_{ij}, \ \vec{x} \leftrightarrow s \in S \tag{36}$$

So in semiclassical reductive quantum gravity, spacetime structure is given by (4.1.1), (5.1.1), (5.1.2), (5.1.3). That structure corresponds to the complementary conditions

$$g_{0\mu} = \delta_{0\mu}, \quad \mu = 0,1,2,3 \tag{37}$$

which fix a solution to the Cauchy problem.

5.2 Creation/Annihilation Operators and the Vacuum State

Consider, e.g., a real scalar field operator $\hat{\varphi}$. We have an expansion

$$\hat{\varphi}(p) = \sum_j \{f_j(p)\hat{a}_j(t) + f_j^*(p)\hat{a}_j^\dagger(t)\}, \ p = (t,s) \tag{38}$$

where f_j is a field mode, $\hat{a}_j/\hat{a}_j^\dagger$ is an annihilation/creation operator,

$$[\hat{a}_j(t),\hat{a}_{j'}^\dagger(t)] = \delta_{jj'}, \ [\hat{a}_j(t),\hat{a}_{j'}(t)] = [\hat{a}_j^\dagger(t),\hat{a}_{j'}^\dagger] = 0 \tag{39}$$

and for a free field

$$\frac{d}{dt}\{\hat{a}_j^\dagger(t)\hat{a}_j(t)\} = 0 \tag{40}$$

The vacuum state is defined by

$$\hat{a}_j(t)\Psi_{\text{vac}} = 0 \tag{41}$$

The uniqueness of the creation/annihilation operators and the vacuum state relies heavily on the spacetime structure (5.1.1), (5.1.2).

5.3 The Energy-Momentum Tensor

The energy-momentum tensor operator is normally ordered:

$$\widehat{T}_{\mu\nu} =: \widehat{T}_{\mu\nu} : \tag{42}$$

The modes f_j and by the same token the expansion (5.2.1) relate to an arbitrarily fixed value of time, so that they depend only on metric, i.e., g_{ij}, but not on the time derivative of the latter. The energy-momentum tensor operator involves at most second derivatives of metric.

The renormalization of $T_{\mu\nu}$ is defined by (5.3.1) rather than (4.2.1), so that the latter equality does not hold:

$$T^{\mu\nu}{}_{;\nu} \neq 0 \tag{43}$$

5.4 Violation of the Semiclassical Einstein Equation

The semiclassical Einstein equation

$$G_{\mu\nu} = 8\pi\varkappa T_{\mu\nu}, \quad T_{\mu\nu} = (\Psi, \widehat{T}_{\mu\nu}\Psi) \tag{44}$$

is violated both by state vector reduction and in view of the inequality (5.3.2). This crucial issue calls for a careful study.

The spacetime structure (5.1.1), (5.1.2) implies a canonical decomposition of the Einstein equation (5.4.1) into space and time/time-space parts:

$$G_{ij} = 8\pi\varkappa T_{ij} \quad \text{(six equations)} \tag{45}$$

$$G_{0\mu} = 8\pi\varkappa T_{0\mu} \quad \text{(four equations)} \tag{46}$$

The space components of the Einstein tensor, G_{ij}, involve the second time derivatives $\ddot{g}_{kl}$ of the metric tensor components g_{kl} [W1], [LL]. It is reasonable to assume that quantum jumps of T_{ij} result in those of $\ddot{g}_{kl}$. That is quite conceivable from the physical point of view: A jump of force (T_{ij}) causes a jump of acceleration ($\ddot{g}_{kl}$).

On the other hand, the time/time-space components, $G_{0\mu}$, involve only the first time derivatives $\dot{g}_{kl}$ [W1], [LL]. The latter should be continuous, not to mention g_{kl} themselves.

In order to compensate for the quantum jumps, it would suffice to retain (5.4.2) and extend (5.4.3). It is this approach that has been exploited in [M1], [M2]. But in view of the inequality (5.3.2), it seems appropriate to extend (5.4.1) as a whole.

5.5 Compensatory Pseudo Energy-Momentum Tensor and Pseudomatter

We extend the semiclassical Einstein equation (5.4.1) by complementing the effective energy-momentum tensor $T_{\mu\nu}$ of matter with a compensatory pseudo energy-momentum tensor $\Theta_{\mu\nu}$:

$$G_{\mu\nu} = 8\pi\varkappa(T_{\mu\nu} + \Theta_{\mu\nu}) \tag{47}$$

The components $G_{0\mu}$ must be continuous, so that the components $\Theta_{0\mu}$ have to compensate for jumps of the four components $T_{0\mu}$. Therefore the compensatory tensor $\Theta_{\mu\nu}$ should involve four degrees of freedom, i.e., four functions on the spacetime manifold M. Equation (5.5.1) implies

$$(T^{\mu\nu} + \Theta^{\mu\nu})_{;\nu} = 0 \tag{48}$$

which is an extension of (1.2.3).

We introduce the concept of pseudomatter by calling $\Theta_{\mu\nu}$ the energy-momentum tensor of the latter.

5.6 Pseudomatter as Pressural Dust

The simplest way to treat pseudomatter is to consider it to be a perfect fluid. Then

$$\Theta_{\mu\nu} = (\sigma + \pi)\upsilon_\mu \upsilon_\nu - \pi g_{\mu\nu} \tag{49}$$

where σ is the density, π is the pressure, and υ_μ is the four-velocity. We have

$$\Theta^{\mu\nu}{}_{;\nu} = \varepsilon(\upsilon^\nu{}_{;\nu}\upsilon^\mu + \upsilon^\mu{}_{;\nu}\upsilon^\nu) + \varepsilon_{,\nu}\upsilon^\nu\upsilon^\mu - \pi^{,\mu} \tag{50}$$

where

$$\varepsilon := \sigma + \pi \tag{51}$$

In the case of

$$\pi = \text{const} \tag{52}$$

(5.6.2) involves σ and π only in the combination (5.6.3), so that we assume that the equation of state for the fluid is (5.6.4), i.e., the fluid is a pressural dust. Then there are four degrees of freedom represented by the functions ε and υ_μ—in view of

$$\upsilon^\mu \upsilon_\mu = \upsilon_0^2 + \upsilon^i \upsilon_i = 1 \tag{53}$$

5.7 Dark Energy, Dark Matter, and the Reductive Semiclassical Einstein Equation

Now we put

$$\pi = -\frac{\Lambda}{8\pi\varkappa} \tag{54}$$

Equation (5.5.1) amounts to

$$G_{\mu\nu} - \Lambda g_{\mu\nu} = 8\pi\varkappa(T_{\mu\nu} + \varepsilon\upsilon_\mu\upsilon_\nu) \tag{55}$$

which includes dark energy, or cosmological constant Λ, and dark matter: $\varepsilon\upsilon_\mu\upsilon_\nu$,

$$\rho_{\text{dark matter}} = \varepsilon \tag{56}$$

We call (5.7.2) the reductive semiclassical Einstein equation.

In the simplest case,

$$|\upsilon_\mu| = \delta_{0\mu} \quad \text{(cold dark matter)} \tag{57}$$

so that (5.7.2) reduces to

$$G_{\mu\nu} - \Lambda g_{\mu\nu} = 8\pi\varkappa(T_{\mu\nu} + \varepsilon\delta_{\mu 0}\delta_{\nu 0}) \tag{58}$$

Thus semiclassical reductive quantum gravity leads naturally to the concepts of dark energy and dark matter. It follows that the presence of dark energy and dark matter in the real world provides an observable evidence of characteristically quantum gravitational effects. That is a challenge to a conventional opinion that "... to day, there is no recognized experimental evidence of characteristically *quantum* gravitational effects" [CH1].

5.8 The Metric Equation

The reductive semiclassical Einstein equation

$$G_{\mu\nu} - \Lambda g_{\mu\nu} = 8\pi\varkappa(T_{\mu\nu} + \varepsilon\upsilon_\mu\upsilon_\nu) \tag{59}$$

represents a system of ten equations for the six components of the metric tensor g_{ij} and the four dark matter variables ε, υ_μ.

The dark matter variables can be eliminated easily. Introducing the notation

$$B_{\mu\nu} := G_{\mu\nu} - \Lambda g_{\mu\nu} - 8\pi\varkappa T_{\mu\nu} \quad \bar{B}_{\mu\nu} := \frac{1}{8\pi\varkappa} B_{\mu\nu} \tag{60}$$

we obtain

$$\bar{B}_{\mu\nu} = \varepsilon\upsilon_\mu\upsilon_\nu \tag{61}$$

We have

$$\bar{B}_{ij} = \varepsilon\upsilon_i\upsilon_j = (\varepsilon\upsilon_i\upsilon_0)(\varepsilon\upsilon_j\upsilon_0)/\varepsilon\upsilon_0^2 \tag{62}$$

Now

$$\varepsilon\upsilon_0^2 = \bar{B}_{00} \tag{63}$$

and

$$\varepsilon\upsilon_i\upsilon_0 = \bar{B}_{0i} \tag{64}$$

Thus we obtain

$$B_{00}B_{ij} - B_{0i}B_{0j} = 0 \tag{65}$$

which represents six equations for the six g_{ij}. We call (5.8.7) the metric equation. It is a dynamical equation for metric. In order that (5.8.7) be linear in $\ddot{g}_{ij}$, $T_{0\mu}$ should not involve those derivatives: the latter are involved in B_{ij}.

Having regard to the spacetime structure given by (4.1.1), (5.1.1), (5.1.2), (5.1.3), the metric equation may be written in an intrinsic, i.e., coordinate-independent form:

$$B_{TT} \otimes B_{SS} - B_{TS} \otimes B_{TS} = 0 \tag{66}$$

where

$$B = G - \Lambda g - 8\pi\varkappa T \tag{67}$$

and

$$B_{TT} = B_{00}\,,\; B_{SS}\,,\; B_{ST} = B_{TS} = B_{0S} \tag{68}$$

are projections on T_t/S_s in (5.1.2).

5.9 The Exact Time of a State Vector Reduction

Let us return to the problem of the time of state vector reduction. In semiclassical reductive quantum gravity, the exact time of a state vector reduction is that of a jump of the classical quantities $\ddot{g}_{ij}$. Note that classical metric is by no means an "apparatus" which causes reduction; on the contrary, it is the latter that produces the jump of $\ddot{g}_{ij}$, which, in turn, manifests itself in the dynamics of metric, matter, and dark matter. It is those classical jumps of $\ddot{g}_{ij}$ that give an operationalistic meaning to the quantum reduction.

5.10 The FLRW Universe

In the Friedmann-Lemaître-Robertson-Walker model,

$$\upsilon_i = 0, \ \upsilon_0^2 = 1 \tag{69}$$

and for a closed universe ($k = 1$), the reductive semiclassical Einstein equation (5.7.2) reduces to

$$2R\ddot{R} + \dot{R}^2 + 1 - \Lambda R^2 = -8\pi\varkappa p R^2 \tag{70}$$

$$3(\dot{R}^2 + 1) - \Lambda R^2 = 8\pi\varkappa(\rho + \varepsilon)R^2 \tag{71}$$

where $R = R(t)$ is the radius of the universe. The matter variables ρ and p (usually $p = p(\rho)$) are determined by matter dynamics, (5.10.2) is the equation of motion for R, and (5.10.3) determines ε.

Let us obtain an equation connecting the change of the total density $\rho + \varepsilon$ with the pressure p. Introducing the notation

$$\tilde{p} := p - \frac{\Lambda}{8\pi\varkappa} \tag{72}$$

$$\tilde{\rho} \overset{\bullet}{:=} \rho + \varepsilon + \frac{\Lambda}{8\pi\varkappa} \tag{73}$$

we have

$$2R\ddot{R} + \dot{R}^2 + 1 = -8\pi\varkappa\tilde{p}R^2 \tag{74}$$

$$3\dot{R}^2 + 3 = 8\pi\varkappa\tilde{\rho}R^2 \tag{75}$$

From (5.10.7) follows

$$6\dot{R}\ddot{R} = 8\pi\varkappa\dot{\tilde{\rho}}R^2 + 16\pi\varkappa\tilde{\rho}R\dot{R} \tag{76}$$

Eliminating $\ddot{R}$ and $\dot{R}^2$ from (5.10.6), (5.10.7), (5.10.8), we obtain

$$\dot{\tilde{\rho}}R^3 + 3(\tilde{p} + \tilde{\rho})R^2\dot{R} = 0 \tag{77}$$

i.e.,

$$\frac{d(\rho + \varepsilon)}{dt}R^3 + 3(p + \rho + \varepsilon)R^2\dot{R} = 0 \tag{78}$$

or

$$\frac{d}{dt}[(\rho + \varepsilon)R^3] = -p\frac{dR^3}{dt} \tag{79}$$

The volume of the universe [LL]

$$V = 2\pi^2 R^3 \tag{80}$$

so that

$$\frac{d}{dt}[(\rho+\varepsilon)V] = -p\frac{dV}{dt} \tag{81}$$

5.11 Basic Dynamical Equations of Semiclassical Reductive Quantum Gravity

Basic dynamical equations of semiclassical reductive quantum gravity as a complete theory are metric and matter equations:

The metric equation:

$$B_{00}B_{ij} - B_{0i}B_{0j} = 0 \tag{82}$$

The matter equations:

The Schrödinger equation

$$\frac{d\Psi}{dt} = -\widehat{H}\Psi, \quad \widehat{H} = \widehat{H}_t \tag{83}$$

State vector reduction equation

quantum jump dynamical equation (84)

The subject of the present study concerns (5.11.1). A version of state vector reduction theory, including (5.11.3), compatible with the present study has been advanced in [M2].

Let

$$\varepsilon\upsilon_0\upsilon_\mu = \bar{B}_{0\mu} \sim \alpha \to 0 \tag{85}$$

then

$$\varepsilon\upsilon_i\upsilon_j = \bar{B}_{ij} = \frac{\bar{B}_{0i}\bar{B}_{0j}}{\bar{B}_{00}} \sim \alpha \to 0 \tag{86}$$

so

$$\varepsilon\upsilon_0\upsilon_\mu = \bar{B}_{0\mu} \sim \alpha \to 0 \Rightarrow \varepsilon\upsilon_\mu\upsilon_\nu = \bar{B}_{\mu\nu} \sim \alpha \to 0 \tag{87}$$

Thus in the absence of dark matter, the metric equation reduces to the semiclassical Einstein equation with the cosmological constant

$$G_{\mu\nu} - \Lambda g_{\mu\nu} = 8\pi\varkappa T_{\mu\nu} \tag{88}$$

But generally, in view of Subsection 5.4, this equation is violated, so that dark matter is indispensable.

5.12 Dark Matter as Pseudodust

Let us return to the reductive semiclassical Einstein equation

$$G_{\mu\nu} - \Lambda g_{\mu\nu} = 8\pi\varkappa(T_{\mu\nu} + \varepsilon\upsilon_\mu\upsilon_\nu) \tag{89}$$

The quantities

$$\varepsilon\upsilon_\mu\upsilon_\nu = \bar{B}_{\mu\nu} \tag{90}$$

are determined from a solution to the metric equation (5.11.1).

We find

$$\varepsilon = \bar{B}^\mu_\mu \tag{91}$$

$$\upsilon_0^2 = \frac{B_{00}}{B^\mu_\mu}, \quad \upsilon_i^2 = \frac{(B_{0i})^2}{B_{00}B^\mu_\mu} \tag{92}$$

For an ordinary dust, for which

$$\upsilon_0^2 \geq 1, \quad \upsilon_i^2 \geq 0 \tag{93}$$

the condition

$$B_{00}B^\mu_\mu \geq 0 \tag{94}$$

should hold. But the metric equation does not guarantee the fulfilment of this condition. Thus we discard the condition (5.12.6) and introduce the term pseudodust. So generally, dark matter is represented by a pseudodust rather than by an ordinary dust. This implies the fruitlessness of efforts to represent dark matter by any kind of ordinary matter whatever.

5.13 Cold Dark Matter

Let us consider the case of cold dark matter in more detail. We define cold dark matter by the following inequalities:

$$|\upsilon^i\upsilon_i| \ll 1 \tag{95}$$

$$|\varepsilon\upsilon_i\upsilon_j| \ll |T_{ij}| \tag{96}$$

From those and (5.6.5), (5.8.3), (5.8.2), (5.12.3), (5.8.1) follows

$$\upsilon_0^2 \approx 1 \tag{97}$$

$$\bar{B}_{00} \approx \varepsilon \tag{98}$$

$$B_{00}B^\mu_\mu \approx (8\pi\varkappa)^2\varepsilon^2 > 0 \qquad \text{see (5.12.6)} \tag{99}$$

$$B_{ij} \approx 0 \tag{100}$$

In the first approximation in υ_i,

$$B_{ij} = 0, \text{ i.e., } G_{ij} - \Lambda g_{ij} = 8\pi\varkappa T_{ij} \tag{101}$$

Now metric g_{kl} is determined by (5.13.7); $B_{0\mu}$ are determined by the g_{kl};

$$\varepsilon = \bar{B}_{00} = \frac{1}{8\pi\varkappa}B_{00} \tag{102}$$

$$\upsilon_i = \frac{\bar{B}_{0i}}{\bar{B}_{00}\upsilon_0} = \frac{B_{0i}}{B_{00}\upsilon_0}, \quad \upsilon_0 = \pm 1 \tag{103}$$

the inequalities (5.13.1), (5.13.2) amount to

$$|B_{0i}B^i_0| \ll B^2_{00} \tag{104}$$

$$|\bar{B}_{0i}\bar{B}_{0j}| \ll |\bar{B}_{00}T_{ij}| \tag{105}$$

respectively.

Conclusion

There are pros and contras of both totally quantum gravity and semiclassical one. None of those pros and contras may be reckoned as decisive. In such a situation, it seems most productive to compare actual constructions. One such a constructions is advanced here—in comparison with some others.

Acknowledgments

I would like to thank Frank Columbus for the invitation to participate in this edition, Alex A. Lisyansky for support, and Stefan V. Mashkevich for helpful discussions.

References

[B] Belinfante, F.J. *General discussion, in: Relativistic theories of gravitation*; ed. L. Infeld; Pergamon Press Ltd, 1964.

[BD] Birrell, N.D.; Davies, P.C.W. *Quantum fields in curved space*; Cambridge University Press, 1982.

[BI] Butterfield, J.; Isham, C. *Spacetime and the philosophical challenge of quantum gravity, in: Physics Meets Philosophy at the Planck Scale: Contemporary Theories in Quantum Gravity*; ed. C. Callender, N. Huggett: Cambridge University Press, 2001.

[BT] Borzeszkowski, H.-H. v.; Treder, H.-J. *The Meaning of Quantum Gravity*; Reidel, 1988.

[C] Carroll, S.M. *Why is the Universe Accelerating?*; arXiv: astro-ph/0310342, 2003.

[CH1] Callender, C.; Huggett, N. *Introduction, in: Physics Meets Philosophy at the Planck Scale: Contemporary Theories in Quantum Gravity*; ed. C. Callender, N. Huggett; Cambridge University Press, 2001.

[CH2] Callender, C.; Huggett, N. (ed.) *Physics Meets Philosophy at the Planck Scale: Contemporary Theories in Quantum Gravity*; Cambridge University Press, 2001.

[D] Dodelson, S. *Modern Cosmology*; Academic Press, 2003.

[DN] Dadhich, N.; Narlicar, J. (ed.) *Gravitation and Relativity: At the turn of the Millenium*; IUCAA, 1998.

[E1] Espagnat, B. d'. *Conceptual Foundations of Quantum Mechanics*; W.A. Benjamin, Inc., 1976.

[E2] Espagnat, B. d'. *Veiled Reality*; Westview Press, 2003.

[F] Ford, L.H. *Quantum Field Theory in Curved Spacetime*; arXiv: gr-qc/9707062, 1997.

[FM] Fisher, A.E.; Marsden, J.E. *Initial data problem and dynamical formalism of general relativity, in: General Relativity* ; ed. S.W. Hawking, W. Israel; Cambridge University Press, 1979.

[Fr] Frampton, P.H. *Dark Energy—A Pedagogic Review*; arXiv: astro-ph/0409166, 2004.

[Fu] Fulling, S.A. *Aspects of Quantum Field Theory in Curved Space-Time*; Cambridge University Press, 1989.

[GSR] Gibbons, G.W.; Shellard, E.P.S.; Rankin, S.J. (ed.) *The Future of Theoretical Physics and Cosmology*; Cambridge University Press, 2003.

[GT] Goldstein, S.; Teufel, S. *Quantum spacetime without observers: Ontological clarity and the conceptual foundations of quantum gravity, in: Physics Meets Philosophy at the Planck Scale: Contemporary Theories in Quantum Gravity*; ed. C. Callender, N. Huggett; Cambridge University Press, 2001.

[HE] Hawking, S.W.; Ellis, G.F.R. *The large scale structure of space-time*; Cambridge University Press, 1973.

[HI] Hawking, S.W.; Israel, W. (ed.) *Three hundred years of gravitation*; Cambridge University Press, 1989.

[I] Israel, W. *General relativity: progress, problems, and prospects*; Canadian Journal of Physics, 63, No 1, 34-43, 1985.

[IPS] Isham, C.J.; Penrose, R.; Sciama, D.W. (ed.) *Quantum Gravity. An Oxford Symposium*; Clarendon Press, 1981.

[K] Kiefer, C. *Quantum Gravity*; Oxford University Press, 2004.

[LL] Landau, L.D.; Lifshitz, E.M. *The Classical Theory of Fields*; Pergamon Press, 1975.

[M1] Mashkevich, V.S. *General Relativity and Quantum Jumps: The Existence of Nondiffeomorphic Solutions to the Cauchy Problem in Nonempty Spacetime and Quantum Jumps as a Provider of Canonical Spacetime Structure*; arXiv: gr-qc/0403056, 2004.

[M2] Mashkevich, V.S. *The Scattering Quantum Jump Universe, in: Progress in General Relativity and Quantum Cosmology*; Nova Science Publishers, Inc., 2005.

[Mø] Møller, C. *The Energy-Momentum Complex in General Relativity and Related Problems, in: Les Theories Relativistes de la Gravitation*; ed. A. Lichnerowicz, M.A. Tonnelat; Paris: CNRS, 1962.

[MR] Maroto, A.L.; Ramírez, J. *A Conceptual Tour About the Standard Cosmological Model*; arXiv: astro-ph/0409280, 2004.

[N] Neumann, J., v. *Mathematical Foundations of Quantum Mechanics*; Princeton University Press, 1955.

[P1] Penrose, R. (ed.) *Quantum Gravity 2: A Second Oxford Symposium*; Clarendon Press, 1981.

[P2] Penrose, R. *Newton, quantum gravity and reality, in: Three hundred years of gravitation*; ed. S. Hawking, W. Israel; Cambridge University Press, 1989.

[R] Rees, M.J. *Dark Matter: Introduction*; arXiv: astro-ph/0402045, 2004.

[Ri] Rich, J. *Fundamentals of Cosmology*; Springer-Verlag, 2001.

[Ro] Rosenfeld, L. *On Quantization of Fields*; Nuclear Physics, 40, 353-6, 1963.

[Rv] Rovelli, C. *"Incerto tempore, incertisque loci": Can we compute the exact time at which a quantum measurement happens?* ; arXiv quantum-ph/9802020 v 3, August 25, 2004.

[S] Stephani, H. *Relativity*; Cambridge University Press, 2004.

[St] Straumann, N. *Dark Energy*; arXiv: gr-qc/0311083, 2003.

[SW] Sachs, R.K.; Wu, H. *General Relativity for Mathematicians*; Springer-Verlag, 1977.

[V] Visser, M. *The quantum physics of chronology protection, in: The Future of Theoretical Physics and Cosmology*; ed. G.W. Gibbons, E.R.S. Shellard, S.J. Rankin; Cambridge University Press, 2004.

[W1] Weinberg, S. *Gravitation and Cosmology*; Wiley, 1973.

[W2] Weinberg, S. *The Quantum Theory of Fields, Vol. I*; Cambridge University Press, 1995.

[Wa] Wald, R.M. *Quantum Field Theory in Curved Spacetime and Black Hole Thermodynamics*; The University of Chicago Press, 1994.

[We] Weinstein, S. *Naive quantum gravity, in: Physics Meets Philosophy at the Planck Scale: Contemporary Theories in Quantum Gravity*; ed. C. Callender, N. Huggett; Cambridge University Press, 2001.

[Wh] Wahsner, R. *Dualism between space-time and matter analyzed from epistemological point of view, in: Advances in the Interplay Between Quantum and Gravity Physics*; ed. P.G. Bergman, V.de Sabbata; Kluwer Academic Publishers, 2002.

In: Trends in Quantum Gravity Research
Editor: David C. Moore, pp. 219-260

ISBN 1-59454-670-3

Chapter 6

QUANTUM RIEMANNIAN GEOMETRY AND BLACK HOLES

Martin Bojowald*
Institute for Gravitational Physics and Geometry,
The Pennsylvania State University, 104 Davey Lab, University Park,
PA 16802, USA; Max-Planck-Institut für Gravitationsphysik,
Albert-Einstein-Institut, Am Mühlenberg 1, D-14476 Potsdam, Germany

Abstract

Black Holes have always played a central role in investigations of quantum gravity. This includes both conceptual issues such as the role of classical singularities and information loss, and technical ones to probe the consistency of candidate theories. Lacking a full theory of quantum gravity, such studies had long been restricted to black hole models which include some aspects of quantization. However, it is then not always clear whether the results are consequences of quantum gravity per se or of the particular steps one had undertaken to bring the system into a treatable form. Over a little more than the last decade loop quantum gravity has emerged as a widely studied candidate for quantum gravity, where it is now possible to introduce black hole models within a quantum theory of gravity. This makes it possible to use only quantum effects which are known to arise also in the full theory, but still work in a rather simple and physically interesting context of black holes. Recent developments have now led to the first physical results about non-rotating quantum black holes obtained in this way. Restricting to the interior inside the Schwarzschild horizon, the resulting quantum model is free of the classical singularity, which is a consequence of discrete quantum geometry taking over for the continuous classical space-time picture. This fact results in a change of paradigm concerning the information loss problem. The horizon itself can also be studied in the quantum theory by imposing horizon conditions at the level of states. Thereby one can illustrate the nature of horizon degrees of freedom and horizon fluctuations. All these developments allow us to study the quantum dynamics explicitly and in detail which provides a rich ground to test the consistency of the full theory.

*E-mail address: bojowald@gravity.psu.edu

1 Introduction

Black holes in classical general relativity are, compared to other astrophysical objects, distinguished by the presence of singularities, where curvature and tidal forces diverge and where space-time stops, and horizons, which can separate off regions from causal contact from another region. Both properties have long been suspected to be changed in a quantum theory of gravity: Singularities denote points where the classical theory breaks down, and at least space-like ones which lie to the past or future of observers are supposed to be removed in a more complete quantum theory. Horizons, on the other hand, are still expected to play an important role also in quantum gravity. The horizon surface should at most be smeared out due to fluctuations in the causal structure on which the concept of horizons relies. For massive black holes (compared to the Planck mass) these horizon fluctuations should be negligible for most purposes such that the classical picture still applies. Instead of modifying the horizon on large scales, quantum gravity is expected to provide a microscopic picture which shows how to build a macroscopic horizon from Planck scale ingredients. If successful, this will then result in a statistical explanation of black hole entropy.

In more detail, the main issues concerning black holes in quantum gravity are as follows:

Singularities: Are they indeed removed and, if yes, what replaces them? There are arguments that not all singularities are equal, with space-like ones to be removed and time-like ones to persist in order to rule out unwanted (such as negative mass) solutions [1]. Also the issue of naked singularities and cosmic censorship arises in this context.

Horizons: First of all, one has to see what an adequate definition of a horizon in quantum gravity could be. The original concept of the event horizon relies on the classical causal structure of all of space-time as well as the presence of singularities in the future. The quasi-local concept of isolated or dynamical horizons [2] uses much weaker assumptions about the structure of space-time such that it is better suited to a quantum treatment at least for large, semiclassical black holes which have only weak curvature at the horizon. For microscopic or primordial black holes, space-time even around the horizon cannot be treated as a smooth classical geometry with a classical causal structure. Here it is not clear if a quantum concept of horizon can even be applied.

If there is an applicable notion of quantum horizon, the issue of black hole entropy can be analyzed. By identifying and counting quantum states uniquely characterizing a horizon of a given area one can compute black hole entropy and compare with the expected semiclassical Bekenstein–Hawking formula. Moreover, detailed pictures of the horizon structure and its fluctuations can be developed which shed more light on quantum gravity in general. If matter fields are present, the horizon should shrink from Hawking radiation which provides insights on how gravity interacts with matter at the quantum level.

Both: Systems such as black holes with singularities as well as horizons have led to much confusion in attempts to guess the outcome of quantum gravity from early glimpses obtained from mainly semiclassical considerations. This is most commonly expressed in the infamous information loss paradox according to which information falling into the singularity implies a non-unitary quantum evolution and thus presumably fundamental limitations to knowledge [3]. These ideas obviously do not take into account what happens to singularities in quantum gravity and thus have to be revisited once a more complete treatment is available.

All these issues probe different aspects of the full theory of quantum gravity and require different techniques. A common feature, except for the entropy counting of isolated horizons, is that they are dynamical aspects such that the Hamiltonian constraint operator in a canonical quantization or an alternative evolution equation is essential. In particular, both black hole singularities as well as their horizons require inhomogeneous situations and an approximation by spatial homogeneity, which works well in cosmological cases, is not sufficient in general to grasp all the important physical aspects. This has the advantage of providing many non-trivial tests of quantum gravity which go beyond what is possible in homogeneous cosmological models.

It certainly also implies that the treatment is more complicated, and indeed progress on the problems listed here has been mixed. The strongest results exist for the counting of black hole entropy of static or isolated horizons which has been derived in different approaches [4, 5, 6, 7, 8]. This has been possible since the isolation (or even extremality in [4]) allows one to ignore the complicated quantum dynamics and still compute the correct number of physical states. Moreover, only the horizon itself is important such that its inhomogeneous neighborhood does not have much influence. This changes if one also wants to study, e.g., horizon fluctuations since they are dynamical and require the neighborhood in which the horizon fluctuates. Thus, both the quantum dynamics and inhomogeneous configurations have to be handled, and there are not many results within a full candidate of quantum gravity so far.

Similarly, the issue of singularities relies on dynamical aspects which for most of the time was too complicated to allow definitive conclusions as to whether or not singularities persist in quantum gravity. In the last few years, there has been progress on the homogeneous situation of cosmological singularities [9, 10, 11] which have been shown to be removed by quantum gravity [12]. Analogous techniques are now also available for some inhomogeneous situations such as the spherically symmetric model [13, 14] which is classically relevant for non-rotating black holes. This has led to an extension of the non-singularity statements from homogeneous models to the spherically symmetric one [15]. Moreover, with new results about quantum horizons a consistent picture of quantum physics of black holes is emerging.

This chapter is also intended as an introduction, by way of examples, to some of the techniques of quantum geometry with an emphasis on aspects which are typical for a loop quantization and essential for physical issues. The main theme will be the understanding of quantum dynamics in inhomogeneous situations and problems surrounding it.

2 Classical Aspects of Spherically Symmetric Systems

A spherically symmetric metric is most easily written in polar coordinates (x, ϑ, φ) and takes the form (with $\mathrm{d}\Omega^2 = \mathrm{d}\vartheta^2 + \sin^2\vartheta \mathrm{d}\varphi^2$)

$$\mathrm{d}s^2 = -N(x,t)^2\mathrm{d}t^2 + q_{xx}(x,t)(\mathrm{d}x + N^x(x,t)\mathrm{d}t)^2 + q_{\varphi\varphi}(x,t)\mathrm{d}\Omega^2 \tag{1}$$

where fields only depend on time t and the coordinate x of the 1-dimensional radial manifold B. This expression makes use of the lapse function $N(x,t)$ and shift vector $N^x(x,t)$ which are prescribed by the slicing of space-time into spatial constant-t slices: coordinate time translations are generated by the vector field

$$\frac{\partial}{\partial t} = Nn + N^x\frac{\partial}{\partial x} \tag{2}$$

with the unit vector field n being normal to the slices. The spatial metric on those slices is then

$$\mathrm{d}q^2 = q_{xx}(x,t)\mathrm{d}x^2 + q_{\varphi\varphi}(x,t)\mathrm{d}\Omega^2 \tag{3}$$

and extrinsic curvature

$$K_{ab} = \tfrac{1}{2}\mathcal{L}_n q_{ab}\,, \tag{4}$$

which determines the conjugate $\pi^{ab} = -\frac{1}{2}\sqrt{\det q}(K^{ab} - q^{ab}K_c^c)$ to the metric in a canonical formulation [16], takes a similar form $K = K_{xx}(x,t)\mathrm{d}x^2 + K_{\varphi\varphi}(x,t)\mathrm{d}\Omega^2$.

A well-known example is obtained by the spherically symmetric vacuum solution to Einstein's field equations, the Schwarzschild metric [17]

$$\mathrm{d}s^2 = -(1-2M/x)\mathrm{d}t^2 + \frac{1}{1-2M/x}\mathrm{d}x^2 + x^2\mathrm{d}\Omega^2 \tag{5}$$

with the mass parameter M. It has the following properties: If we first restrict our attention to larger $x > 2M$, the metric is static since its coefficients do not depend on time and $N^x = 0$. When x becomes large compared to the mass, i.e. if we approach the asymptotic regime far away from the black hole, the metric becomes asymptotically flat. The black hole region is characterized by the horizon which appears at $x = 2M$ as a coordinate singularity in the Schwarzschild metric and can be defined in a coordinate independent manner as the outer boundary of a region where trapped surfaces, i.e. envelops of light rays which cannot expand outwards to infinity, occur. If we enter the black hole region through the horizon we notice that now t becomes a space-like coordinate since the tt component changes sign. The role of coordinate time is then played by x on which the metric coefficients depend. Thus, the interior is not static, but since the metric components now do not depend on the spatial coordinate t it is homogeneous (of Kantowski–Sachs form).

2.1 Metric and Triad

The metric components $q_{\varphi\varphi} = x^2$, $q_{xx} = (1-2M/x)^{-1}$ and $N^2 = -g_{tt} = 1 - 2M/x$ can be used to characterize the three different regimes of a massive black hole with mass $M \gg 1$: At asymptotic infinity we have $x \gg 2M \gg 1$ and thus

$$q_{\varphi\varphi} \gg 1 \qquad q_{xx} \sim 1\,.$$

At the horizon we have $x \sim 2M$ and

$$q_{\varphi\varphi} \gg 1 \qquad q_{xx} \gg 1$$

while at the singularity we have $0 \sim x \ll 2M$ and

$$q_{\varphi\varphi} \ll 1 \qquad |q_{xx}| \ll 1 \qquad N \gg 1\,.$$

In the latter case, q_{xx} is relevant only if we approach the singularity on slices with t constant which are time-like inside the horizon. The lapse function, on the other hand, is the relevant metric component if we approach the singularity on slices which are space-like inside.

These regimes of metric components can be used for a first glimpse on how a quantization may deal with the singularity or horizon. From cosmological models it is known that expressions for, e.g., curvature components can be modified when they become large, cutting off classical divergences (in isotropic cosmology they are all inverse powers of the scale factor [18], or spin connection components in anisotropic models [11]). Similarly here, some spin connection components contain information about intrinsic curvature. Their form can be obtained from the general expression (see, e.g., [19])

$$\Gamma_a^i = -\varepsilon^{ijk} e_j^b (\partial_{[a} e_{b]}^k + \tfrac{1}{2} e_k^c e_a^l \partial_{[c} e_{b]}^l) \tag{6}$$

where e_a^i are components of the co-triad (i.e. $e_a^i e_b^i = q_{ab}$) and e_j^b of its inverse. In spherical symmetry, co-triads take a special form just as the metric (3) does. Since it does not matter how a triad is rotated, it need not be exactly invariant under the rotation group acting on space, but it is enough for it to be invariant up to a gauge rotation. This is realized for co-triads of the form

$$\begin{aligned} e_a^i \tau_i \mathrm{d}x^a &= e_x(x)\tau_3 \mathrm{d}x + (e_1(x)\tau_1 + e_2(x)\tau_2)\mathrm{d}\vartheta + (e_1(x)\tau_2 - e_2(x)\tau_1)\sin\vartheta \mathrm{d}\varphi \\ &=: e_x(x)\tau_3 \mathrm{d}x + e_\varphi(x)\bar{\Lambda}(x)\mathrm{d}\vartheta + e_\varphi(x)\Lambda(x)\sin\vartheta \mathrm{d}\varphi \end{aligned} \tag{7}$$

where we use SU(2) generators $\tau_j = -\frac{i}{2}\sigma_j$ with Pauli matrices σ_j, and $\Lambda =: \cos\eta\tau_2 + \sin\eta\tau_1$ and $\bar{\Lambda} := \exp(-\frac{\pi}{2}\tau_3)\Lambda\exp(\frac{\pi}{2}\tau_3)$ are defined to have unit norm in su(2), i.e. $\cos\eta = e_1/e_\varphi$ and $\sin\eta = -e_2/e_\varphi$ with $e_\varphi^2 = e_1^2 + e_2^2$. Infinitesimal rotations of space now act by Lie derivatives on e with respect to superpositions of vector fields $X = \sin\varphi\partial_\vartheta + \cot\vartheta\cos\varphi\partial_\varphi$, $Y = -\cos\varphi\partial_\vartheta + \cot\vartheta\sin\varphi\partial_\varphi$, and $Z = \partial_\varphi$, while gauge rotations of the triad act by conjugation in su(2). We thus obtain explicitly

$$\begin{aligned} \mathcal{L}_X e &= (e_1\tau_1 + e_2\tau_2)\cos\varphi \mathrm{d}\varphi - (-e_2\tau_1 + e_1\tau_2)\frac{\cos\varphi}{\sin\vartheta}\mathrm{d}\vartheta = \left[e, \frac{\cos\varphi}{\sin\vartheta}\tau_3\right] \\ \mathcal{L}_Y e &= (e_1\tau_1 + e_2\tau_2)\sin\varphi \mathrm{d}\varphi - (-e_2\tau_1 + e_1\tau_2)\frac{\sin\varphi}{\sin\vartheta}\mathrm{d}\vartheta = \left[e, \frac{\sin\varphi}{\sin\vartheta}\tau_3\right] \\ \mathcal{L}_Z e &= 0 \end{aligned}$$

showing that any rotation in space simply amounts to a gauge rotation of the triad. The corresponding metric is thus invariant under rotations, and indeed a co-triad (7) implies a metric of the form (3) with

$$q_{xx} = e_x^2 \quad , \quad q_{\varphi\varphi} = e_1^2 + e_2^2 = e_\varphi^2\,. \tag{8}$$

A spherically symmetric spin connection takes the form

$$\Gamma^i_a \tau_i dx^a = \Gamma_x \tau_3 dx + \Gamma_\varphi \bar{\Lambda}^\Gamma d\vartheta + \Gamma_\varphi \Lambda^\Gamma \sin\vartheta d\varphi + \tau_3 \cos\vartheta d\varphi \tag{9}$$

where the last term must be added since a connection transforms differently from a co-triad under gauge transformations. Indeed,

$$\begin{aligned} \mathcal{L}_X(\tau_3\cos\vartheta d\varphi) &= -\tau_3\left(\frac{\sin\varphi}{\sin\vartheta}d\varphi + \frac{\cos\vartheta\cos\varphi}{\sin^2\vartheta}d\vartheta\right) = d\left(\frac{\cos\varphi}{\sin\vartheta}\tau_3\right) \\ \mathcal{L}_Y(\tau_3\cos\vartheta d\varphi) &= -\tau_3\left(\frac{\cos\varphi}{\sin\vartheta}d\varphi - \frac{\cos\vartheta\sin\varphi}{\sin^2\vartheta}d\vartheta\right) = d\left(\frac{\sin\varphi}{\sin\vartheta}\tau_3\right) \end{aligned}$$

and $\mathcal{L}_Z(\tau_3\cos\vartheta d\varphi) = 0$ such that we have the correct transformation of Γ with the same gauge rotation as above.

The explicit formula (6) applied to a spherically symmetric co-triad shows that

$$\Gamma_x = -\eta' \quad , \quad \Gamma_\varphi = -e'_\varphi/e_x \quad , \quad \Lambda^\Gamma = \bar{\Lambda}, \tag{10}$$

with $\bar{\Lambda}$ as defined for the co-triad, such that the φ-component Γ_φ is gauge invariant while Γ_x is pure gauge. Modifications to classical behavior similar to those in cosmological models can now be expected, e.g., from the spin connection component $\Gamma_\varphi = -\sqrt{q_{\varphi\varphi}}'/\sqrt{q_{xx}}$ when metric components become small. Classically, this expression diverges at small q_{xx}, which can be changed in a quantum theory for the corresponding operator. Since, as we will see later, Γ_φ appears in the equations of motion, a modification here would change the behavior of solutions. Horizons of massive black holes would remain unmodified since there both metric components are large. The singularity, however, looks less clear: only for time-like slices does q_{xx} become small, indicating a modification and the possibility of removal of the singularity. But if we approach the singularity on space-like slices with x constant, in the interior N^2 (playing then the role of q_{xx}) remains large which does not suggest modifications. Indeed, the slices then are homogeneous and Γ_φ vanishes identically which means that we will need another measure for the removal of singularities in this case.

At asymptotic infinity, however, we would encounter severe problems since q_{xx} is close to one at which point modifications can already be noticeable, spoiling the classical limit of the theory. This is a sign of using the wrong variables since the modification is a consequence of quantum effects, and the success of a quantization can depend significantly on the choice of fundamental variables. Indeed, there are variables better suited to a demarkation of the different regimes than the metric. This is in particular the case for the densitized triad defined by $E^a_i = e^a_i|\det(e^j_b)|$ where e^a_i is the inverse of the co-triad e compatible with the metric. A spherically symmetric densitized triad is of the general form

$$E = E^x(x)\tau_3\sin\vartheta\frac{\partial}{\partial x} + (E^1(x)\tau_1 + E^2(x)\tau_2)\sin\vartheta\frac{\partial}{\partial\vartheta} + (E^1(x)\tau_2 - E^2(x)\tau_1)\frac{\partial}{\partial\varphi} \tag{11}$$

written down as an su(2) valued densitized vector field. The gauge invariant components are E^x and $(E^\varphi)^2 = (E^1)^2 + (E^2)^2$ whose relation with the metric components is

$$|E^x| = q_{\varphi\varphi} \qquad E^\varphi = \sqrt{q_{xx}q_{\varphi\varphi}} \tag{12}$$

(note that E^x can be positive or negative depending on the orientation $\mathrm{sgn}\det E = \mathrm{sgn} E^x (E^\varphi)^2$ of the triad). The angular components have the same internal directions Λ and $\bar{\Lambda}$ as the co-triad.

For the Schwarzschild solution with $|E^x| = x^2$ and $E^\varphi = x/\sqrt{1-2M/x}$ we now have the following behavior: At asymptotic infinity

$$|E^x| \gg 1 \qquad E^\varphi \gg 1\,,$$

at the horizon

$$|E^x| \gg 1 \qquad E^\varphi \gg 1$$

and at the singularity

$$|E^x| \ll 1 \qquad E^\varphi \ll 1\,.$$

Thus, irrespective of the approach to the singularity, the behavior is just as needed for unmodified classical behavior far away from the black hole all the way up to the horizon, while inverse triad components, such as the spin connection component

$$\Gamma_\varphi = -(E^x)'/2E^\varphi \tag{13}$$

will be modified at the singularity with small E^φ.

2.2 Basic Variables

For detecting the classical singularity it seems much more reliable to use the densitized triad rather than the metric, which is also the case in homogeneous models with an explicit impact on the removal of singularities [10]. Indeed, the densitized triad as a basic variable is important in other ways, too: it arises naturally when one attempts to quantize gravity in a background independent manner. These two issues, the fate of classical singularities and background independence, are superficially quite different but turn out to be deeply related.

Most recent progress in a background independent quantization of general relativity has come after a reformulation in terms of Ashtekar variables [20, 21] where the densitized triad E^a_i plays the role of a momentum canonically conjugate to the Ashtekar connection $A^i_a = \Gamma^i_a - \gamma K^i_a$ with the spin connection Γ^i_a as a function of E^a_i via (6), extrinsic curvature $K^i_a = e^b_i K_{ab}$ and the Barbero–Immirzi parameter $\gamma > 0$ [22]. The extrinsic curvature components here make A^i_a canonically conjugate to E^a_i, while the spin connection provides A^i_a with the transformation properties of a connection. This reformulation thus casts general relativity as a gauge theory and does not only bring it formally closer to other interactions but also leads to a direct way for a background independent quantization.

Usually, a field theory would be quantized by smearing the fields with test functions over 3-dimensional regions so as to make their classical Poisson *-algebra well defined. For instance, a scalar ϕ with Lagrangian $\sqrt{\det q}(\frac{1}{2}\dot{\phi}^2 + \frac{1}{2}q^{ab}\partial_a\phi\partial_b\phi + V(\phi))$ on a background metric q_{ab} (assuming lapse function $N = 1$ and shift vector $N^a = 0$) has momentum $p_\phi = \sqrt{\det q}\dot{\phi}$ which transforms as a density (which is often ignored when the background metric is fixed as, e.g., Minkowski space). This has the singular Poisson relations $\{\phi(x), p_\phi(y)\} = \delta(x,y)$. However, if we smear the fields with test functions f and g on space to obtain $\phi[f] := \int \sqrt{\det q} f(x)\phi(x)\mathrm{d}^3x$ and $p_\phi[g] := \int g(x) p_\phi(x)\mathrm{d}^3x$ we obtain the well-defined Poisson algebra $\{\phi[f], p_\phi[g]\} = \int \sqrt{\det q} f(x)g(x)\mathrm{d}^3x$. This does not contain δ-functions, but does depend on the background metric q which is not available for a background independent formulation of gravity. The very first step of a background independent quantization of general relativity, therefore, has to face the problem that the physical fields, with the metric or densitized triad among them, need to be smeared for a well-defined algebra to be represented on a Hilbert space, but that a background metric must not be introduced.

For a scalar, there is a simple way out: as is easily verified, we still obtain a well-defined algebra if we only smear p_ϕ for which we do not need a background metric since it is already a density. Similarly, in the case of gravity we can evade the problem in Ashtekar's formulation since with connections and densitized vector fields as basic variables there is a natural, background independent smearing leading to a well-defined algebra: Instead of 3-dimensional smearings for all basic fields we use a 1-dimensional smearing of the connection and a 2-dimensional one for the densitized triad, giving rise to holonomies

$$h_e(A) = \mathcal{P}\exp\int_e \tau_i A_a^i \dot{e}^a \mathrm{d}t \tag{14}$$

along edges e in space, and fluxes

$$F_S(E) = \int_S \tau^i E_i^a n_a \mathrm{d}^2y \tag{15}$$

through surfaces S. (We use the tangent vector $\dot{e}^a$ to the curve e and the co-normal n_a to the surface S, both of which are defined without reference to a metric.)

It turns out that this smearing is sufficient for a well-defined classical Poisson algebra which even has a unique diffeomorphism invariant representation [23, 24, 25, 26, 27]. This representation defines the basic framework of loop quantum gravity [28, 29, 30, 31]. States are represented usually in the connection representation $\psi[A]$ on which holonomies act as multiplication operators and fluxes as derivative operators. This can all be done rigorously thanks to a rich structure on the infinite dimensional space of connections which is under much better control than the space of metrics. As a consequence, flux operators have discrete spectra implying a discrete structure of spatial geometry [32, 33, 34] which is also realized in symmetric models [35, 36]. Moreover, since flux spectra are discrete and contain zero, there are no densely defined inverse operators. Instead there are techniques [37] which allow one to quantize co-triad or other inverse components of the basic E_i^a by operators which reduce to the inverse in a classical regime but modify the classical divergence at small values. This has already been described and used above for the spherically symmetric spin connection component. Here, it is important that those expressions are taken as functions of the densitized triad components and not metric components. These effects come from properties of flux operators as basic operators in a background independent formulation which relies on the densitized triad as basic variable and so far is not known in a metric formulation. Indeed, as observed before, the densitized triad is much better suited to separate the classical singularity from other regimes such that modifications are only expected there.

2.3 Dynamics

Up until now we have discussed kinematical properties of the spherically symmetric system. The dynamical behavior of triad and connection (or extrinsic curvature) components is dictated by the Hamiltonian constraint

$$H[N] = (2G)^{-1}\int_B \mathrm{d}x N(x)|E^x|^{-1/2}\left((K_\varphi^2 E^\varphi + 2K_\varphi K_x E^x) + (1-\Gamma_\varphi^2)E^\varphi + 2\Gamma_\varphi' E^x\right) \quad (16)$$

in terms of the spin connection component Γ_φ as before and the extrinsic curvature components in

$$K = K_x(x)\tau_3 \mathrm{d}x + (K_1(x)\tau_1 + K_2(x)\tau_2)\mathrm{d}\vartheta + (K_1(x)\tau_2 - K_2(x)\tau_1)\sin\vartheta \mathrm{d}\varphi \quad (17)$$

where again only K_x and $K_\varphi^2 = K_1^2 + K_2^2$ are gauge invariant. In addition, there is the diffeomorphism constraint

$$D[N^x] = (2G)^{-1}\int_B N^x(x)(-2E^\varphi K_\varphi' + K_x E^{x\prime}). \quad (18)$$

Physical fields $(K_x, E^x; K_\varphi, E^\varphi)$ have to solve the constraint equations $H[N] = 0 = D[N^x]$ for all functions N and N^x on B (except for possible boundary conditions which we ignore here) and evolve in coordinate time according to Hamiltonian equations of motion $\dot{E}^x = \{E^x, H[N] + D[N^x]\}$, etc. to be computed with the Poisson relations $\{K_x(x_1), E^x(x_2)\} = -2G\delta(x_1, x_2)$, $\{K_\varphi(x_1), E^\varphi(x_2) = -G\delta(x_1, x_2)$. For the triad components this gives

$$\begin{aligned} \dot{E}^x &= 2NK_\varphi\sqrt{|E^x|} + N^x E^{x\prime} && (19)\\ \dot{E}^\varphi &= N(K_\varphi E^\varphi + K_x E^x)|E^x|^{-1/2} + (N^x E^\varphi)' && (20) \end{aligned}$$

which, when solved for the extrinsic curvature components, agrees with their geometrical definition via

$$K_a^i = e_i^b K_{ab} = (2N)^{-1} e_i^b \mathcal{L}_{\partial_t - N^x\partial_x} e_a^j e_b^j \quad (21)$$

from (4) and (2). Evaluating this for a spherically symmetric co-triad (7) or densitized triad (11) indeed gives spherically symmetric components

$$K_x = N^{-1}(\dot{e}_x - (N^x e_x)') \quad , \quad K_\varphi = N^{-1}(\dot{e}_\varphi - N^x e_\varphi')$$

and the same internal directions $\Lambda^K = \Lambda$, $\bar{\Lambda}^K = \bar{\Lambda}$ as those of the triad. The extrinsic curvature components then have Hamiltonian equations of motion

$$\begin{aligned} \dot{K}_x &= -NK_\varphi K_x|E^x|^{-1/2} + \tfrac{1}{2}NK_\varphi^2 E^\varphi|E^x|^{-3/2} + (N^x K_x)' \\ &\quad + \tfrac{1}{2}N|E^x|^{-1/2}\left(E^\varphi|E^x|^{-1} - \tfrac{1}{4}(E^{x\prime})^2(|E^x|E^\varphi)^{-1} - E^{x\prime}E^{\varphi\prime}(E^\varphi)^{-2} + E^{x\prime\prime}(E^\varphi)^{-1}\right) \\ &\quad + \tfrac{1}{2}N'\left(E^{x\prime}(E^\varphi)^{-1}|E^x|^{-1/2} - 2\sqrt{|E^x|}E^{\varphi\prime}(E^\varphi)^{-2}\right) + N''\sqrt{|E^x|}(E^\varphi)^{-1} && (22)\\ \dot{K}_\varphi &= -\tfrac{1}{2}NK_\varphi^2|E^x|^{-1/2} + N^x K_\varphi' \\ &\quad + \tfrac{1}{2}N|E^x|^{-1/2}(\tfrac{1}{4}(E^{x\prime})^2(E^\varphi)^{-2} - 1) + \tfrac{1}{2}N'\sqrt{|E^x|}E^{x\prime}(E^\varphi)^{-2}. && (23) \end{aligned}$$

These coupled non-linear equations are difficult to solve in general, but the Schwarzschild solution can easily be reproduced by assuming staticity: $K_x = K_\varphi = N^x = 0$ which already implies that the diffeomorphism constraint is satisfied. Equations (22) and (23) then assume the form of consistency conditions for the lapse function in order to ensure the existence of a static slicing. Both conditions are identically satisfied for a lapse function

$$N \propto E^{x\prime}/E^\varphi \tag{24}$$

using that E^φ and E^x are subject to the constraint equation

$$(\Gamma_\varphi^2 - 1)E^\varphi - 2\Gamma'_\varphi E^x = (\tfrac{1}{4}(E^{x\prime})^2/(E^\varphi)^2 - 1)E^\varphi + (E^{x\prime}/E^\varphi)' E^x = 0$$

following from (16) with $K_x = 0 = K_\varphi$ and Γ_φ from (13).

It remains to solve this constraint for E^x and E^φ. If we choose our radial coordinate such that $|E^x| = x^2$, this simplifies to a differential equation

$$-2x^3 E^{\varphi\prime} + 3x^2 E^\varphi - (E^\varphi)^3 = 0$$

whose solution $E^\varphi(x) = x(1 + c/x)^{-1/2}$ is the Schwarzschild component for E^φ with $c = -2M$, which then also reproduces the correct lapse function from (24).

This shows how the dynamical equations appear in a canonical formalism, and also how special the simplicity of the static Schwarzschild solution is. With slight modifications to the equations, e.g. coming from quantum modifications, the assumption of staticity will no longer be consistent since two conditions $\dot{K}_x = 0 = \dot{K}_\varphi$ have to be satisfied by only one function N. Thus, quantum corrections are expected to change the static behavior of the classical solution, even though it would come from only small changes outside the horizons of massive black holes. What this means for the inside where quantum effects dominate around the singularity has to be analyzed by direct methods from quantum gravity.

3 Quantization: Overview

Even though the vacuum spherically symmetric system has only a finite number of physical degrees of freedom given by the black hole ADM mass and its conjugate momentum [38, 39, 40], a Dirac quantization requires field theory techniques in order to deal with infinitely many kinematical degrees of freedom. Almost all of these degrees of freedom will then be removed by the Hamiltonian constraint which acts as a functional differential or difference operator. Thus, many of the field theoretic aspects of the full theory can be probed here which also implies a corresponding level of complication. So far, the system is not fully understood in a loop quantization even in the vacuum case, and other techniques which can be applied more easily to this system do not allow definitive conclusions about the singularity. It is therefore necessary at this stage to refer to approximation methods. These methods allow different glimpses which one can then try to bring together for a consistent picture. Here, we briefly collect different classes of approximations, which will be described in more detail in the following sections.

3.1 Homogeneous Techniques

Currently, loop techniques for homogeneous geometries, following techniques introduced in [41], are fully developed to a degree that one can analyze properties of physical solutions. (The main open issue is the physical inner product, about which not much is known even in the simplest cases [42, 43].) There are explicit expressions for the most important operators such as the volume operator [35], matter Hamiltonians or the Hamiltonian constraint [9, 10, 11] which is a big advantage compared to the full theory where the ubiquitous volume operator cannot be diagonalized even in principle. The constraint equation takes the form of a difference equation for the wave function in the triad representation which explicitly shows how one can evolve through the classical singularity. Moreover, one can define effective classical equations with diverse correction terms [44, 45, 46, 47, 48, 49, 50]. They capture the main quantum effects [44, 51, 52, 53, 54, 55, 56,] and can be analyzed more easily than the quantum difference equation directly (see, e.g., [59, 60, 61, 62, 63, 64, 65]).

In some cases these effective classical equations provide an intuitive explanation for the removal of singularities since they display bouncing behavior of a cosmological solution. This can also be used to model the case of matter collapsing into a black hole. As a model, the ball of matter can be assumed to be homogeneous such that the collapse of the outer shell radius is described by effective equations for an isotropic system. These equations are modified at small scales, i.e. when the ball collapses to a certain size. In the modified regime there are matter systems which show a bounce, which now can be interpreted as the collapsing matter parts repelling each other and bouncing back after maximal contraction. So far, this is not much different from a bouncing universe and indeed described by the same equations. The difference is that the matter ball does not present the full system, but that there is also the outside. Without specifying the matter content there, one can try to match the interior to a generalized Vaidya metric outside allowing for matter radiated away. This allows to study the formation or disappearance of horizons which may or may not shield the bounce replacing the classical singularity [66].

Limitations of these techniques are that only the interior carries quantum effects, while the outside is described by a generalized Vaidya metric of general relativity. Some quantum effects are transported to the outside by matching to the effective interior, which then enter the Vaidya solution effectively through a non-standard energy momentum tensor. This still shows possible changes in the behavior of horizons, but is of course more indirect than a complete inhomogeneous analysis.

A different approach using homogeneous techniques only applies to the Schwarzschild solution which is homogeneous inside the horizon. One can then describe the interior by a quantum equation which as in cosmological cases, is a difference equation not breaking down at the classical singularity. Also here we thus obtain a mechanism to evolve through a classical singularity, and there are many more non-trivial aspects which only arise in a loop quantization and show its consistency [67]. In particular, the singularity is removed, but the horizon which presents another boundary to the classical interior remains.

3.2 Extrapolation

The previous analysis provides a picture of a non-zero Schwarzschild black hole interior which one can now extrapolate in two ways: The non-singular interior first has to be embedded in a full space-time which can happen in several different ways. Moreover, for a realistic black hole this must be generalized to the presence of matter. While there are many gaps to be filled in by detailed constructions and calculations, one can already see different implications for the issue of information loss [68].

3.3 Inhomogeneous Techniques

Operators for the spherically symmetric system (with or without matter) are now available explicitly at a level similar to that in homogeneous models [13, 14]. In particular, there is a similar simplification in the volume operator which translates to matrix elements of the Hamiltonian constraint also being known explicitly. However, the constraint is much more difficult to analyze since it now presents a functional difference equation in infinitely many kinematical variables. The construction and regularization of the constraint is more subtle compared to homogeneous cases, but similar to the full theory where there are different versions. These can then be studied explicitly and their physical implications analyzed, leading possibly to conclusions as to which operator is most suited for the full theory.

Even though the singularity issue is not yet solved in generality, there are indications that a mechanism similar to that in homogeneous models is at work. This will then provide a large class of systems where one and the same mechanism, derived from basic loop properties, provides a removal of singularities in non-trivial ways.

There are regimes where the constraint operator can be approximated by a simpler expression. Interestingly, this is true in particular in the neighborhood of isolated or slowly evolving horizons [69] such that horizon properties such as fluctuations and its growth from infalling matter or shrinking from Hawking radiation can be analyzed. Moreover, the regime provides perturbation techniques which allow us to study general properties of the constraint operator and matter Hamiltonians.

3.4 Full Theory

The full theory has a rigorous quantum representation [70] and well-defined candidates for the Hamiltonian constraint [71]. Understanding the dynamics in general is certainly very complicated, and even computing matrix elements of the constraint is involved. Most full results which contribute to the physical picture are thus non-dynamical: Spatial geometry is discrete [32, 33, 34] as a characteristic of the full quantum representation, and there are well-defined quantum matter Hamiltonians [37]. Black hole (and cosmological) horizons can be introduced as a boundary provided they are isolated [2]. This condition ensures that the dynamics at the boundary is not essential and allows the correct counting of black hole entropy [7, 8].

4 Homogeneous Techniques

In the Schwarzschild interior $r < 2M$ one can choose a homogeneous slicing such that the metric is of the Kantowski–Sachs form

$$ds^2 = -N(T)^2 dT^2 + (2M/T - 1) dR^2 + T^2 d\Omega^2 \qquad (25)$$

with $T = r$, $R = t$ and a lapse function $N(T)^2 = T/(2M - T)$. The spatial metric is then related to a homogeneous triad of the form (11) where E^x and E^φ are constants on spatial slices. Their conjugates are given by Ashtekar connection components of the general spherically symmetric and homogeneous form

$$A_a^i \tau_i dx^a = A_x \tau_3 dx + A_\varphi \bar{\Lambda}^A d\vartheta + A_\varphi \Lambda^A \sin\vartheta d\varphi + \tau_3 \cos\vartheta d\varphi \tag{26}$$

which in this case are simply proportional to extrinsic curvature components $K_x = -A_x/\gamma$ and $K_\varphi = -A_\varphi/\gamma$ since $\Gamma = \tau_3 \cos\vartheta d\varphi$ from (9) and (10) with homogeneity. Moreover, $\Lambda^A = \Lambda$ (as defined for the triad) follows from the Gauss constraint. Since Λ is constant in a homogeneous model and subject to gauge rotations, we will fix it to $\Lambda = \tau_2$ in this section, such that $\bar{\Lambda} = \tau_1$. The symplectic structure for the 4-dimensional phase space is determined by $\{K_x, E^x\} = -2G$, $\{K_\varphi, E^\varphi\} = -G$.

4.1 Quantum Representation

Loop quantum gravity is based on spin network states which are generated by holonomies as multiplication operators. Similarly, homogeneous models in loop quantum cosmology are based on a representation [72] which emerges from holonomies of homogeneous connections and which turns out to be inequivalent to the usual Schrödinger representation used in a Wheeler–DeWitt like quantization. For the Kantowski–Sachs model an orthonormal basis of states is given by the family

$$\langle K_\varphi, K_x | \mu, \nu \rangle = \exp(-\tfrac{i}{2}\gamma(\mu K_\varphi + \nu K_x)) \qquad \mu, \nu \in \mathbb{R}, \mu \geq 0 \tag{27}$$

such that the kinematical Hilbert space is non-separable. (There are arguments to reduce this to a separable Hilbert space as in [73] using properties of observables [74].) One can see one of the basic loop properties that only exponentials of connection or extrinsic curvature components are represented directly, but not the components themselves: It is clear that, e.g., $\exp(-i\gamma K_x/2)$ acts directly as a shift operator

$$\widehat{\exp(-i\gamma\kappa K_x)} |\mu, \nu\rangle = |\mu, \nu + 2\kappa\rangle \tag{28}$$

but since this operator family is not represented continuously, this does not allow us to obtain an operator for K_x by differentiation. Indeed,

$$\langle \mu, \nu | \widehat{\exp(-i\gamma\kappa K_x)} | \mu, \nu \rangle = \langle \mu, \nu | \mu, \nu + 2\kappa \rangle = \delta_{0,\kappa}$$

is not continuous at $\kappa = 0$. This is different from a Wheeler–DeWitt quantization where extrinsic curvature components would be basic operators represented directly. Instead, here we have to express those components through holonomies such as $\exp(\gamma K_x \tau_3) = \cos(\frac{1}{2}\gamma K_x) + 2\tau_3 \sin(\frac{1}{2}\gamma K_x)$ and use the action

$$\cos(\tfrac{1}{2}\gamma K_x)|\mu, \nu\rangle = \tfrac{1}{2}(|\mu, \nu + 1\rangle + |\mu, \nu - 1\rangle) \tag{29}$$

$$\sin(\tfrac{1}{2}\gamma K_x)|\mu, \nu\rangle = \tfrac{i}{2}(|\mu, \nu - 1\rangle - |\mu, \nu + 1\rangle). \tag{30}$$

Another difference to the Wheeler–DeWitt representation arises for triad operators which in the Wheeler–DeWitt case would be simply multiplication operators on a wave function in the metric representation and thus have continuous spectra. In the loop case, however, the triad operators

$$\hat{E}^x = i\frac{\ell_{\rm P}^2}{4\pi}\frac{\partial}{\partial K_x} \qquad \hat{E}^\varphi = i\frac{\ell_{\rm P}^2}{8\pi}\frac{\partial}{\partial K_\varphi} \tag{31}$$

with the Planck length $\ell_{\rm P} = \sqrt{8\pi G\hbar}$ have the basis states (27) as eigenstates

$$\hat{E}^x|\mu,\nu\rangle = \tfrac{1}{8\pi}\gamma\ell_{\rm P}^2\nu|\mu,\nu\rangle \qquad \hat{E}^\varphi|\mu,\nu\rangle = \tfrac{1}{16\pi}\gamma\ell_{\rm P}^2\mu|\mu,\nu\rangle \tag{32}$$

and thus discrete spectra (i.e., normalizable eigenstates). Again, this is different from the Wheeler–DeWitt quantization but directly analogous to full loop quantum gravity. In particular the volume $V = 4\pi E^\varphi\sqrt{|E^x|}$ has a quantization with discrete spectrum with eigenvalues

$$V_{\mu,\nu} = 2\pi(\gamma\ell_{\rm P}^2/8\pi)^{3/2}\mu\sqrt{|\nu|}\,. \tag{33}$$

4.2 Inverse Triad Components

It is often necessary to quantize inverse powers of densitized triad components, for instance for matter Hamiltonians or curvature components. Since the basic triad operators have discrete spectra containing zero, they do not have densely defined inverses which could otherwise be used for this purpose. Nevertheless one can proceed, and in the end have regular properties, by rewriting the classical inverse in an equivalent way and quantizing the new expression [37]. We demonstrate this for the spatial curvature given by ${}^3R = 2/|E^x|$ for which we need an inverse of E^x. This can be taken as a measure for the classical singularity where it diverges. Since there is no direct way of quantizing this expression via an inverse of $\hat{E}^x$ we first write

$$\frac{E^\varphi\,{\rm sgn}\,E^x}{2\sqrt{|E^x|}} = \frac{-1}{8\pi G}\{K_x,V\} = \frac{1}{4\pi\gamma G}\,{\rm tr}\,\tau_3 e^{-\gamma K_x\tau_3}\{e^{\gamma K_x\tau_3},V\}$$

where the first step replaces the inverse power of E^x by only positive powers occurring in V at the expense of introducing K_x for which we do not have a direct quantization. Nevertheless, in the next step we obtain an equivalent expression which only contains exponentials of K_x which we can quantize directly. Using the volume operator and turning the Poisson bracket into a commutator then yields a densely defined operator

$$\begin{aligned}\widehat{\frac{E^\varphi\,{\rm sgn}\,E^x}{\sqrt{|E^x|}}} &= \frac{-i}{2\pi\gamma G\hbar}{\rm tr}\,\tau_3 e^{-\gamma K_x\tau_3}[e^{\gamma K_x\tau_3},\hat{V}] \\ &= \frac{4i}{\gamma\ell_{\rm P}^2}(\sin(\tfrac{1}{2}\gamma K_x)\hat{V}\cos(\tfrac{1}{2}\gamma K_x)-\cos(\tfrac{1}{2}\gamma K_x)\hat{V}\sin(\tfrac{1}{2}\gamma K_x))\end{aligned} \tag{34}$$

with eigenvalues

$$\frac{2}{\gamma\ell_{\rm P}^2}(V_{\mu,\nu+1}-V_{\mu,\nu-1}) = \tfrac{1}{2}\sqrt{\frac{\gamma\ell_{\rm P}^2}{8\pi}}\mu(\sqrt{|\nu+1|}-\sqrt{|\nu-1|})\,. \tag{35}$$

Since $\hat{E}^\varphi$ has eigenvalues $\gamma\ell_P^2\mu/16\pi$, we can write

$$\frac{\widehat{\operatorname{sgn} E^x}}{\sqrt{|E^x|}}|\mu,\nu\rangle = (\gamma\ell_P^2/8\pi)^{-1/2}(\sqrt{|\nu+1|}-\sqrt{|\nu-1|}) \quad (36)$$

which has the expected behavior for $|\nu| \gg 1$ but behaves very differently from the classical expectation for small $|\nu|$.

Taking this as a measure for the singularity indicates that it is removed in quantum gravity since the eigenvalues remain finite even when $\nu = 0$ at the classical singularity. Nevertheless, a final confirmation of an absence of singularities can only come from considerations of the dynamics which must allow us to evolve further even when we reach a point corresponding to the classical singularity. Only then can we conclude that the singularity as a boundary of space-time has been removed.

4.3 Dynamics

The spherically symmetric Hamiltonian constraint (16) can be used to find the expression for the homogeneous Kantowski–Sachs interior

$$H[N] = (2G)^{-1}N|E^x|^{-1/2}\left((K_\varphi^2+1)E^\varphi + 2K_\varphi K_x E^x\right)\quad (37)$$

where we used $\Gamma_\varphi = 0$ with homogeneity in (13). There are different terms in this expression, those quadratic in K and the K-independent one which comes from the spin connection in the curvature of the Ashtekar connection. In the full theory there would only be curvature components of A_a^i in the Euclidean part $\varepsilon_{ijk}F_{ab}^iE_j^aE_k^b$ of the constraint, which can be represented via holonomies by using

$$s_1^a s_2^b F_{ab}^i(x)\tau_i = (h_\alpha - 1)/\Delta + O(\Delta)$$

where α is a small loop of coordinate area Δ and with tangent vectors s_1 and s_2. For small Δ in a limit removing a regulator one can use h_α as an excellent approximation for the curvature components, and stick this together with quantizations of the triad components to obtain a quantization of the constraint [71]. This is different in a homogeneous context (or in any symmetric model where some directions are homogeneous) because we have only exponentials of connection components, but not holonomies with an adjustable edge length that shrinks in a continuum limit. Nevertheless, since the constraint operator in the full theory is based on holonomies quantizing the F-components, this has to be the case also for symmetric models related to the full theory. The only possibility to use h_α as a good approximation is then given when the arguments of the exponentials are small in semiclassical regimes where the classical constraint is to be reproduced. In other regimes, one does not expect the classical constraint to be of any value for guidance and in fact usually obtains strong quantum corrections.

In a semiclassical regime one has small curvature such that the extrinsic curvature components can be assumed to be small when checking the classical limit of the constraint. However, Ashtekar connection components are not necessarily small since for them also the spin connection plays a role. Here, another difference to the full theory arises: while in general spin connection components do not have coordinate independent meaning and in fact can be made arbitrarily small in any neighborhood, some of the components (such as (13) in the spherically symmetric model) obtain invariant meaning in a symmetric context where only transformations respecting the symmetry are allowed. Usually, unless the model has flat symmetry orbits such that the spin connection vanishes, one cannot expect the components to be small even in semiclassical regimes. This requires a special treatment of the spin connection in symmetric models, which is possible in a general manner [11, 14]. For this reason we have started the quantization in this model with extrinsic curvature components and will also use them instead of Ashtekar connection components in holonomies when constructing the Hamiltonian constraint.

One may ask what the relation to the full theory then is where holonomies of the Ashtekar connection are basic, while holonomies of a tensor such as extrinsic curvature cannot even be defined. The arguments presented before explain why extrinsic curvature is important to analyze the classical limit, but this does not show the contact to the full theory. This will be much clearer in inhomogeneous models which are in between homogeneous ones and the full theory. Here, we will have directions along symmetry orbits, for which the techniques just described will apply, and inhomogeneous directions for which we will use holonomies of the Ashtekar connection as in the full theory. As we will discuss in more detail in the quantization of the spherically symmetric model, all this fits into a general scheme which allows to derive expressions in all different classes of models.

We can now express the terms quadratic in curvature components via holonomies, such as

$$K_\varphi^2+1=-\frac{2}{\gamma^2\delta^2}\operatorname{tr}\tau_3(e^{-\delta\gamma K_\varphi\tau_1}e^{-\delta\gamma K_\varphi\tau_2}e^{\delta\gamma K_\varphi\tau_1}e^{\delta\gamma K_\varphi\tau_2}+\gamma^2\delta^2\tau_3)+O(\delta)$$

and

$$K_xK_\varphi=\frac{2}{\gamma^2\delta^2}\operatorname{tr}\tau_1(e^{-\delta\gamma K_x\tau_2}e^{-\delta\gamma K_\varphi\tau_3}e^{\delta\gamma K_x\tau_2}e^{\delta\gamma K_\varphi\tau_3})+O(\delta)\,.$$

Triad components, together with Pauli matrices in the traces, can be obtained in the right combinations from the Poisson brackets

$$\tau_3\frac{E^\varphi}{\sqrt{|E^x|}}=-\frac{1}{4\pi\gamma\delta G}e^{-\delta\gamma K_x\tau_3}\{e^{\delta\gamma K_x\tau_3},V\}$$

as already used for (34), and

$$\tau_1\sqrt{|E^x|}=-\frac{1}{4\pi\gamma\delta G}e^{-\delta\gamma K_\varphi\tau_1}\{e^{\delta\gamma K_\varphi\tau_1},V\}\,.$$

In all expressions, besides the volume $V=4\pi\int\mathrm{d}x\sqrt{|E^x|}E^\varphi$ only holonomies $h_x^{(\delta)}:=e^{-\gamma\delta K_x\tau_3}$, $h_\vartheta^{(\delta)}:=e^{-\gamma\delta K_\varphi\tau_1}$ and $h_\varphi^{(\delta)}:=e^{-\gamma\delta K_\varphi\tau_2}$ of the symmetric Ashtekar connection (26), expressed through extrinsic curvature components, occur which can be quantized directly.

In a more symmetric form, which as we will see later also applies in general, we write

$$\begin{aligned}\frac{(K_\varphi^2+1)E^\varphi+2K_xK_\varphi E^x}{\sqrt{|E^x|}}&\sim\frac{1}{2\pi\gamma^3\delta^3G}\operatorname{tr}((h_\vartheta h_\varphi h_\vartheta^{-1}h_\varphi^{-1}+\gamma^2\delta^2\tau_3)h_x\{h_x^{-1},V\}\\&\quad+h_xh_\vartheta h_x^{-1}h_\vartheta^{-1}h_\varphi\{h_\varphi^{-1},V\}+h_\varphi h_xh_\varphi^{-1}h_x^{-1}h_\vartheta\{h_\vartheta^{-1},V\})\\&=\frac{1}{4\pi\gamma^3\delta^3G}\sum_{IJK}\varepsilon^{IJK}\operatorname{tr}\left((h_Ih_Jh_I^{-1}h_J^{-1}-\gamma^2\delta^2F(\Gamma)_{IJ})h_K\{h_K^{-1},V\}\right)\end{aligned}$$

with the curvature components $F(\Gamma)_{IJ}$ of the spin connection, i.e. here

$$F(\Gamma) = \mathrm{d}\Gamma = -\tau_3 \sin\vartheta \mathrm{d}\vartheta \wedge \mathrm{d}\varphi$$

such that only $F(\Gamma)_{\vartheta\varphi} := i_{X_\varphi} i_{X_\vartheta} F(\Gamma) = -\tau_3$ appears, with the symmetry generators $X_\vartheta = \partial_\vartheta$ and $X_\varphi = (\sin\vartheta)^{-1}\partial_\varphi$.

Quantizing and evaluating the action explicitly through the action of basic operators leads to a constraint operator of the form

$$\begin{aligned}
\hat{H}^{(\delta)} &= -iN(\gamma^3\delta^3 G\ell_{\mathrm{P}}^2)^{-1} \sum_{IJK} \varepsilon^{IJK} \operatorname{tr}\left((h_I^{(\delta)} h_J^{(\delta)} h_I^{(\delta)-1} h_J^{(\delta)-1} - \gamma^2\delta^2 F(\Gamma)_{IJ}) h_K^{(\delta)} [h_K^{(\delta)-1}, \hat{V}] \right) \\
&= -2iN(\gamma^3\delta^3 G\ell_{\mathrm{P}}^2)^{-1} \left(8 \sin\frac{\delta\gamma K_\varphi}{2} \cos\frac{\delta\gamma K_\varphi}{2} \sin\frac{\delta\gamma K_x}{2} \cos\frac{\delta\gamma K_x}{2} \right. \\
&\quad \times \left(\sin\frac{\delta\gamma K_\varphi}{2} \hat{V} \cos\frac{\delta\gamma K_\varphi}{2} - \cos\frac{\delta\gamma K_\varphi}{2} \hat{V} \sin\frac{\delta\gamma K_\varphi}{2} \right) \\
&\quad + \left(4 \sin^2\frac{\delta\gamma K_\varphi}{2} \cos^2\frac{\delta\gamma K_\varphi}{2} + \gamma^2\delta^2 \right) \\
&\quad \left. \times \left(\sin\frac{\delta\gamma K_x}{2} \hat{V} \cos\frac{\delta\gamma K_x}{2} - \cos\frac{\delta\gamma K_x}{2} \hat{V} \sin\frac{\delta\gamma K_x}{2} \right) \right)
\end{aligned} \tag{38}$$

where $\delta > 0$ is regarded as a parameter analogous to the edge length in the full theory. From the holonomy operators one obtains shifts in the labels when acting on a state $|\mu, \nu\rangle$ which in the triad representation given by the coefficients $\psi_{\mu,\nu}$ in a decomposition $|\psi\rangle = \sum_{\mu,\nu} \psi_{\mu,\nu} |\mu, \nu\rangle$ leads to the difference equation

$$\begin{aligned}
0 = (\hat{H}^{(\delta)}\psi)_{\mu,\nu} &= 2\delta\sqrt{|\nu + 2\delta|}(\psi_{\mu+2\delta,\nu+2\delta} - \psi_{\mu-2\delta,\nu+2\delta}) \\
&+ \tfrac{1}{2}(\sqrt{|\nu+\delta|} - \sqrt{|\nu-\delta|})\left((\mu+4\delta)\psi_{\mu+4\delta,\nu} - 2(1+\gamma^2\delta^2)\mu\psi_{\mu,\nu} + (\mu-4\delta)\psi_{\mu-4\delta,\nu}\right) \\
&- 2\delta\sqrt{|\nu - 2\delta|}(\psi_{\mu+2\delta,\nu-2\delta} - \psi_{\mu-2\delta,\nu-2\delta}) \,.
\end{aligned} \tag{39}$$

We are now in a position to analyze whether or not there is a singularity in the quantum theory. There are a few key differences to the usual classical formulation, the first one coming from the fact that we are using triad variables. Compared to a metric formulation, this provides us with an additional sign factor $\operatorname{sgn} E^x$ determining the orientation of space. Accordingly, there are two regions of minisuperspace separated by the line $E^x = 0$ where the classical singularity would be. We have already seen that the classical divergence of inverse powers of E^x does not occur in a loop quantization, but the real test of a singularity can only come from the dynamics: starting with initial values in one region of minisuperspace we need to find out whether we can uniquely evolve to the other side, through the classical singularity. In the quantum theory this is done for the wave function which we can prescribe for sufficiently many initial values at some $\nu > 0$ and additional boundary values at $\mu = 0$ so as to provide a good initial value formulation for the difference equation (39) as described in detail in [10]. One can then see by direct inspection that indeed this will uniquely fix the wave function not just at positive ν where we started, but also at negative ν, at the other side of the classical singularity. Thus, quantum geometry automatically allows us to evolve through the classical singularity which therefore is removed from quantum gravity.

Intuitively, the region of negative ν corresponds to a region of a space-time diagram at the other side of the singularity, as sketched in Fig. 1, which therefore is no longer a boundary but a region of high curvature where the classical theory and its smooth space-time picture break down [67].

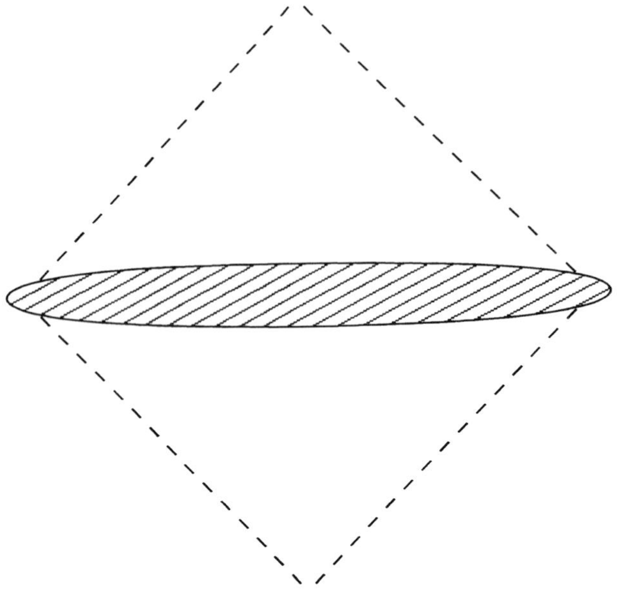

Figure 1: Interior of a Schwarzschild black hole with the quantum region (hatched) replacing the classical singularity. This allows to extend space-time to the new upper region. How these regions are embedded in a full space-time is left unspecified here.

There are many basic aspects which are playing together in just the right way for this result to hold true. They all come directly from the loop quantization and are not put in by hand; in fact, they had been recognized as essential for a background independent quantization a few years before their role in removing classical singularities emerged. The loop representation is important in two ways since via discrete triad spectra it leads to the kinematical results of non-diverging inverse powers of densitized triad components, and through the representation of holonomies to the dynamical constraint as a difference operator. Moreover, the theory is based on densitized triads which, as discussed before, has consequences for the position of classical singularities in minisuperspace important for how one can evolve through them. This automatically provides us with the sign factor from orientation and thus a region beyond the classical singularity. Still, also the dynamical law has to be of the right form for an evolution to this other side of minisuperspace to be possible.

Thus, we have a few essential effects which automatically come from a loop quantization. Once recognized and identified, they can easily be copied in other quantization schemes inspired by loop quantum gravity and cosmology. However, in such a case one has to guess anew in each model what the relevant basic properties would be since there is no underlying scheme for guidance. With the loop quantization we have such a general scheme which just needs to be evaluated in different models. Only then can the results be regarded as reliable expectations for quantum gravity, rather than possibly artificial consequences of choices made. The sign of triad components, for instance, appears automatically and then gives rise to the additional side of the classical singularity to which we can evolve. In loop inspired approaches without a link to the full theory, however, the sign is introduced by hand by extending the range of metric variables to negative values. While this leads to similar results in isotropic models [75], except that the geometrical meaning of the sign remains unclear, there are differences in the black hole interior [76]. In particular, this approach would suggest that even the horizon can be penetrated by the homogeneous quantum evolution despite the fact that space-time becomes inhomogeneous outside. This problem does not occur in the quantization described here since the horizon remains a boundary (corresponding to $\mu = 0$).

4.4 Effective Dynamics

The non-singular quantum dynamics is obviously very different from the classical one even though they can be shown to agree in classical regions at large densitized triad components and small curvature [77]. In between, there is a regime where equations of motion of the classical type, i.e. ordinary differential equations in coordinate time, should be able to describe the system even though quantum effects are already at work. One can think of these equations as describing the position of wave packets which spread only slightly in semiclassical regimes [45, 46, 47]. Quantum effects will then provide modifications, e.g. where inverse powers of densitized triad components occur in a matter Hamiltonian which are replaced by regular expressions in quantum geometry. This provides different means to calculate implications of quantum effects which can so far be done in homogeneous situations.

For instance if we assume the distribution of a matter system collapsing into a black hole to be isotropic, its outer radius $Ra(t)$ is described by a solution $a(t)$ to the Friedmann equation, with R being the coordinate radius where we cut off spatial slices from a closed FRW model. If we choose a scalar ϕ with potential $V(\phi)$ and momentum p_ϕ, we have the Friedmann equation

$$a(\dot{a}^2+1) = \frac{8\pi G}{3}\left(\tfrac{1}{2}a^{-3}p_\phi^2 + a^3 V(\phi)\right) \tag{40}$$

which develops a singularity corresponding to the part of the final black hole singularity covered by matter.

The corresponding effective classical equations are modified by replacing the classically diverging a^{-3} in the matter Hamiltonian with a regular function $d(a)$ derived from finite inverse scale factor operators such as (34) [78, 79]. Including two ambiguity parameters j (a half integer) and $0 < l < 1$, this can be parameterized as

$$d(a)_{j,l} := a^{-3} p_l(3a^2/\gamma j \ell_{\mathrm{P}}^2)^{3/(2-2l)} \tag{41}$$

with

$$\begin{aligned} p_l(q) \;=\; & \frac{3}{2l}q^{1-l}\left(\frac{1}{l+2}\left((q+1)^{l+2}-|q-1|^{l+2}\right)\right. \\ & \left. -\frac{1}{l+1}q\left((q+1)^{l+1}-\mathrm{sgn}(q-1)|q-1|^{l+1}\right)\right). \end{aligned} \tag{42}$$

The essential property of $d(a)_{j,l}$ is that it is increasing from zero for $a^2 < \frac{1}{3}\gamma j\ell_P^2$ which through the Friedmann equation implies a different dynamical behavior at small volume. This model then provides an intuitive explanation for the removal of classical singularities even at the effective level since the equations lead to a bounce at non-zero a: due to the modified density the kinetic term is negligible at small a, and matter evolution equations from the modified matter Hamiltonian imply friction of ϕ [80]. The potential term is then dominating and almost constant which means that the bounce is approximately of de Sitter form [52]. In this interpretation of collapsing matter this means that it does not collapse completely but rebounds after a point of minimal contraction is reached.

So far, we had only access to the inside of the matter contribution which we assumed to be isotropic. The solution can now be matched to a suitable solution describing the outside, which would be able to tell us, for instance, whether horizons form. For pressure-less matter one can match to the static Schwarzschild solution as in the Oppenheimer–Snyder model [81]. This is the case classically only for dust, which however can develop pressure if quantum modifications come into play. (This is in agreement with our earlier observations that quantum effects will not allow the presence of a static vacuum solution.) We have thus chosen the more general scalar matter which has pressure even classically. Physically, pressure leads to shock waves at the outer boundary giving rise to a non-static exterior. This can be described by a generalized Vaidya metric

$$ds^2 = -(1-2M(\chi,v)/\chi)dv^2 + 2dvd\chi + \chi^2 d\Omega^2 \tag{43}$$

which we can match to the interior (Fig. 2) by requiring equal induced metric and extrinsic curvature at the time-like matching surface Σ defined by $r = R$ inside and $\chi = \chi(v)$ outside.

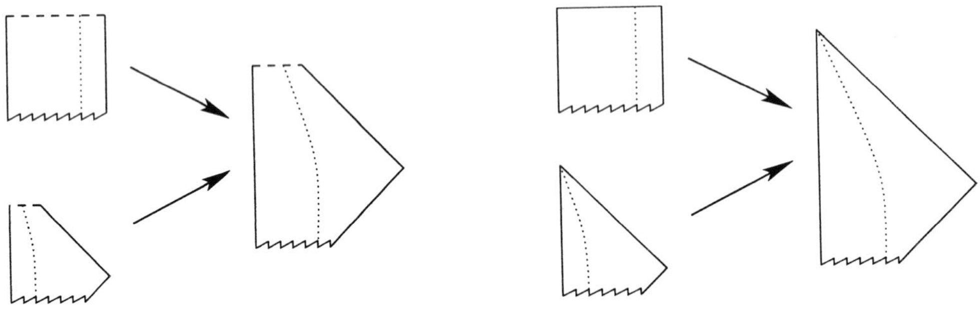

Figure 2: A closed Friedmann–Robertson–Walker model and a generalized Vaidya metric are matched to form a collapse model. Singularities are indicated by dashed lines and the matching surface by dotted lines. The bottom parts of the diagrams are cut off since they depend on details of the solutions. The left hand side shows the classical case with a future singularity, while the right hand side shows the singularity-free effective case.

In this way we can see what the collapsing matter implies for the outside space-time at least in a neighborhood [66]. To have access to the full outside all the way up to an asymptotic observer we would need to specify the matter content outside. While this is possible e.g. as the same scalar matter as inside, only inhomogeneous, it would not necessarily be correct physically. In fact, we have modified only the classical equations describing the interior, while we did not use effective equations outside. When the matter distribution extends over a large region in the early stages, one does not expect strong modifications, but this is not clear close to the bounce. In this region, the interior equations are strongly modified, and this is transferred to the outside via the matching conditions in a rather indirect way: We use the classical generalized Vaidya metric, but did not specify the matter content. It is effectively the energy momentum tensor which carries quantum effects from the interior to the outside via the matching. Prescribing the outside matter content would remove this transfer and stop us from seeing possible quantum effects outside.

We write the interior metric as

$$ds^2 = -dt^2 + X(r,t)^2 dr^2 + Y(r,t)^2 d\Omega^2$$

with $X(r,t) = a(t)/(1+r^2/4)$ and $Y(r,t) = rX(r,t)$. On the matching surface $r = R$ of the interior and $\chi = \chi(v)$ of the generalized Vaidya exterior the metric and extrinsic curvature have to agree. From the metrics we obtain

$$\chi|_\Sigma = Y|_\Sigma \tag{44}$$

and

$$\left.\frac{dv}{dt}\right|_\Sigma = \left.(1-2M/\chi - 2d\chi/dv)^{-1/2}\right|_\Sigma \tag{45}$$

while the extrinsic curvature, computed again from (21), gives us

$$\left.\frac{YY'}{X}\right|_\Sigma = \left.\chi\frac{1-2M/\chi - d\chi/dv}{\sqrt{1-2M/\chi-2d\chi/dv}}\right|_\Sigma \tag{46}$$

and

$$0 = \partial_v M + \frac{d^2\chi}{dv^2} + \left(1-\frac{2M}{\chi}-3\frac{d\chi}{dv}\right)\left(\frac{M}{\chi}-\partial_\chi M\right) \tag{47}$$

which yields a condition for $\partial M/\partial\chi$ at constant v.

With $d\chi/dv|_\Sigma = \dot{\chi}|_\Sigma/\dot{v}|_\Sigma$ and (44) we use (46) to write the square root in (45) in terms of Y' and $\dot{Y}$ which leads to

$$\left.\frac{dv}{dt}\right|_\Sigma = \left.\frac{(Y'/X+\dot{Y})}{1-2M/Y}\right|_\Sigma . \tag{48}$$

Using (44) and defining $c := Y'/X$, (46) becomes

$$c^2(1-2M/\chi + 2d\chi/dv) = (1-2M/\chi - d\chi/dv)^2$$

which with

$$(d\chi/dv)^2 = \dot{Y}^2(1-2M/\chi - 2d\chi/dv)$$

(following from $d\chi/dv = \dot{\chi}/\dot{v}$ and (45)) gives $c^2 = 1-2M/\chi + \dot{Y}^2$. Thus,

$$2M|_\Sigma = (Y\dot{Y}^2 + Y(1-c^2))|_\Sigma . \tag{49}$$

A trapped surface forms in a generalized Vaidya metric when $2M = \chi$, which lies on the matching surface if $2M = Y$. From (49) this yields the simple condition

$$|\dot{Y}| = c = Y'/X \tag{50}$$

which for FRW reduces to

$$|\dot{a}| = (1 - R^2/4)/R\,. \tag{51}$$

Assuming, for now, that this condition will be satisfied at a time $t(R)$ during collapse, we obtain a horizon covering the bounce (Fig. 3). The squared norm of its normal is given by $\partial_v M(1 - 2\partial_\chi M)$ which can be computed from (47) using $\mathrm{d}M/\mathrm{d}v = \partial_v M + \partial_\chi M \mathrm{d}\chi/\mathrm{d}v$ and turns out to be zero if the horizon condition (51) is satisfied. The horizon is thus always null when it first intersects the matching surface.

After the first trapped surface forms on the matching surface, $|\dot{a}|$ continues to increase before it turns around when the peak in $d(a)_{j,l}$ is reached. From then on, $|\dot{a}|$ decreases and reaches $\dot{a} = 0$ at the bounce. In between, the trapped surface condition (51) will be satisfied a second time at an inner horizon. Unlike the outer one, it lies in the modified regime where energy conditions are effectively violated and $\ddot{a} > 0$ [44]. It is also null at the matching surface but can become time-like soon and evaporate later. Similarly, the outer horizon can become time-like when matter having experienced the quantum modifications starts to propagate through it. Thus, also the outer horizon can evaporate and shrink toward the matching surface at later times, when the inner matter is already expanding.

The horizon thus evaporates and the bouncing matter has a chance to reappear. Indeed, the condition (51) will be satisfied also at a time after the bounce where $\dot{a}$ is now positive. Thus, the horizon will intersect again with the matter shells and disappear. At such a point, however, the matching breaks down since $\mathrm{d}v/\mathrm{d}t$ diverges when $2M = Y$ and $\dot{Y} > 0$. From a single matching of the interior we obtain only a part of the collapse before a horizon disappears. At the endpoint of the horizon the interior coordinate t ceases to be good, and we have to match to another patch (Fig. 3).

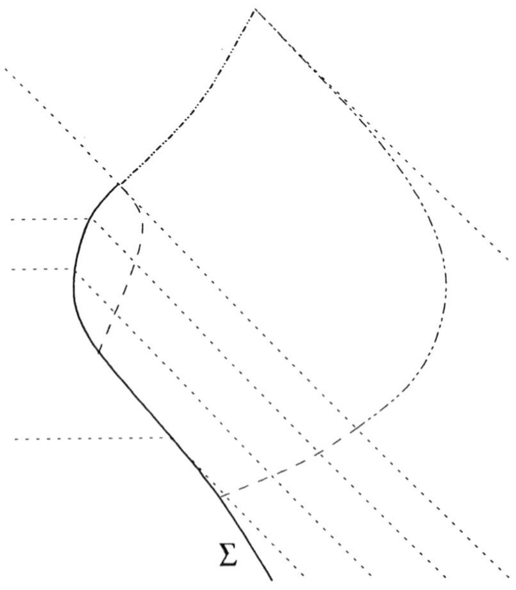

Figure 3: Sketch of the bouncing effective interior, covered by evaporating horizons (dashed). Constant v (outside) and t (inside) slices are dotted. The matching described in the text only refers to the part before the inner horizon collides with the expanding matter after the bounce.

The precise position of the horizon only follows when we specify the matter content and field equations (i.e. Einstein's equations or modified ones) and integrate with M and $\partial M/\partial\chi$ as boundary conditions at the matching surface. Since we leave this open, we do not get precise information on the horizon but only qualitative properties. After some time into the collapse, the horizon is expected to shrink since modifications in the interior imply small violations of energy conditions (which also allow the bounce to take place). Radiation of negative energy implies, analogously to Hawking radiation, that the horizon becomes time-like and shrinks. Later, it can meet the matching surface again at which point the matter becomes visible from behind the horizon. If the initial mass was large, it takes a long time for the bounce to occur and the matter to reemerge such that for most of the time the system looks like a classical black hole to an outside observer.

It is not guaranteed that a horizon forms at all since fulfillment of the horizon condition depends on initial values. In particular, once we choose R to specify the matching surface in the interior, Eq. (51) fixes the value for $\dot{a}$ which needs to be reached for a horizon to occur. Classically, $\dot{a}$ is unbounded as we approach the singularity such that the condition will always be true at one point and there is always a horizon covering the classical singularity in this model. With the effective equations, however, $\dot{a}$ is bounded for given initial conditions, and depending on the value of R it can happen that the horizon condition is never fulfilled. In this case, there would not be a black hole but only a matter distribution collapsing to high densities and rebounding. This rules out black holes of a certain type, in particular those of small mass: Starting with a configuration such that a horizon forms, we can decrease R toward zero without changing $a(t)$. The right hand side of (51) then increases and at one point the condition can no longer be fulfilled. Since with decreasing R we carve out a smaller piece of the homogeneous interior, the total initial mass is smaller, giving us a lower bound for the mass of black holes in this model. Precise values have to be derived from more detailed models, but this argument shows that large, astrophysical black holes will be unaffected while primordial ones of small masses do not form [66].

5 Extrapolation

We have seen two results from homogeneous techniques employed in the preceding section:

- At the fully quantum level of the Kantowski–Sachs model describing the Schwarzschild black hole interior the singularity is absent (Fig. 1).
- Matter systems allow effective classical equations for their collapse such that the classical singularity is replaced by a bounce sometimes shrouded by a horizon (Fig. 3).

Both results have been arrived at with very different techniques, and have different physical meaning. The first one only applies to the vacuum case but provides us with a strict result as to how the classical singularity is replaced in quantum gravity. It directly shows that general relativity is singular because it relies on the smooth classical space-time picture. This picture breaks down at high curvature and has to be replaced by discrete quantum geometry, providing a non-singular evolution.

The second result works with matter but is more intuitive, only giving a picture from effective classical equations. It provides a physical, rather than geometrical explanation for the failure of general relativity in strong curvature regimes. Singularities in general relativity can be understood as a consequence of the always attractive nature of classical gravity: Once matter collapses to a sufficiently high density, be it an isolated part or the whole universe, there is nothing to prevent total collapse into a singularity. Viewing the Friedmann equation, e.g. with scalar matter, as describing a mechanics system with the matter energy density serving as potential shows this by the fact that the energy density decreases as a function of a at fixed ϕ and p_ϕ, in particular the kinetic term $\frac{1}{2}a^{-3}p_\phi^2$. Thus, there is an attractive force driving the system toward $a = 0$ (or, as usually expressed in cosmology, positive pressure which thermodynamically is defined as the negative change of energy with volume). The modification of a^{-3} by the regular function $d(a)$ in (41), which turns around at a peak value and then approaches zero rather than infinity at $a = 0$, implies that now the energy density increases as a function of volume at small scales. This can be interpreted as quantum gravity becoming repulsive at small scales, which can then easily prevent total collapse into a singularity. Moreover, at non-zero but small scales this repulsive component is still active and leads to modified behavior. For instance, in an expanding universe it implies that the expansion is accelerated leading directly to inflation [44]. In the interpretation of collapsing matter, the same effect makes the horizon shrink after the bounce such that only the strong quantum region is covered.

Since we have used approximations, the question arises how these partial results can fit into a full picture of quantum black holes. The first result indicates that space-time can be extended through classical singularities, but since it gives us access only to the interior, it is not clear if the new region we reach can also be accessed from an outside observer. (If not, the black hole would appear as a wormhole through which one can travel into a new region of the universe.) The second result now indicates that we can in fact access the new region since there is only one matching region outside the collapsing matter, suggesting a picture as in Fig. 4.

However, here it is important to bear in mind that we only effectively described matter outside falling into the collapsing shells. In a more realistic model, there would be such inhomogeneous matter colliding with the homogeneous core and making it more heavy. This can then lead to singularities forming in the outside region. For their resolution we would again have to use quantum geometry and face the same problem as to whether or not this will lead to a new region accessible from the outside.

It is clear that a decisive answer can only be obtained with inhomogeneous techniques, which we are going to describe in the next section. Still, even at this level one can see that there are only a few possible scenarios which can be distinguished by using inhomogeneous properties of quantum geometry. Irrespective of which outcome inhomogeneous models will show us, one can already see special features of quantum geometry leading to a new paradigm about black hole evaporation. For the first time, this takes into account a resolution of the classical singularity with implications for apparent loss of information [68]. In fact, while Hawking radiation still emerges in a neighborhood of the dynamical horizon and is still approximately thermal, this is by no means everything coming out of the black hole at late times. Infalling matter now evolves through the quantum region of high curvature and reappears later, restoring correlations which are not recovered by Hawking radiation alone. In particular, there is no reason for the final state measured on all of future null infinity to be mixed if we started with a pure initial state. In the usual picture one would

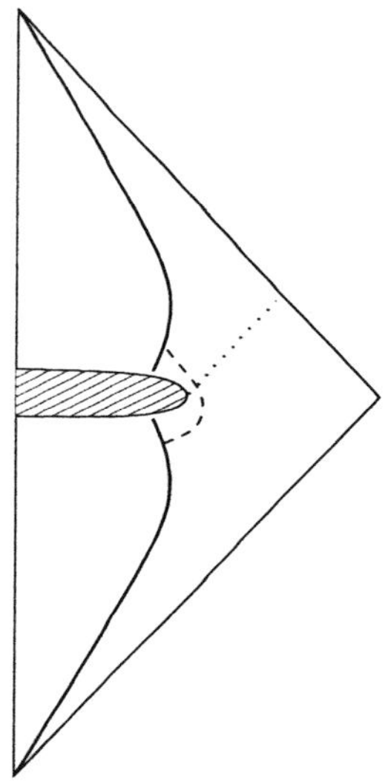

Figure 4: Combination of Figs. 1 and 3 where the quantum region around the classical singularity is shrouded by an evaporating horizon (dashed). The dotted line marks the boundary of the part of space-time unaffected by the strong curvature inside.

cut out the quantum region (or the place of the classical singularity) and consider the future space-time without allowing penetration through that region. Future null infinity then stops at the intersection with the dotted line in Fig. 4, and a state retrieved at this part of null infinity is indeed mixed since it is obtained by averaging over the rest to the future. In this way, both the singularity problem and the information loss paradox are resolved by loop quantum gravity.

6 Inhomogeneous Techniques

For the spherically symmetric model we need to perform the loop quantization for inhomogeneous configurations (17) and (11) such that the basic fields now depend on the radial coordinate x. Instead of using states such as (27) with a finite number of labels, we now have a field theory with infinitely many kinematical degrees of freedom. An orthonormal basis of states is given by [13]

$$\langle A_x, K_\varphi | \ldots, k_n, \mu_n, k_{n+1}, \mu_{n+1}, \ldots \rangle = \prod_n \exp(\tfrac{1}{2} i k_n \int_{e_n} A_x \mathrm{d}x) \exp(-i\mu_n \gamma K_\varphi(v_n)) \qquad (52)$$

with countably many labels $k_n \in \mathbb{Z}$ and $0 \leq \mu_n \in \mathbb{R}$ labeling edges e_n and vertices v_n, respectively, of a 1-dimensional graph in the radial line. Note that, as already indicated before, we are using exponentials of the extrinsic curvature component K_φ along homogeneous directions but holonomies of the connection component A_x along the inhomogeneous direction. Both exponentials are represented as multiplication operators.

Spatial geometry is encoded in densitized triad operators acting by

$$\hat{E}^x(x)|\ldots,k_n,\mu_n,\ldots\rangle = \frac{\gamma\ell_{\mathrm{P}}^2}{8\pi}\frac{k_{n^+(x)}+k_{n^-(x)}}{2}|\ldots,k_n,\mu_n,\ldots\rangle \tag{53}$$

$$\int_I \hat{E}^{\varphi}|\ldots,k_n,\mu_n,\ldots\rangle = \frac{\gamma\ell_{\mathrm{P}}^2}{8\pi}\sum_{v_n\in I}\mu_n|\ldots,k_n,\mu_n,\ldots\rangle \tag{54}$$

where $n^{\pm}(x)$ is the edge label to the right (left) of x, and I is an interval on the radial line (over which we need to integrate E^{φ} since it is a density). As before in the homogeneous case we also obtain densely defined operators for inverse powers of the triad components, which can in particular be done for the inverse of E^{φ} in the spin connection component (13).

6.1 Hamiltonian Constraint

For the Hamiltonian constraint (16) we again have to obtain curvature components from holonomies, now using holonomies of A_x for the inhomogeneous radial direction and exponentials of K_{φ} for homogeneous directions along symmetry orbits. Terms containing the spin connection component Γ_{φ} belonging to homogeneous directions will be quantized separately. One may wonder if this procedure will easily give the right components in the Hamiltonian constraint, given that it has the rather simple expression (16) in terms of extrinsic curvature components while we are using the Ashtekar connection component A_x. As we will see, the general scheme will yield automatically the right combination of components by a straightforward construction of loops to be used in holonomies.

To see this in detail, we first note the difference between A_x and K_x, which is given by the x-component of the spin connection. For a general spherically symmetric triad it takes the form

$$\Gamma = -\eta'\tau^3 dx + \frac{E^{x\prime}}{2E^{\varphi}}\Lambda d\vartheta - \frac{E^{x\prime}}{2E^{\varphi}}\bar{\Lambda}\sin\vartheta d\varphi + \tau_3\cos\vartheta d\varphi \tag{55}$$

as in (9) with (10). Here, we recognize (13) as used before as the component along homogeneous directions. This component is a scalar (noting that both $E^{x\prime}$ and E^{φ} are densities of weight one), while the x-component $\Gamma_x = -\eta'$ does not have covariant meaning and indeed can be made arbitrarily small locally by a suitable gauge transformation. This is analogous to the situation in the full theory, while the gauge invariant meaning of Γ_{φ} mimics homogeneous models.

Even though Γ_x can be made arbitrarily small locally by a gauge transformation, we cannot assume this when constructing a suitable Hamiltonian constraint operator. Thus, it must be built into the construction so as to combine with A_x from radial holonomies to give K_x as in the expression for the constraint. This $K_{\varphi}K_x$-term in the constraint can, according to the general construction where connection or extrinsic curvature components derive from closed holonomies, only come from a loop which has one edge along a symmetry orbit and one in the radial direction. Starting in a point v, such a holonomy is of the form $h_x^{(\delta_x)}h_{\varphi}^{(\delta)}(v_+)(h_x^{(\delta_x)})^{-1}(h_{\varphi}^{(\delta)}(v))^{-1}$ with a new vertex v_+ displaced from v by a coordinate distance δ_x of the radial edge. (We distinguish between δ_x for the radial direction and δ for the angular directions since the continuum limit is technically different in both cases.) This term appears together with $\bar{\Lambda}(v)$ (coming from quantizing the triad components) in a trace whose expansion in δ

$$\begin{aligned}
&-2\,\mathrm{tr}(h_x h_\varphi(v_+) h_x^{-1} h_\varphi(v)^{-1}\bar{\Lambda}(v))\\
&\sim -2\gamma\delta(K_\varphi(v_+)\,\mathrm{tr}(\Lambda(v_+)\bar{\Lambda}(v)) + 2K_\varphi(v_+)\textstyle\int A_x \mathrm{d}x\,\mathrm{tr}(\tau_3\Lambda(v_+)\bar{\Lambda}(v)))\\
&= \gamma\delta(K_\varphi(v_+)\sin(\eta_+ - \eta) + K_\varphi(v_+)\textstyle\int A_x \mathrm{d}x \cos(\eta_+ - \eta))\\
&= \delta\delta_x\gamma K_\varphi(v)(A_x(v) + \eta'(v)) + O(\delta^2)
\end{aligned} \tag{56}$$

has all the right terms, with A_x coming directly from the radial holonomy and $\eta(v_+)$ in $\delta\eta' \sim \eta(v_+) - \eta(v)$ from the internal direction $\Lambda(v_+)$ at the new vertex. The other term of the form K_φ^2 is obtained from angular holonomies only, as in the homogeneous case,

$$-2\,\mathrm{tr}(h_\vartheta h_\varphi h_\vartheta^{-1} h_\varphi^{-1}\tau_3) \sim \delta^2\gamma^2 K_\varphi^2\,. \tag{57}$$

The matrices τ_3 and $\bar{\Lambda}(v)$ in the traces come from Poisson brackets expressing triad components as in the homogeneous constraint. Moreover, the spin connection components in (16) are again expressed through the curvature

$$F(\Gamma) = -\Gamma_\varphi'\Lambda \mathrm{d}x\wedge \mathrm{d}\vartheta + \Gamma_\varphi'\bar{\Lambda}\sin\vartheta \mathrm{d}x\wedge \mathrm{d}\varphi + (\Gamma_\varphi^2 - 1)\tau_3 \sin\vartheta \mathrm{d}\vartheta\wedge \mathrm{d}\varphi$$

of the spin connection. Through

$$\sum_{IJK} \varepsilon^{IJK}\,\mathrm{tr}(\delta_I\delta_J F(\Gamma)_{IJ} h_K\{h_K^{-1}, V\}) \propto (\Gamma_\varphi^2 - 1)\{A_x, V\} - 2\Gamma_\varphi'\{-\gamma K_\varphi, V\})$$

we obtain the additional terms of the constraint, where we included the length parameters δ_I (i.e. δ_x for $I = x$ and δ for ϑ or φ).

This demonstrates how the general procedure works without additional input: We use exponentials of extrinsic curvature components for homogeneous directions and holonomies of Ashtekar connection components for inhomogeneous ones as dictated by the background independent representation. Spin connection components for inhomogeneous directions then come in the right form to combine with extrinsic curvature components, while those in homogeneous directions are split off and quantized separately. This is possible because those components, in contrast to the inhomogeneous ones, do have covariant meaning. This ties together the constructions in homogeneous models and the full theory, and at the same time opens a direct route to effective classical equations: Homogeneous spin connection components usually contain inverse powers of the densitized triad, such as (13). When they are quantized, the classical divergence will be removed implying modifications at small scales. in homogeneous models this has been used, e.g., in the Bianchi IX case where it has been shown to remove the classical chaos [82, 83]. Similarly, one can use this mechanism to derive effective classical equations for the spherically symmetric model and find possible consequences.

Before trusting those effective equations in the inhomogeneous case one needs to make sure that there is a well-defined Hamiltonian constraint operator emerging from the procedure described here. So far, we have only discussed those holonomies and spin connection components which give us the contributions to the constraint, but they must now be stuck together with quantizations of triad components so as to build a well-defined operator for the whole expression. Moreover, since the expression (16) is an integrated density, one has to discretize the integration first and then, after quantizing the individual terms, perform the continuum limit removing the regulator. The discretization had already been understood above, with v and v_+ being the endpoints of a discrete interval of size δ_x, and we convinced ourselves that the continuum limit of the discretization will yield the correct result. In more detail, one writes

$$H[N] = \int \mathrm{d}x N(x)\mathcal{H}(x) \sim \sum_n \delta_x^{(n)} N(v_n)\mathcal{H}(v_n)$$

where we discretized the radial line into intervals of coordinate length $\delta_x^{(n)}$, each one containing the point v_n. Classically, both expressions can be made to agree for any subdivision by choosing points v_n in the intervals according to the mid point theorem. Alternatively, if one wants to fix the v_n to be endpoints of the intervals, the discretization agrees with the classical constraint in the continuum limit in which $n \to \infty$ and $\delta_x^{(n)} \to 0$ for all n.

For the Hamiltonian constraint in general each term in the sum then has contributions of the form

$$\delta_x \mathcal{H}(v) \propto \sum_{I,J,K} \varepsilon^{IJK} \mathrm{tr}(h_{IJ} h_K [h_K^{-1}, \hat{V}]) \tag{58}$$

where we sum over triples (I,J,K) of independent directions which in symmetric models are given by generators of the symmetry transformations ((ϑ,φ) in the spherically symmetric case), and in the full theory or inhomogeneous directions of a symmetric model by edges of a graph (x in spherical symmetry). The holonomies h_{IJ} are formed according to the symmetry: if both directions I and J are inhomogeneous, h_{IJ} is a holonomy along a closed loop α_{IJ} constructed from edges in the IJ-plane of a graph to act on; if at least one of the two directions is homogeneous, $h_{IJ} = h_I h_J h_I^{-1} h_J^{-1} - \gamma^2 \delta_I \delta_J \hat{F}_{IJ}$ with h_I being exponentials of the (su(2)-valued) extrinsic curvature components belonging to the I-direction for a homogeneous direction I or holonomies along an inhomogeneous direction. (The appearance of $F(\Gamma)$ can be understood as a correction term since loops made from holonomies along vector fields generating symmetries do not close if orbits have non-zero curvature [84].) The size of loops α_{IJ} or single holonomies is determined by the size δ_x of the discretization. The final holonomy h_K either belongs to an edge transversal to both directions I and J, or again to exponentiated extrinsic curvature components if K is homogeneous. These combinations are chosen in such a way that h_{IJ} yields the correct curvature components, the commutator gives the necessary triad components, and both terms together provide just the right product of lattice sizes such as δ_x in order for the sum to take the form of a Riemann summation of the original integral. In the full theory, this procedure only results in the so-called Euclidean part of the constraint which can be used to construct the Lorentzian constraint [71]. In the models used here, however, the prescription (58) is sufficient even for Lorentzian signature.

This can now be illustrated and applied in the spherically symmetric model where the first case above cannot appear since there is only one inhomogeneous direction. We have thus two cases, one in which direction I or J is radial, resulting in the first product of holonomies discussed above, combined with a commutator

$$\begin{aligned} h_\varphi [h_\varphi^{-1}, \hat{V}] &= \hat{V} - \cos\tfrac{1}{2}\gamma K_\varphi \hat{V} \cos\tfrac{1}{2}\gamma K_\varphi - \sin\tfrac{1}{2}\gamma K_\varphi \hat{V} \sin\tfrac{1}{2}\gamma K_\varphi \\ &\quad -2\Lambda(\cos\tfrac{1}{2}\gamma K_\varphi \hat{V} \sin\tfrac{1}{2}\gamma K_\varphi - \sin\tfrac{1}{2}\gamma K_\varphi \hat{V} \cos\tfrac{1}{2}\gamma K_\varphi)\,. \end{aligned}$$

In the second case we have the other product of holonomies and a commutator

$$\begin{aligned} h_x[h_x^{-1},\hat{V}] &= \hat{V}-\cos\tfrac{1}{2}\textstyle\int A_x\hat{V}\cos\tfrac{1}{2}\int A_x-\sin\tfrac{1}{2}\int A_x\hat{V}\sin\tfrac{1}{2}\int A_x \\ &\quad +2\tau_3(\cos\tfrac{1}{2}\textstyle\int A_x\hat{V}\sin\tfrac{1}{2}\int A_x-\sin\tfrac{1}{2}\int A_x\hat{V}\cos\tfrac{1}{2}\int A_x)\,. \end{aligned}$$

The integration here is over an interval of size δ_x such that in the limit of a fine discretization this term is of order δ_x as needed for the Riemann sum.

What we did not specify yet is how the discretization is adapted to a graph the constructed operator is supposed to act on, i.e. whether v and maybe $v_\pm$ are already vertices of the graph or arbitrary points. At this point, choices need to be made which lead to different versions of the constraint. The same choices arise in the full theory [85, 86, 87], but they can be studied much more easily in the spherically symmetric model such that it may be possible to rule out some versions.

It is already non-trivial to check that different versions lead to well-defined operators at all. For this, the action after performing the continuum limit, in which the number of discretization points becomes infinite, must be finite. If the action were non-zero at each discretization point, there would not be a well-defined operator in the limit and the regulator could not be removed. One would then only deal with a lattice regulated theory rather than a quantization of the continuum theory. In the full theory, there is a well-defined operator because the action of the constraint is zero unless a discretization point is already a vertex. Starting with states with finitely many vertices then leads to a densely defined operator. This comes about in the full theory because the constraint contains the volume operator in such a way that it acts only on planar vertices if there is no vertex already present in the graph. Since the full volume operator annihilates all planar vertices, there are only finitely many contributions from the vertices already present.

In spherical symmetry, however, all vertices are planar since graphs are just 1-dimensional. This simple general argument is thus not available and it is not obvious that the same construction scheme will result in a well-defined operator. It turns out, however, that this is the case as a consequence of how triad components in the constraint are quantized: a discretization point which is not already a vertex of the graph to act on will be annihilated such that only finitely many contributions from the vertices remain. One can thus use the same type of operator, just with adaptations to the symmetric situation.

Nevertheless, one can also choose different constructions where the discretization is given directly by the graph, i.e. discretization intervals would be complete edges of the graph. Since endpoints of discretization intervals are always vertices of a state, the continuum limit would then require also states to change and become finer and finer. In this picture, the continuum limit of the constraint operator can only be tested on states which are suited to the continuum behavior, while there are also other states where discreteness is essential and where the classical constraint would be corrected from quantum effects. Moreover, in the continuum limit the number of vertices diverges and the constraint operator becomes ill-defined just as the usual Wheeler–DeWitt operator is. Both schemes result in well-defined operators, but they lead to quite different equations of motion and require different conceptual viewpoints about the continuum limit. When the continuum limit is to be ensured for each state, one requires in a sense that the classical equations are sensible at arbitrarily small scales, and corrections could only come from quantum uncertainties. In the second picture, on the other hand, the classical continuum picture arises only after a certain coarse graining, or by working only with states which are not sensitive to the microscopic structure. If one chooses a state which is sensitive to small scales, or looks very closely at small scales of even a semiclassical state, then corrections to the classical expressions arise, for instance as a consequence of the underlying discreteness. This second viewpoint has been taken successfully in cosmological models, and is, as we will see, also fruitful in black hole models.

6.2 Dynamics

Since the constraint operator is again constructed from holonomies which act by shifting the labels, it implies difference equations for states in a triad representation. (Note that also in the spherically symmetric model the triad representation exists, unlike in the full theory where flux operators do not commute [88].) These equations are now not only partial difference equations but also have many independent variables. Interestingly, the type of difference equation is very different for the two versions of the constraint operator: in the second case the number of edges and vertices of the original state is unchanged and the operator only acts on the labels. This results in difference equations with independent variables k_e for each edge and μ_v for each vertex. Since the operator does not change the number of edges and vertices, one obtains coupled difference equations in a fixed number of variables for each sector given by the number of vertices.

In the first version of the constraint, however, the situation is very different. Now, new vertices are created and edges split in each action of the operator. Thus, for a triad representation it is not enough to work with a fixed number of vertices. Rather, all graphs have to be taken into account for the equation, which implies that one has to deal with infinitely many independent variables and thus functional difference equations.

We thus return to the simpler type of difference equation implied by the other version of the constraint and discuss what one can already say about the singularity issue. First of all, one will have to identify classical singularities on minisuperspace in order to study the constraint equation in a neighborhood. In an isotropic model this is simple since the only way is for the volume to go to zero [10]. Similarly, one can identify the classical singularity on an anisotropic minisuperspace where all densitized triad components would go to zero. The situation is not so clear in midisuperspaces such as spherical symmetry since there are more possibilities for a singularity to develop. Even though this has not been settled in general, there are many cases where a singularity is characterized by E^x approaching zero (which on classical solutions such as Schwarzschild implies that E^φ becomes zero, too). This is also in agreement with the general mechanism removing singularities seen so far in homogeneous models: there it is the sign coming from orientation which leads to different regions of minisuperspace separated by the classical singularity. The quantum evolution, as we have seen in the Kantowski–Sachs model, can then allow us to evolve between the two regions, thus removing the classical singularity as a boundary.

The role of orientation is now played by $\operatorname{sgn} E^x$ since $\det q = E^x(E^\varphi)^2$. Since E^x depends on the radial position, or the edge after quantization, the boundary of our midisuperspace has many components which we can identify with an inhomogeneous classical singularity, corresponding to the fact that inhomogeneous singularities behave differently in their different points. For the states this means that we encounter a section of a classical singularity each time an edge label k_e becomes zero. As in homogeneous models, we can then use the evolution equation in the triad representation in order to see if an evolution through this part of the boundary is possible. Since the structure of the difference equation for a given edge label is very similar to the homogeneous equation, one can expect that the boundary indeed disappears and that the quantum evolution connects regions of midisuperspace corresponding to different local orientations. There would thus be no singular boundary, and the same mechanism as in homogeneous models could also remove spherically symmetric singularities.

This scenario has been verified in [15], noting a crucial difference to homogeneous models wich require a symmetric ordering of the constraint. Thus, quantization choices are reduced by looking at less symmetric models, so far in such a way which maintains the validity of the general picture. Yet, there are also open issues left for a general understanding. For instance, while the results are independent of the matter Hamiltonian and can be extended to cylindrical gravitational wave models, thus also allowing local degrees of freedom, they are so far based only one type of the constraint which leads to difference equations easier to deal with. The behavior with the other version is not easy to see, but if it is singularity free, too, the mechanism is likely to be different. Most importantly, the kind of initial/boundary value problem suitable for the constraint equations needs to be analyzed in more detail to guarantee that there are suitable and sufficiently many solutions with the correct classical limit. At this point the anomaly issue, i.e. whether two constraint operators with different lapse functions have the correct commutator, becomes important. These considerations thus provide a promising and treatable way to distinguish different versions of the quantization by their physical implications, which can then be extrapolated to the full theory.

6.3 Horizons

A feature of black holes which is new compared to cosmological models, and which requires inhomogeneous situations, is given by the presence of horizons. Global concepts such as the event horizon are, of course, not helpful in our case since we would need to solve the Hamiltonian constraint completely before being able to discuss this issue. There are more practical definitions such as apparent horizons which, however, are much more general and do not distinguish between fully dynamical situations and almost static systems. The quantum behavior would be most easy to analyze if we can define horizons locally and in a controlled manner which does not require the full dynamics at once. Such a concept is presented by isolated [2] or slowly evolving dynamical horizons [89], which even quite unexpectedly simplify the spherically symmetric Hamiltonian constraint in their neighborhood.

6.3.1 Definition

There are three main parts to the definition of an isolated horizon Δ with spatial sections $S \cong S^2$ of given area a_0, embedded into the space manifold Σ by $\iota\colon S \to \Sigma$ [90, 30]:

(i) The canonical fields (A_a^i, E_i^a) on the horizon are completely described by a single field $W = \frac{1}{2}\iota^* A^i r_i$ on S which is a U(1)-connection obtained from the pull-back of the

Ashtekar connection to S. Here, r_i is an internal direction on the horizon chosen such that W is a connection in the spin bundle on S^2 and $r^i E^a_i = \sqrt{\det q}\, r^a$ on the horizon with the internal metric q on S and the outward normal r^a to S in Σ.

(ii) The intrinsic horizon geometry, given by the pull-back of the 2-form $\Sigma^i_{ab} := \varepsilon_{abc} E^c_i$ to S, is determined by the curvature $F = \mathrm{d}W$ of W by

$$F = -\frac{2\pi}{a_0} \iota^* \Sigma^i r_i . \tag{59}$$

(iii) The constraints hold on S.

A further consequence of the isolated horizon conditions [90] is that the curvature $\mathcal{F}$ of the pull-back of A^i_a to S has to equal the curvature of W: $r_i \mathcal{F}(\iota^* A^i) = 2\mathrm{d}W$. This can be seen as one of the distinguishing features of an isolated horizon since even slowly evolving horizons at rate ε (related to the expansion of horizon cross sections [89]) will break it, though just by an amount of the order ε^2.

When the horizon is introduced as a boundary, condition (i) is used to identify the horizon degrees of freedom represented by the field W. Condition (ii) then shows that these degrees of freedom are fields of a Chern–Simons theory on the horizon. It is the main condition since it relates the horizon degrees of freedom to the bulk geometry, which after quantization selects the relevant quantum states to be counted. Condition (iii), on the other hand, does not play a big role since an isolated horizon as boundary implies a vanishing lapse function on S for the Hamiltonian constraint which then is to be imposed only in the bulk.

Thus, when computing black hole entropy in this way, as we will describe later, the Hamiltonian constraint does not play any role since it does not act at the boundary, and the Hamiltonian generating evolution along the horizon need not be considered. In the spherically symmetric model one can hope that the constraint is simple enough for an application in this case, either to generate evolution or to impose the horizon not as a boundary but inside space such that the constraint would have to be imposed. In the latter case, moreover, we will not be able to have an independent boundary theory which is then matched to the bulk, but would have to find the relevant degrees of freedom within the original quantum theory.

6.3.2 Spherical Symmetry

We can now evaluate the conditions for spherically symmetric connections of the form

$$A = A_x(x)\tau_3 \mathrm{d}x + A_\varphi(x)\bar{\Lambda}^A(x)\mathrm{d}\vartheta + A_\varphi(x)\Lambda^A(x)\sin\vartheta\, \mathrm{d}\varphi + \tau_3 \cos\vartheta\, \mathrm{d}\varphi \tag{60}$$

and densitized triads (11), where in general the internal directions Λ^A and Λ are different. The connection component A_x has been discussed before, while the relation $A_\varphi \Lambda^A = \Gamma_\varphi \bar{\Lambda} - \gamma K_\varphi \Lambda$, following from the definition of the Ashtekar connection together with (9), (10) and (17), implies $A_\varphi^2 = \Gamma_\varphi^2 + \gamma^2 K_\varphi^2$.

We choose $r_i := \mathrm{sgn}(E^x)\delta_{i,3}$ such that in fact $r^i E^a_i = |E^x| \sin\vartheta\, \partial_x$ with the intrinsic horizon area element $|E^x|\sin\vartheta$ of a metric $|E^x|\mathrm{d}\Omega^2$. Thus, $W = \frac{1}{2} r_i \iota^* A^i = \frac{1}{2}\,\mathrm{sgn}(E^x)\cos\vartheta\, \mathrm{d}\varphi$

whose integrated curvature given by $\oint_S \mathrm{d}W = -2\pi\,\mathrm{sgn}(E^x(x))$ agrees with the Chern number of the spin bundle, depending on the orientation given by $\mathrm{sgn}(E^x)$.

Evaluating (59) first shows that in the spherically symmetric context it is not restrictive since we have $a_0 = 4\pi|E^x(S)|$ and the right hand side given by $-\frac{1}{2}\,\mathrm{sgn}(E^x(S))$ equals F for all E. This is not surprising since the spherically symmetric intrinsic geometry of S is already given by the total area which is fixed from the outset. (What is free is the sign of $E^x(S)$, or orientation, which confirms ideas of [91].) Now the first condition plays a major role, which we evaluate in the form $r_i\mathcal{F}(\iota^* A^i) = 2\mathrm{d}W$ [90]. Since $\mathcal{F}(\iota^* A) = (A_\varphi^2 - 1)\tau_3 \sin\vartheta \mathrm{d}\vartheta \wedge \mathrm{d}\varphi$, the condition requires $A_\varphi = 0$ which will be the main restriction we have to impose on quantum states in addition to the constraints. This condition $A_\varphi = 0$ selects 2-spheres in a spherically symmetric space-time corresponding to cross-sections of a horizon. Indeed, for the Schwarzschild solution we have $A_\varphi = \Gamma_\varphi$ since the extrinsic curvature vanishes. With (13) and the Schwarzschild triad we obtain the correct condition $x = 2M$ for the horizon. In general, $A_\varphi^2 = \Gamma_\varphi^2 + \gamma^2 K_\varphi^2 = 0$ implies $\Gamma_\varphi = 0$ and $K_\varphi = 0$.

A slowly evolving horizon at rate ε satisfies the condition $r_i\mathcal{F}(\iota^* A^i) = 2\mathrm{d}W$ only up to terms of the order ε^2. Thus, A_φ is not exactly zero but must be small of order ε, which then is true also for Γ_φ and K_φ.

6.3.3 Dynamics

In spherical symmetry we can locate a horizon on a state [69], which must be at a vertex in order for $K_\varphi = 0$ to be sharp enough. The condition that K_φ be zero for an isolated horizon or small for a slowly evolving horizon then leads to important simplifications which allow a perturbative treatment of the dynamics around the horizon. Indeed, when acting with the constraint at the horizon both terms made from holonomies h_{IJ} contain factors of $\sin\frac{1}{2}\gamma K_\varphi$ at the horizon vertex or a neighboring one which must then be small. Ignoring those terms in an approximation leads to an operator which is diagonal on the spin network states (52) and thus easy to solve as a constraint or to use for generating time evolution at a boundary. The additional terms ignored in this approximation can then be included in a perturbative treatment of the near horizon dynamics.

Without many calculations this already shows how the horizon fluctuates dynamically. Classically an isolated horizon has constant area which thus commutes with the Hamiltonian constraint. This is also true at the quantum level to leading order of the above approximation since the area operator $\hat{A}(S) = 4\pi|\hat{E}^x(S)|$ has the same eigenstates (52) as the leading order constraint. Thus, at this level the horizon area is an observable not just when the horizon is treated as a boundary, but also if its full neighborhood is quantized. However, there are additional terms which arise in higher orders of the perturbation scheme. There are two reasons for horizon area fluctuations even in the isolated case: While classically $K_\varphi = 0$ exactly at the horizon and only this value is important, the quantization does not allow this to hold arbitrarily sharply. Otherwise, the volume of a shell around the horizon, which depends on the conjugate momentum E^φ of K_φ could not be sharp independently of the mass which would contradict semiclassical properties to hold true at least for massive black holes. Secondly, the constraint operator acting at the horizon itself depends on neighboring values of K_φ through $h_\varphi(v_+)$ in (56). This would give non-zero contributions even if K_φ at the horizon would be zero exactly.

Both terms lead to small dynamical changes in the horizon area coming from typical quantum gravity properties. The first reason is quantum uncertainty which does not allow a sharp condition $K_\varphi = 0$, and the second space-discreteness and non-locality which implies that not only K_φ at the horizon itself is relevant but also the values in neighboring vertices which are not necessarily zero. For large black holes, the correction terms are expected to be small: uncertainty will not change the horizon area much compared to its already large size, and in neighboring vertices of a semiclassical state K_φ will still be extremely small. Thus, for large black holes the horizon area is an excellent observable, while for microscopic black holes large fluctuations are expected which may even prevent horizons as they are known classically. This agrees with the picture we have obtained from effective equations and matching techniques before.

7 Full Theory

The methods developed so far in symmetric models mimic those of the full theory, with some adaptations to preserve the symmetry. In this section, for completeness, we describe what this looks like in the full theory and discuss applications which work without assuming symmetries.

7.1 Representation

As discussed before, the full theory of loop quantum gravity is based on holonomies for arbitrary edges in space and fluxes for surfaces, forming the basic classical Poisson algebra. In the connection representation, states are functionals on the infinite dimensional space of connections [92, 93, 70] through holonomies, and a dense subspace of the Hilbert space is spanned by cylindrical functions

$$\psi(A) = f_\gamma(h_{e_1}(A), \ldots, h_{e_n}(A)) \tag{61}$$

which depend on only finitely many holonomies. Since there is now no symmetry requirement, the edges can be arbitrary curves in space and form a graph γ with vertices at their intersection points. The inner product for two states associated with the same graph is given by

$$\langle f_\gamma | g_\gamma \rangle = \int_{SU(2)^n} \prod_{e\in\gamma} \mathrm{d}\mu_H(h_e) f_\gamma(h_1, \ldots, h_n)^* g_\gamma(h_1, \ldots, h_n) \tag{62}$$

with the Haar measure $\mathrm{d}\mu_H$ on the structure group SU(2). For two functions with different graphs, they need to be extended to a bigger one which is always possible by cutting edges or inserting new ones on which the extended state depends trivially. An orthonormal basis is given by spin network states [94, 95], associated with graphs labeled by irreducible SU(2) representations j_e at edges and contraction matrices C_v at vertices, of the form

$$T_{\gamma,j,C}(A) = \prod_{v\in\gamma} C_v \cdot \prod_{e\in\gamma} \rho_{j_e}(h_e(A)) \tag{63}$$

where the representation matrices $\rho_{j_e}(h_e(A))$ evaluated in edge holonomies are multiplied together in vertices according to the symbols C_v.

Fluxes are quantized as derivative operators in the connection representation since the densitized triad is conjugate to the Ashtekar connection. Replacing the triad components in (15) by functional derivatives and acting on a cylindrical function, we obtain

$$\begin{aligned}\hat{F}_S f_\gamma &= -8\pi i\gamma\hbar G\int_S \mathrm{d}^2y\tau^i n_a \frac{\delta}{\delta A_a^i(y)} f_\gamma(h(A)) \\ &= -i\gamma\ell_{\mathrm{P}}^2 \sum_{e\in\gamma}\int_S \mathrm{d}^2y\tau^i n_a \frac{\delta h_e}{\delta A_a^i(y)}\frac{\mathrm{d}f_\gamma(h)}{\mathrm{d}h_e}\end{aligned}$$

which has contributions only from intersection points y of the surface S of the flux with the graph γ associated with the state. Moreover, each derivative operator for an intersection point can be seen to be equivalent to an angular momentum operator such that its spectrum is discrete and equidistant. Since there is a finite sum over all such contributions, the spectrum of flux operators is discrete, too. Not all the angular momentum operators involved necessarily commute, and so triad operators do not always commute with each other such that a triad representation does not exist [88] (unlike in the symmetric models studied before).

The densitized triad describes spatial geometry, and spatial quantum geometry is encoded in flux operators. From the basic ones one can construct geometrical operators such as the area [32, 33] or volume operator [34] which also have discrete spectra. Thus, quantum spatial geometry is discrete in a precise way, given by the spectra of geometric operators. The area spectrum is known completely, but for the volume operator this is impossible to compute explicitly since arbitrarily large matrices would have to be diagonalized. Spatial geometry at the quantum level is thus rather complicated in general if explicit calculations need to be done.

This translates to the Hamiltonian constraint and other operators, for which the volume operator plays a crucial role. Classically, the Hamiltonian constraint is given by [21]

$$H[N] = (8\pi G)^{-1}\int \mathrm{d}^3x N(x)|\det E|^{-1/2}(F_{ab}^i E_j^a E_k^b \varepsilon_{ijk} - 2(1+\gamma^2)K_{[a}^i K_{b]}^j E_i^a E_j^b) \qquad (64)$$

with the curvature F_{ab}^i of the Ashtekar connection, and the extrinsic curvature $K_a^i = \gamma^{-1}(\Gamma_a^i - A_a^i)$ a function of the basic variables through (6). Both parts of the constraint can be quantized using building blocks similar to (58), resulting in a well-defined operator [71] even when matter Hamiltonians are included [37]. Edges for the holonomies have to be chosen, which can be done in a diffeomorphism invariant manner and even in such a way that the quantization is anomaly free at least on states satisfying the diffeomorphism constraint [96].

Which version of the quantization is the correct one, however, is still an open issue since in particular the classical limit and that of perturbations on a classical background ("gravitons") are difficult to analyze. Moreover, finding and interpreting solutions in full generality is complicated by technical and conceptual problems.

It is thus important to devise approximation schemes, other than symmetry reduction as employed before, in order to shed light on physical properties of the full theory. One powerful possibility consists in imposing an isolated horizon as a boundary [90] since boundary conditions imply that the constraint is not to be imposed there (a constraint has lapse function going to zero at the isolated horizon). Thus, also at the quantum level the constraint operator can be ignored and aspects of the basic quantum representation receive physical meaning. Indeed, boundary degrees of freedom are obtained from intersections of spin network states with the horizon surface, and flux operators are important to select physical states corresponding to an isolated horizon. By counting those states and comparing with the Bekenstein–Hawking expectation the theory can be tested.

7.2 Black Hole Entropy

An isolated horizon S with prescribed area a_0 as a boundary leads to an additional boundary term in the symplectic structure [90],

$$\Omega = (8\pi\gamma G)^{-1}\int \mathrm{d}^3x \partial A_a^i \wedge \partial E_i^a + \frac{a_0}{2\pi}(16\pi\gamma G)^{-1}\int_S \mathrm{d}^2y r_i r_j \partial A_a^i \wedge \partial A_b^j \varepsilon^{ab} \qquad (65)$$

where we denote differentials on field space by ∂, ε^{ab} is the anti-symmetric tensor on the boundary surface, and r_i the internal vector as in the definition of an isolated horizon. The boundary term to the symplectic structure can be recognized as that of a U(1) Chern–Simons theory which thus describes the horizon degrees of freedom by the U(1) connection $W_a = \frac{1}{2}r_i A_a^i$.

We quantize the full system by using quantum geometry in the bulk and quantum Chern–Simons theory on the horizon. Doing this results in the curvature $F = \mathrm{d}W$ becoming an operator with equidistant spectrum which, via (59) needs to be matched to the flux $r_i\Sigma^i$ through the horizon. As shown before, quantum geometry indeed implies a flux operator with equidistant spectrum such that the matching is possible at the quantum level. Since also the pre-factors match, there are always solutions to the horizon condition which can now be counted, for a given area a_0, to compute the entropy as the logarithm of the number of states.

This results in an expression for entropy which is proportional to the horizon area [7, 8], confirming expectations from semiclassical considerations. Intuitively, entropy counts the number of ways that one can construct a macroscopic horizon of area a_0 from elementary discrete parts [97] (which is generalized in this picture since there is not just one elementary type but different ones given by the spin label of an intersection point with a spin network). Since the discreteness scale is set by the Barbero–Immirzi parameter γ, the number of possible such configurations and thus entropy must depend on γ. Indeed, γ appears in the constant of proportionality between entropy and area which allows us to fix γ by requiring the Bekenstein–Hawking law. Moreover, since there are different types of black holes — charged, distorted, rotating or with non-standard matter couplings — and the value is already fixed by the simplest case of a Schwarzschild black hole, one can test the theory since now entropy must result in the right way without any further parameter to tune. This is indeed the case [98, 99], providing a non-trivial test of the theory.

The scale of discreteness is then fixed which, since it must be small enough, can already be confronted with observations. It turns out that $\gamma = 0.2735$ [100, 101] is of the order one such that the discreteness lies around $\sqrt{\gamma}\ell_P \approx \frac{1}{2}\ell_P$ and is thus much too small to be observable directly. Indeed there have long been reasons to expect a scale of discreteness around the Planck length which is now confirmed by detailed calculations in loop quantum gravity. It is not at all obvious that this comes about since there are many non-trivial steps in the derivation, and mistakes in the foundations of the theory could easily lead to larger values which could already be in conflict with observations.

At this point it is important to consider the physical meaning of γ. It can be seen as a fundamental parameter setting the scale of discreteness which is thus characteristic of quantum gravity. (In fact, one can express the continuum limit as a limit $\gamma \to 0$ [77].) In usual arguments, this is expected to be done by ℓ_P, which has to appear anyway just for dimensional reasons. However, in ℓ_P only the gravitational constant G and Planck's constant $\hbar$ enter such that the Planck length is already fixed by classical gravity and quantum mechanics alone. Since these theories are unrelated to full quantum gravity, there is no reason for γ to equal one even though one can expect a value of the order one from dimensional arguments. A precise value for the scale of discreteness can only come from a detailed quantum theory of gravity and calculations which are sensitive to the underlying discrete structure, as realized by loop quantum gravity.

8 Conclusion

In the preceding sections we described the current status of what black holes look like from the viewpoint of non-perturbative, background independent quantum gravity. There are results obtained with different approximations to the full theory which provide a consistent picture of black holes without pathologies or puzzles, such as the singularity problem or the information loss paradox, perceived earlier from general relativity alone or from combinations of classical gravity and quantum field theory on a background.

The main type of approximation used here is that of a symmetry reduction as often employed in classical or quantum physics. This allows to study the background independent quantum dynamics and its characteristic features in different explicit ways. Compared to the full theory, there are several technical simplifications for instance from a volume operator with explicitly known spectrum. But also at a conceptual level, the interpretation of solutions or physical situations is simplified.

Even though special properties of a given symmetric model, such as simplifying coordinate or field transformations, have not been made use of and essential ingredients have rather been modeled on the full theory, the question arises what one could do without symmetry assumptions. For a fair judgment one has to bear in mind that background independence in quantum field theory is a new concept, which is introduced non-perturbatively. There are hardly any comparable results in other realistic quantum field theories, and quantum gravity introduces its own conceptual issues to the theory. Moreover, the fact that common perturbative approximation schemes are not available is a consequence of the property of gravity that a split into a free field theory plus perturbations is not possible. One thus has to deal with the fully non-linear framework which otherwise is usually avoided in quantum field theory. Loop quantum gravity provides a framework in which these hard questions, which sooner or later will have to be faced by any approach to quantum gravity, are being confronted directly.

A consequence of the non-linearity is that operators, even if they can be defined in a well-defined manner, are by no means unique since there are often factor ordering or regularization choices. Loop quantum gravity, nevertheless, has succeeded in finding characteristic effects from a background independent quantization. Details, of course, depend on several quantization choices, but one can directly investigate the robustness of results to ambiguities. As described here, this allows one to solve conceptual problems in the physics of black holes, and also in cosmology as detailed elsewhere [102, 79, 103], while parameters can be fixed in detail by consistency conditions or phenomenology.

Acknowledgements

The author is grateful to Abhay Ashtekar, Rituparno Goswami, Roy Maartens, Parampreet Singh and Rafal Swiderski for collaboration on some of the results described here.

References

[1] Horowitz, G. T.; Myers, R. C. *Gen. Rel. Grav.* 1995, 27, 915–919.

[2] Ashtekar, A.; Krishnan, B. *Liv. Rev. Rel.* 2004, 7, 1–77.

[3] Hawking, S. W. *Phys. Rev. D* 1976, 14, 2460–2473.

[4] Strominger, A.; Vafa, C. *Phys. Lett. B* 1996, 379, 99–104.

[5] Rovelli, C. *Phys. Rev. Lett.* 1996, 14, 3288–3291.

[6] Krasnov, K. *Phys. Rev. D* 1997, 55, 3505–3513.

[7] Ashtekar, A.; Baez, J. C.; Corichi, A.; Krasnov, K. *Phys. Rev. Lett.* 1998, 80, 904–907.

[8] Ashtekar, A.; Baez, J. C.; Krasnov, K. *Adv. Theor. Math. Phys.* 2001, 4, 1–94.

[9] Bojowald, M. *Class. Quantum Grav.* 2002, 19, 2717–2741.

[10] Bojowald, M. *Class. Quantum Grav.* 2003, 20, 2595–2615.

[11] Bojowald, M.; Date, G.; Vandersloot, K. *Class. Quantum Grav.* 2004, 21, 1253–1278.

[12] Bojowald, M. *Phys. Rev. Lett.* 2001, 86, 5227–5230.

[13] Bojowald, M. *Class. Quantum Grav.* 2004, 21, 3733–3753.

[14] Bojowald, M.; Swiderski, R. Spherically Symmetric Quantum Geometry: Hamiltonian Constraint, gr-qc/0511108 (at press in Class. Quantum Grav.)

[15] Bojowald, M. *Phys. Rev. Lett.* 2005, 95, 061301.

[16] Arnowitt, R.; Deser, S.; Misner, C. W. In Gravitation: *An Introduction to Current Research*; Witten, L.; Ed.; Wiley, New York, 1962.

[17] Schwarzschild, K. Sitzber. Deut. Akad. Wiss. Berlin, *Phys.-Math.* Klasse 1916, 189–196; [english translation: physics/9905030].

[18] Bojowald, M. *Phys. Rev. D* 2001, 64, 084018.

[19] Thiemann, T. *Class. Quantum Grav.* 1995, 12, 181–198.

[20] Ashtekar, A. *Phys. Rev. D* 1987, 36, 1587–1602.

[21] Barbero G., J. F. *Phys. Rev. D* 1995, 51, 5507–5510.

[22] Immirzi, G. *Class. Quantum Grav.* 1997, 14, L177–L181.

[23] Sahlmann, H. (2002). Some Comments on the Representation Theory of the Algebra Underlying Loop Quantum Gravity [gr-qc/0207111]

[24] Sahlmann, H. (2002). When Do Measures on the Space of Connections Support the Triad Operators of Loop Quantum Gravity? [gr-qc/0207112]

[25] Okolow, A.; Lewandowski, J. *Class. Quantum Grav.* 2003, 20, 3543–3568.

[26] Sahlmann, H.; Thiemann, T. (2003). On the superselection theory of the Weyl algebra for diffeomorphism invariant quantum gauge theories [gr-qc/0302090]

[27] Fleischhack, C. (2004). Representations of the Weyl Algebra in Quantum Geometry [math-ph/0407006]

[28] Rovelli, C. Living Reviews in Relativity 1998, 1, 1, `http://www.livingreviews.org/Articles/Volume1/1998-1rovelli.`

[29] Thiemann, T. (2001). Introduction to Modern Canonical Quantum General Relativity [gr-qc/0110034]

[30] Ashtekar, A.; Lewandowski, J. *Class. Quantum Grav.* 2004, 21, R53–R152.

[31] Rovelli, C. Quantum Gravity; Cambridge University Press, Cambridge, UK, 2004.

[32] Rovelli, C.; Smolin, L. *Nucl. Phys. B* 1995, 442, 593–619, *Erratum: Nucl. Phys. B* 1995, 456, 753.

[33] Ashtekar, A.; Lewandowski, J. *Class. Quantum Grav.* 1997, 14, A55–A82.

[34] Ashtekar, A.; Lewandowski, J. *Adv. Theor. Math. Phys.* 1997, 1, 388–429.

[35] Bojowald, M. *Class. Quantum Grav.* 2000, 17, 1509–1526.

[36] Bojowald, M.; Swiderski, R. *Class. Quantum Grav.* 2004, 21, 4881–4900.

[37] Thiemann, T. *Class. Quantum Grav.* 1998, 15, 1281–1314.

[38] Thiemann, T.; Kastrup, H. A. *Nucl. Phys. B* 1993, 399, 211–258.

[39] Kastrup, H. A.; Thiemann, T. *Nucl. Phys. B* 1994, 425, 665–686.

[40] Kuchař, K. V. *Phys. Rev. D* 1994, 50, 3961–3981.

[41] Bojowald, M.; Kastrup, H. A. *Class. Quantum Grav.* 2000, 17, 3009–3043.

[42] Hossain, G. M. *Class. Quantum Grav.* 2004, 21, 179–196.

[43] Noui, K.; Perez, A.; Vandersloot, K. *Phys. Rev. D* 2005, 71, 044025.

[44] Bojowald, M. *Phys. Rev. Lett.* 2002, 89, 261301.

[45] Bojowald, M.; Singh, P.; Skirzewski, A. *Phys. Rev. D* 2004, 70, 124022.

[46] Bojowald, M.; Skirzewski, A. (2005). Effective Equations of Motion for Quantum Systems [math-ph/0511043]

[47] Ashtekar, A.; Bojowald, M.; Willis, J. in preparation

[48] Willis, J. On the Low-Energy Ramifications and a Mathematical Extension of Loop Quantum Gravity; PhD thesis, The Pennsylvania State University, 2004.

[49] Date, G.; Hossain, G. M. *Class. Quantum Grav.* 2004, 21, 4941–4953.

[50] Banerjee, K.; Date, S. *Class. Quant. Grav.* 2005, 22, 2017–2033.

[51] Date, G.; Hossain, G. M. *Phys. Rev. Lett.* 2005, 94, 011301.

[52] Singh, P.; Toporensky, A. *Phys. Rev. D* 2004, 69, 104008.

[53] Vereshchagin, G. V. *JCAP* 2004, 07, 013.

[54] Date, G.; Hossain, G. M. *Phys. Rev. Lett.* 2005, 94, 011302.

[55] Lidsey, J. E.; Mulryne, D. J.; Nunes, N. J.; Tavakol, R. *Phys. Rev. D* 2004, 70, 063521.

[56] Bojowald, M.; Maartens, R.; Singh, P. *Phys. Rev. D* 2004, 70, 083517.

[57] Mulryne, D. J.; Nunes, N. J.; Tavakol, R.; Lidsey, J. *Int. J. Mod. Phys. A* 2005, 20, 2347–2357.

[58] Lidsey, J. E. *JCAP* 2004, 0412, 007.

[59] Bojowald, M. *Phys. Rev. Lett.* 2001, 87, 121301.

[60] Bojowald, M.; Hinterleitner, F. *Phys. Rev. D* 2002, 66, 104003.

[61] Hinterleitner, F.; Major, S. *Phys. Rev. D* 2003, 68, 124023.

[62] Green, D.; Unruh, W. *Phys. Rev. D* 2004, 70, 103502.

[63] Bojowald, M.; Date, G. *Class. Quantum Grav.* 2004, 21, 121–143.

[64] Cartin, D.; Khanna, G.; Bojowald, M. *Class. Quantum Grav.* 2004, 21, 4495–4509.

[65] Cartin, D.; Khanna, G. *Phys. Rev. Lett.* 2005, 94, 111302.

[66] Bojowald, M.; Goswami, R.; Maartens, R.; Singh, P. *Phys. Rev. Lett.* 2005, 95, 091302.

[67] Ashtekar, A.; Bojowald, M. *Class. Quantum Grav.* 2006, 23, 391–411.

[68] Ashtekar, A.; Bojowald, M. *Class. Quantum Grav.* 2005, 22, 3349–3362.

[69] Bojowald, M.; Swiderski, R. *Phys. Rev. D* 2005, 71, 081501(R).

[70] Ashtekar, A.; Lewandowski, J.; Marolf, D.; Mourão, J.; Thiemann, T. *J. Math. Phys.* 1995, 36, 6456–6493.

[71] Thiemann, T. *Class. Quantum Grav.* 1998, 15, 839–873.

[72] Ashtekar, A.; Bojowald, M.; Lewandowski, J. *Adv. Theor. Math. Phys.* 2003, 7, 233–268.

[73] Bojowald, M. *Class. Quantum Grav.* 2000, 17, 1489–1508.

[74] Velhinho, J. M. *Class. Quantum Grav.* 2004, 21, L109–L113.

[75] Husain, V.; Winkler, O. *Phys. Rev. D* 2004, 69, 084016.

[76] Modesto, L. (2004). The Kantowski-Sachs Space-Time in Loop Quantum Gravity [gr-qc/0411032]

[77] Bojowald, M. *Class. Quantum Grav.* 2001, 18, L109–L116.

[78] Bojowald, M. Class. *Quantum Grav.* 2002, 19, 5113–5130.

[79] Bojowald, M. *Pramana* 2004, 63, 765–776.

[80] Bojowald, M.; Vandersloot, K. *Phys. Rev. D* 2003, 67, 124023.

[81] Oppenheimer, J. R.; Snyder, H. *Phys. Rev.* 1939, 56, 455–459.

[82] Bojowald, M.; Date, G. *Phys. Rev. Lett.* 2004, 92, 071302.

[83] Bojowald, M.; Date, G.; Hossain, G. M. *Class. Quantum Grav.* 2004, 21, 3541–3569.

[84] Bojowald, M. *Class. Quantum Grav.* 2001, 18, 1055–1070.

[85] Lewandowski, J.; Marolf, D. *Int. J. Mod. Phys. D* 1998, 7, 299–330.

[86] Gambini, R.; Lewandowski, J.; Marolf, D.; Pullin, J. *Int. J. Mod. Phys. D* 1998, 7, 97–109.

[87] Thiemann, T. (2003). The Phoenix Project: Master Constraint Programme for Loop Quantum Gravity [gr-qc/0305080]

[88] Ashtekar, A.; Corichi, A.; Zapata, J. *Class. Quantum Grav.* 1998, 15, 2955–2972.

[89] Booth, I.; Fairhurst, S. *Phys. Rev. Lett.* 2004, 92, 011102.

[90] Ashtekar, A.; Corichi, A.; Krasnov, K. *Adv. Theor. Math. Phys.* 2000, 3, 419–478.

[91] Kastrup, H. A. *Ann. Phys.* (Leipzig) 2000, 9, 503–522.

[92] Ashtekar, A.; Lewandowski, J. *J. Geom. Phys.* 1995, 17, 191–230.

[93] Ashtekar, A.; Lewandowski, J. *J. Math. Phys.* 1995, 36, 2170–2191.

[94] Rovelli, C.; Smolin, L. *Phys. Rev. D* 1995, 52, 5743–5759.

[95] Baez, J. C. In The Interface of Knots and Physics; Kauffman, L.; Ed.; American Mathematical Society, Providence, RI, 1996; pp 167–203.

[96] Thiemann, T. *Phys. Lett. B* 1996, 380, 257–264

[97] Bekenstein, J. D.; Mukhanov, V. F. *Phys. Lett. B* 1995, 360, 7–12.

[98] Ashtekar, A.; Engle, J.; van den Broeck, C. *Class. Quantum Grav.* 2005, 22, L27.

[99] Ashtekar, A.; Corichi, A. *Class. Quantum Grav.* 2003, 20, 4473–4484.

[100] Domagala, M.; Lewandowski, J. *Class. Quantum Grav.* 2004, 21, 5233–5243.

[101] Meissner, K. A. *Class. Quantum Grav.* 2004, 21, 5245–5251.

[102] Bojowald, M.; Morales-Técotl, H. A. In Proceedings of the Fifth Mexican School (DGFM): The Early Universe and Observational Cosmology; *Lect. Notes Phys.* **646**; Springer-Verlag, Berlin, 2004; pp 421–462.

[103] Bojowald, M. *Living Rev. Relativity* 2005, 8, 11, `http://relativity.livingreviews.org/Articles/lrr-2005-11/`.

INDEX

A

B

C

D

E

F

G

H

T

U

V

W

Y